Nanofluids Technology for Thermal Sciences and Engineering

This text highlights how nanofluids can be used in thermal solutions across multiple industries, including electronics, energy, and manufacturing. It emphasizes the enhanced heat transfer properties of nanofluids and their potential to significantly improve the efficiency of heat exchange processes. This book discusses topics such as nanoparticle synthesis, nanofluid testing, performance enhancement using nanofluids, thermal behavior of hybrid nanofluids, Brinkman equation in nanofluids and safety considerations in nano fluid-based systems.

This book:

- Discusses the recent innovation, technological development of nanofluids and explores nanoparticle synthesis and characterization for nanofluid development.
- Offers a comprehensive understanding of nanofluid technology and nanofluid for aerospace application, covering diverse topics from fundamental properties to advanced research frontiers in nanofluids for thermal engineering.
- Includes real-world case studies and practical techniques that will help the readers to apply nanofluid technology in various thermal engineering scenarios.
- Covers heat exchanger performance improvement with nanofluids, hybrid nanofluids, Flow of Newtonian and Non-Newtonian hybrid Nanofluid, and oil-based Tri-hybrid Nanofluid.
- Explains experimental techniques for nanofluid testing and validation and presents safety and environmental considerations in nanofluid-based systems.

It is primarily written for senior undergraduates, graduate students, and academic researchers in the fields of manufacturing engineering, industrial engineering, production engineering, mechanical engineering, automotive engineering, and aerospace engineering.

Advances in Manufacturing, Design and Computational Intelligence Techniques

The book series editor is inviting edited, reference and text book proposal submission in the book series. The main objective of this book series is to provide researchers a platform to present state-of-the-art innovations, research related to advanced materials applications, cutting-edge manufacturing techniques, innovative design and computational intelligence methods used for solving nonlinear problems of engineering. The series includes a comprehensive range of topics and its application in engineering areas such as additive manufacturing, nanomanufacturing, micromachining, biodegradable composites, material synthesis and processing, energy materials, polymers and soft matter, nonlinear dynamics, dynamics of complex systems, MEMS, green and sustainable technologies, vibration control, AI in power station, analog-digital hybrid modulation, advancement in inverter technology, adaptive piezoelectric energy harvesting circuit, contactless energy transfer system, energy efficient motors, bioinformatics, computer-aided inspection planning, hybrid electrical vehicle, autonomous vehicle, object identification, machine intelligence, deep learning, control-robotics-automation, knowledge-based simulation, biomedical imaging, image processing and visualization. This book series compiled all aspects of manufacturing, design and computational intelligence techniques from fundamental principles to current advanced concepts.

https://www.routledge.com/Advances-in-Manufacturing-Design-and-Computational-Intelligence-Techniques/book-series/CRCAIMDCIT

Series Editor: Dr. Ashwani Kumar

Laser-Based Technologies for Sustainable Manufacturing
Edited by Avinash Kumar, Ashwani Kumar, Abhishek Kumar

Thermal Energy Systems: Design, Computational Techniques and Applications
Edited by Ashwani Kumar, Varun Pratap Singh, Chandan Swaroop Meena, Nitesh Dutt

Additive Manufacturing in Industry 4.0: Methods, Techniques, Modeling, and Nano Aspects
Edited by Vipin Kumar Sharma, Ashwani Kumar, Manoj Gupta, Vinod Kumar, Dinesh Kumar Sharma, Subodh Kumar Sharma

Advanced Materials for Biomedical Applications
Edited by Ashwani Kumar, Yatika Gori, Avinash Kumar, Chandan Swaroop Meena, Nitesh Dutt

Applications of Computational Intelligence Techniques in Communications
Edited by Mridul Gupta, Pawan Kumar Verma, Rajesh Verma and Dharmendra Kr. Upadhya

Nanofluids Technology for Thermal Sciences and Engineering

Research, Development, and Applications

Edited by
Mukesh Kumar Awasthi, Nitesh Dutt, and Ashwani Kumar

CRC Press is an imprint of the
Taylor & Francis Group, an **informa** business

Designed cover image: shutterstock

First edition published 2025
by CRC Press
2385 NW Executive Center Drive, Suite 320, Boca Raton FL 33431

and by CRC Press
4 Park Square, Milton Park, Abingdon, Oxon, OX14 4RN

CRC Press is an imprint of Taylor & Francis Group, LLC

ISBN: 9781032799117 (hbk)
ISBN: 9781032799131 (pbk)
ISBN: 9781003494454 (ebk)

DOI: 10.1201/9781003494454

Typeset in Sabon
by codeMantra

Contents

Aim and Scope

The aim and scope of this book *Nanofluids Technology for Thermal Sciences and Engineering: Research, Development, and Applications* are multifaceted, reflecting its commitment to advancing knowledge and practical applications in the field of nanofluid technology for thermal management. One of the primary aims of this book is to provide readers with a comprehensive understanding of the fundamental principles governing nanofluids. It aims to elucidate the composition, properties, and synthesis methods of nanofluids, ensuring that readers have a solid foundation upon which to build their knowledge.

This book begins by offering readers a solid foundation in the fundamental concepts of nanofluids, including their composition, properties, and synthesis methods. It then moves on to discuss the diverse applications of nanofluids in thermal solutions, ranging from electronics cooling to renewable energy systems. The authors highlight the superior thermal conductivity and heat transfer properties of nanofluids, showcasing their potential to enhance the efficiency of heat exchange processes in various engineering applications.

The scope of this book extends to showcasing the vast range of practical applications for nanofluid technology. It delves into how nanofluids can be used in thermal solutions across multiple industries, including electronics, energy, and manufacturing. Practical insights are offered through case studies that illustrate successful implementations, enabling readers to appreciate the real-world impact of this technology. This book emphasizes the enhanced heat transfer properties of nanofluids and their potential to significantly improve the efficiency of heat exchange processes. By exploring the mechanisms behind these enhancements, the aim is to inspire researchers and engineers to harness nanofluids to address thermal challenges in a more energy-efficient manner.

Another essential aspect of this book's scope is to address the challenges associated with nanofluid technology. It explores issues such as stability, safety, and scalability, equipping readers with the knowledge needed to navigate potential obstacles. Furthermore, this book offers insights into emerging solutions and strategies to overcome these challenges effectively.

This book does not merely dwell on the present state of the field but also looks towards the future. The scope extends to discussing emerging trends and innovations in nanofluid research, including the integration of nanofluids into advanced technologies such as 3D printing and microfluidics.

This book's scope recognizes the cross-disciplinary nature of nanofluid technology. It caters to a diverse readership, including researchers, engineers, and professionals from various backgrounds interested in thermal management and nanotechnology. Its comprehensive coverage aims to bridge the gap between theoretical knowledge and practical applications, making it a valuable resource for anyone seeking to harness nanofluid technology for thermal solutions.

In summary, this book aims to provide a holistic understanding of nanofluid technology while exploring its practical applications, addressing challenges, and envisioning its future potential. Its scope encompasses both the foundational aspects and the evolving landscape of nanofluid technology, making it an indispensable reference for those seeking to innovate and optimize thermal solutions across industries. The top-tier research content contained within this book will serve as a valuable reference guide for a diverse audience, including postgraduate students, Ph.D. researchers, fluid dynamics scientists, mechanical engineers, aspiring researchers, experts in heat transfer, manufacturing engineers, design engineers, mechatronics students, and professionals specializing in energy studies. The relevance of this book extends across a wide spectrum of industries, encompassing fields such as fluid dynamics, heat transfer, manufacturing, design engineering, research and development, automotive manufacturing, aviation, electronics, civil engineering, and heat transfer research. This resource will be of great assistance to individuals engaged in research and innovation within these various sectors.

Preface

This book titled *Nanofluids Technology for Thermal Sciences and Engineering: Research, Development, and Applications* serves as a comprehensive guide to the intriguing realm of nanofluids and their transformative influence on thermal engineering and heat transfer applications. Within the pages of this book, we navigate through the innovative domains of nanotechnology, shedding light on how nanofluids are poised to revolutionize various industries, shaping the future of thermal management.

Chapter 1 initiates our exploration by laying the foundation, introducing readers to the fundamental concepts of nanofluids—remarkable liquids infused with nanoparticles. Readers gain an understanding of their unique properties, explore a wide range of applications, and confront the challenges faced by researchers and engineers in harnessing their potential. Chapter 2 provides a comprehensive examination of various methods used for producing and analyzing nanoparticles and nanofluids. The top-down strategy involves methods such as chemical etching, laser ablation, and mechanical milling, while the bottom-up approach encompasses a different set of techniques. Chapter 3 focuses on Nanofluid Testing and Validation, delving into experimental techniques and the detailed target properties of nanofluids. Examples include validating fluidic properties using a rheometer and measuring thermal conductivity using methods such as hot disc, laser flash, and transient hot wire techniques.

Chapter 4 explores nanofluid applications in renewable thermal energy storage, highlighting the potential transformation of the thermal energy storage field with the incorporation of nanotechnology. Advances in nanomaterial production, system design, and optimization are expected to enhance heat transfer, energy storage capacities, efficiency, and dependability. Chapter 5 provides a brief summary of key emerging fields of nanofluid applications in thermal engineering, such as NEPCMs, graphite-based nanofluids, AI-integrated nanofluids, and nanofluids in biomedical applications. The advantages, challenges, and applications of each field are highlighted, pointing out their potential future scope. Chapter 6 critically explores safety and environmental factors related to nanofluid-based systems, emphasizing a comprehensive hazard assessment and rigorous risk management strategies to mitigate potential adverse effects on occupational safety and the environment.

Chapter 7 highlights the role of nanofluids in Heat Exchanger Design and Performance Improvement, emphasizing the enhancement of thermophysical qualities for improved heat transfer capabilities. Chapter 8 discusses the improved heat transmission and decreased thermal resistance of nanofluids in comparison to conventional coolants for automobile applications, assessing various nanoparticles for suitability in automotive cooling. Chapters 9–14 delve into specific studies on the flow of nanofluids between rotating disks, magnetohydrodynamic flows, hybrid nanofluid dynamics, and heat transfer properties, providing detailed insights into each area.

Chapter 15 explores the transformative role of nanofluids in aerospace thermal management, offering enhanced thermal conductivity for addressing extreme heat challenges in propulsion, avionics, and atmospheric re-entry. Chapter 16 explores the key components and technologies involved in the production of concrete, emphasizing its potential to reduce carbon emissions and resource depletion compared to traditional concrete.

In conclusion, this book serves as a passport to the forefront of nanofluid technology, inviting students, researchers, engineers, and anyone curious about thermal solutions to embark on an enlightening journey. Together, readers will explore, innovate, and shape the future of thermal engineering with nanofluids as a guiding star.

Editors
Mukesh Kumar Awasthi, Nitesh Dutt, Ashwani Kumar

Acknowledgment

We extend our heartfelt gratitude to CRC Press (Taylor & Francis) and the dedicated editorial team for their pivotal role in bringing this book, *Nanofluids Technology for Thermal Sciences and Engineering: Research, Development, and Applications*, to fruition. Their invaluable suggestions and unwavering support have been instrumental. Their insightful guidance and unwavering commitment to excellence significantly enhanced the overall quality of our work.

Additionally, we express sincere appreciation to the numerous contributors and reviewers whose illuminating perspectives enriched every chapter of this comprehensive resource. Their collective expertise undeniably contributed to the depth and breadth of the content. The book begins by providing readers with a solid foundation in the fundamental concepts of nanofluids, exploring into their composition, properties, and synthesis methods.

The narrative then seamlessly transitions into an exploration of the diverse applications of nanofluids in thermal solutions, spanning from electronics cooling to renewable energy systems. The comprehensive coverage ensures that readers gain a holistic understanding of the subject matter, thanks to the collaborative efforts of the contributors and reviewers.

This book is dedicated to the passionate and dedicated individuals whose commitment drives advancements in Nanofluids Technology. As we acknowledge their vital contributions, our hope is that this work becomes a source of inspiration and a valuable reference for the scholarly community. We recognize and honor the collaborative spirit that unites us all in the pursuit of excellence and progress within the dynamic intersection of engineering and technology.

Editors
Mukesh Kumar Awasthi, Nitesh Dutt and Ashwani Kumar

About the Editors

Mukesh Kumar Awasthi did his PhD on the topic "Viscous Correction for the Potential Flow Analysis of Capillary and Kelvin-Helmholtz instability". He is an assistant professor in the Department of Mathematics at Babasaheb Bhimrao Ambedkar University, Lucknow. Dr. Awasthi specialized in the mathematical modeling of flow problems. He has taught courses of Fluid Mechanics, Discrete Mathematics, Partial differential equations, Abstract Algebra, Mathematical Methods, and Measure theory to postgraduate students. He has excellent knowledge in the mathematical modeling of flow problems, and he can solve these problems analytically as well as numerically. He has a good grasp of the subjects like viscous potential flow, electro-hydrodynamics, magneto-hydrodynamics, heat, and mass transfer. Dr. Awasthi qualified for the National Eligibility Test (NET) conducted on all India level in 2008 by the Council of Scientific and Industrial Research (CSIR) and got a Junior Research Fellowship (JRF) and Senior Research Fellowship (SRF) for doing research. He has published 115 plus research publications (journal articles/books/book chapters/conference articles) in many national and international journals and conferences. Also, he has published eight books. He has attended many symposia, workshops, and conferences in mathematics as well as fluid mechanics. He got "Research Awards" four times consecutively from 2013 to 2016 from the University of Petroleum and Energy Studies, Dehradun, India. He has also received start-up research funds for his project "Nonlinear study of the interface in multilayer fluid system" from UGC, New Delhi. He is also listed in the top 2% of influential researchers in the world prepared by Stanford University based on Scopus data in 2022.

Dr. Nitesh Dutt is an associate professor in the Department of Mechanical Engineering, COER University Roorkee, Uttarakhand, India. He has more than 7 years of teaching experience. He did his bachelor's degree in Mechanical Engineering, and has a Masters and PhD from IIT Roorkee. He has published more than 11 research articles in international journals and conferences. His main areas of research include nuclear engineering, heat and mass transfer, thermodynamics, fluid mechanics, refrigeration and air conditioning, computational fluid dynamics (CFD).

Dr. Ashwani Kumar holds a PhD in Mechanical Engineering with a specialization in Mechanical Vibration and Design. Currently, he is serving as a senior lecturer in Mechanical Engineering (Gazetted Officer Group B) at the Technical Education Department Uttar Pradesh Kanpur, India (under the Government of Uttar Pradesh). He has been serving the department since December 2013. Previously, Dr. Kumar worked as an assistant professor in the Department of Mechanical Engineering at Graphic Era University, Dehradun, India (NIRF Ranking 55) from July 2010 to November 2013. With over 13 years of experience in research, academia, and administration, he has taken on various roles, including Coordinator for AICTE-Extension of Approval, Nodal officer for PMKVY-TI Scheme (Government of India), internal coordinator-CDTP scheme (Government of Uttar Pradesh), Industry Academia relation officer, Assistant Centre Superintendent (ACS)-Institute Examination Cell, and Zonal Officer to conduct Joint Entrance Examination (JEE-Diploma). He also served as a Sector Magistrate for Lok Sabha-Vidhan Sabha Election. As an academician and researcher, Dr. Kumar holds the position of Series Editor for five book series published by CRC Press (Taylor & Francis USA) and Wiley Scrivener Publishing USA. The book series include Advances in Manufacturing, Design and Computational Intelligence Techniques, Renewable and Sustainable Energy Developments, Smart Innovations and Technological Advancements in Mechanical and Materials Engineering, Solar Thermal Energy Systems: Advancements in Engineering, Ergonomics, and Sustainable Development, and Computational Intelligence and Biomedical Engineering. Additionally, he serves as the guest editor of a special issue titled "Sustainable Buildings, Resilient Cities, and Infrastructure Systems" for Buildings (ISSN: 2075-5309, I.F. 3.8). Dr. Kumar is also the Editor-in-Chief for the *International Journal of Materials, Manufacturing, and Sustainable Technologies* (IJMMST, ISSN: 2583-6625) and the Editor of the *International Journal of Energy Resources Applications* (IJERA, ISSN: 2583-6617). His contributions extend to being a guest editor and editorial board member for eight international journals, as well as a review board member for 20 prestigious international journals indexed in SCI/SCIE/Scopus, including *Applied Acoustics*, *Measurement*, *JESTEC*, *AJSE*, *SV-JME*, and *LAJSS*. He has authored and co-authored over 100 research articles in journals, book chapters, and conferences and has published three patents. Recognized for his excellence in academic and research endeavors, he has received the Best Teacher award. He has successfully supervised 15 B. Tech., M.Tech, and PhD theses and serves as an external doctoral committee member at SRM University, New Delhi. His current research interests encompass AI & ML in mechanical engineering, smart materials & manufacturing techniques, thermal energy storage, building efficiency, renewable energy harvesting, sustainable transportation, and heavy vehicle dynamics.

Contributors

Pratibha S. Agrawal
Department of Applied Chemistry
Laxminarayan Innovation Technological University
Nagpur, India

Fasaha Ahmad
Department of Applied Chemistry
Laxminarayan Innovation Technological University
Nagpur, India

Khulood Al Ghafri
Department of Mechanical Engineering
Global College of Engineering and Technology (GCET)
Muscat, Sultanate of Oman

Mukesh Kumar Awasthi
Department of Mathematics
Babasaheb Bhimarao Ambedkar University
Lucknow, India

M. Baskaran
Department of Mechatronics Engineering
K. S. Rangasamy College of Technology
Tiruchengode, India

Ashtashil Vrushketu Bhambulkar
Department of Civil Engineering
JSPM's Imperial College of Engineering and Research
Pune, India

Aditi Bonde
Department of Applied Chemistry
Laxminarayan Innovation Technological University
Nagpur, India

Ruchi Chandrakar
Civil Engineering Department, Faculty of Technology
Kalinga University
Naya Raipur, India

Mayur Chaware
Department of Applied Chemistry
Laxminarayan Innovation Technological University
Nagpur, India

Vinodhini Chinnathambi
Department of Mechatronics Engineering
Kongu Engineering College
Perundurai, India

Kalidas Das
Department of Mathematics
Krishnagar Government College
Nadia, India

Krishna Priyadarshini Das
Department of Mechanical Engineering
Indian Institute of Technology Delhi
New Delhi, India

Tusar Kanti Das
Department of Mathematics
Dudhnoi College
Dudhnoi, India

Satya Deo
Department of Mathematics
University of Allahabad
Prayagraj, India

Bamdeb Dey
Department of Mathematics
Assam Don Bosco University
Guwahati, India

Shweta Dode
Department of Applied Chemistry
Laxminarayan Innovation Technological University
Nagpur, India

Pinaki Ranjan Duari
Department of Basic Science & Humanities (Mathematics Section)
Asansol Engineering College
West Bengal, India

Nitesh Dutt
Department of Mechanical Engineering
COER University Roorkee
Uttarakhand, India

R. Gangadevi
Department of Mechatronics Engineering
SRM Institute of Science and Technology
Kanchipuram, India

Raja Gunasekaran
Department of Mechanical Engineering
Velalar College of Engineering and Technology
Erode, India

Reshu Gupta
Applied Science Cluster (Department of Mathematics)
UPES
Dehradun, India

Srijan Gupta
Department of Applied Chemistry
Laxminarayan Innovation Technological University
Nagpur, India

Milad Heidari
Department of Mechanical Engineering
Global College of Engineering and Technology (GCET)
Muscat, Sultanate of Oman

Gobinath Velu Kaliyannan
Department of Mechatronics Engineering
Kongu Engineering College
Perundurai, India

Suganeswaran Kandasamy
Department of Mechatronics Engineering
Kongu Engineering College
Perundurai, India

Ashwani Kumar
Department of Mechanical Engineering
Technical Education Department Uttar Pradesh
Kanpur, India

Ashwini Kumar
Department of Mechanical Engineering
Arka Jain University
Jharkhand, India

Deepak Kumar Maurya
Department of Mathematics
Rajendra Singh (Rajju Bhaiya) Institute of Physical Sciences for Study and Research
Veer Bahadur Singh Purvanchal University
Jaunpur, India

Rahul Mishra
Department of Mechanical Engineering
Kalinga University
Naya Raipur, India

Nithyavathy Nagarajan
Department of Social Science
Kumaraguru Institute of Agriculture
Erode, India

Jintu Mani Nath
Department of Mathematics
Mangaldai College
Mangaldai, India

Bhagyashri Patgiri
Department of Mathematics
Cotton University
Guwahati, India

Abhijit Pattnayak
Department of Mechanical Engineering
Indian Institute of Technology Delhi
New Delhi, India

Ashish Paul
Department of Mathematics
Cotton University
Guwahati, India

K. Peera
Department of Zoology
Sri Venkateswara University
Tirupati, India

Shailandra Kumar Prasad
Mechanical Engineering Department
R.V.S. College of Engineering & Technology
Jamshedpur, India

Uttam Roy
Department of Administration
Jadavpur University
Kolkata, India

Neelav Sarma
Department of Mathematics
Cotton University
Guwahati, India

S. Senthilraja
Department of Mechatronics Engineering
SRM Institute of Science and Technology
Kanchipuram, India

Varun Pratap Singh
Department of Mechanical Engineering, School of Advanced Engineering
UPES
Dehradun, India

Sivasakthivel Thangavel
Department of Mechanical Engineering
Global College of Engineering and Technology (GCET)
Muscat, Sultanate of Oman

Chapter 1

Recent innovation, technological development, and futuristic application of nanofluids

Mukesh Kumar Awasthi,
Ashwani Kumar, and Nitesh Dutt

1.1 INTRODUCTION

1.1.1 Definition of nanofluids

Nanofluids represent a class of engineered fluids that consist of a base fluid, typically a traditional heat transfer fluid like water or oil, with dispersed nanoscale particles. The term "nanofluid" itself signifies the nanometer-sized particles that are suspended within the fluid. These nanoparticles, often ranging from 1 to 100 nm in size, significantly alter the thermophysical properties of the base fluid, leading to enhanced heat transfer capabilities and other unique characteristics. The field of nanofluid research is interdisciplinary, drawing from physics, chemistry, and engineering, and has gained considerable attention due to its promising applications in various industries.

1.1.2 Historical background

The exploration of nanofluids dates back to the early 1990s when Choi and Eastman introduced the concept of nanofluids and began experimenting with the addition of nanoparticles to enhance thermal conductivity. Their seminal paper, published in 1995, laid the groundwork for subsequent research in the field [1]. Since then, the field has evolved, with researchers investigating various nanoparticles, base fluids, and applications. The initial focus on thermal conductivity enhancement has expanded to include studies on viscosity, stability, and other properties critical for practical applications.

1.1.3 Definition and classification

Nanofluids are typically classified based on the type of nanoparticles dispersed in the base fluid. Common nanoparticles include metallic oxides (such as alumina and titania), metallic nanoparticles (such as copper and silver), and carbon-based nanoparticles (such as graphene and carbon

DOI: 10.1201/9781003494454-1

nanotubes). The unique properties of nanofluids make them promising candidates for improving the efficiency of heat exchangers, enhancing cooling systems, and advancing other thermal management applications.

The definition of nanofluids encompasses the incorporation of nanoscale particles into traditional fluids, leading to altered thermophysical properties. The historical background reveals a trajectory of research that started with the pursuit of enhanced thermal conductivity and has since evolved into a diverse field with broad applications. As the understanding of nanofluids continues to deepen, so too does their potential to revolutionize heat transfer and other industrial processes.

1.1.4 Significance and relevance in modern research

In contemporary scientific research, nanofluids have emerged as a significant and dynamic field with wide-ranging applications, demonstrating their relevance across diverse industries. The incorporation of nanoscale particles into conventional fluids brings about distinct improvements in thermal, mechanical, and rheological properties, thus addressing challenges faced in various technological domains.

One of the key areas where nanofluids have demonstrated immense significance is in thermal management and heat transfer applications. The enhanced thermal conductivity of nanofluids, attributed to the increased surface area and unique thermal properties of nanoparticles, offers unprecedented opportunities for improving the efficiency of heat exchangers and cooling systems. This has direct implications for industries ranging from electronics, where efficient thermal dissipation is crucial, to energy production, where optimizing heat transfer processes can lead to increased efficiency.

Moreover, the biomedical field has witnessed a surge in interest in nanofluids due to their potential applications in hyperthermia treatment and drug delivery. Nanofluids can be engineered to carry therapeutic agents and nanoparticles to specific targets within the body, offering a level of precision that conventional drug delivery methods struggle to achieve. The controlled release of drugs from nanofluid carriers holds promise for personalized medicine and targeted therapies.

In the realm of materials science and manufacturing, nanofluids play a pivotal role in lubrication and cutting-edge machining processes. The unique properties of nanofluids, such as their ability to reduce friction and wear, have led to advancements in the development of high-performance lubricants. Additionally, in machining applications, nanofluids are explored for their cooling and lubricating capabilities, improving tool life and machining efficiency.

The versatility of nanofluids extends to environmental applications as well. Researchers are investigating their potential in water treatment and purification, leveraging the properties of nanoparticles to enhance filtration processes and remove contaminants. The use of nanofluids in environmental remediation showcases the interdisciplinary nature of nanofluid research, where principles from chemistry, physics, and engineering converge to address pressing global challenges.

Despite the immense promise, the research on nanofluids also confronts challenges that necessitate thorough investigation. Issues related to stability, agglomeration, and long-term effects on human health and the environment require careful consideration. As the field progresses, researchers are actively working on standardizing methods for synthesizing and characterizing nanofluids, ensuring reproducibility and comparability of results.

The significance of nanofluids in modern research is evident across multiple disciplines, from enhancing thermal management in electronics to revolutionizing drug delivery in medicine. The ability to tailor nanofluids for specific applications underscores their versatility and potential impact on various industries. As nanofluid research continues to evolve, addressing challenges and exploring new frontiers, the transformative potential of these engineered fluids in modern technology becomes increasingly apparent.

1.2 PROPERTIES OF NANOFLUIDS

1.2.1 Size and composition of nanoparticles

The size of nanoparticles plays a pivotal role in determining their thermal and rheological characteristics. Smaller nanoparticles generally lead to higher surface area, which, in turn, enhances the interaction with the base fluid. This increased interaction often results in improved thermal conductivity and heat transfer capabilities of nanofluids. Figure 1.1 illustrates the relationship between nanoparticle size and thermal conductivity for copper thin films [2].

The composition of nanoparticles in nanofluids is diverse, encompassing various materials such as metallic oxides, metallic nanoparticles, and carbon-based nanoparticles. Table 1.1 provides an overview of commonly used nanoparticle materials and their properties.

The choice of nanoparticle material depends on the specific application requirements. For instance, metallic nanoparticles with high thermal conductivity may be preferred in heat transfer applications, while carbon-based nanoparticles might find use in lubrication due to their moderate to high thermal conductivity and unique lubricating properties.

It is essential to consider not only the average size of nanoparticles but also their size distribution within nanofluids. Figure 1.2 illustrates a typical size distribution curve, showcasing the range of nanoparticle sizes present. Achieving a narrow and well-controlled size distribution is critical to ensuring consistent and predictable nanofluid properties.

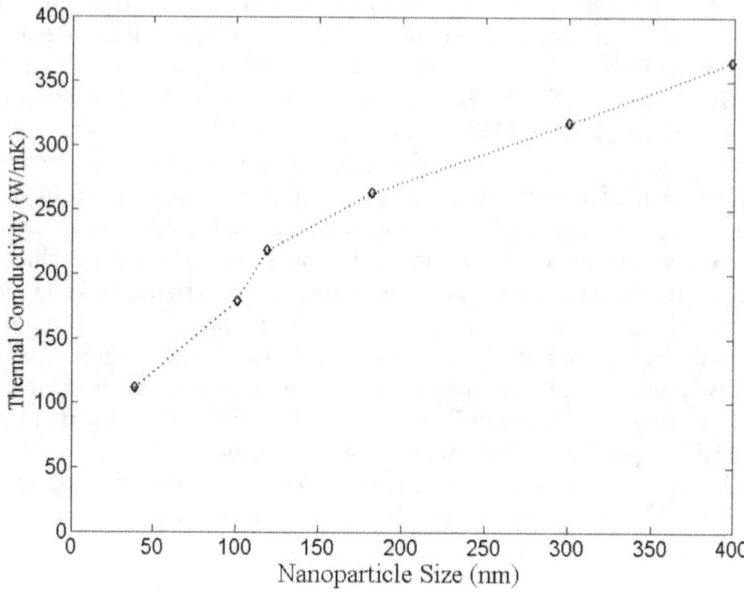

Figure 1.1 Relationship between nanoparticle size and thermal conductivity.

Table 1.1 Common nanoparticle materials and properties

Nanoparticle material	*Thermal conductivity (W/mK)*	*Melting point (°C)*	*Applications*
Metallic oxides	High	Varies	Heat exchangers, electronics
Metallic nanoparticles	Very high	Depends on metal	Biomedical, catalysis
Carbon-based	Moderate to high	Varies	Electronics, lubrication

In conclusion, the size and composition of nanoparticles are pivotal factors influencing the behavior and performance of nanofluids. Proper understanding and control of these parameters are crucial for tailoring nanofluids to specific applications, ensuring optimal thermal and rheological properties. Figures 1.1, and 1.2, along with Table 1.1, provide visual and tabular aids for comprehending the intricate relationships within the realm of nanoparticle characteristics in nanofluids.

1.2.1.1 Thermal conductivity enhancement

The quest for improved thermal conductivity in fluids has led to the emergence of nanofluids as a revolutionary solution. Nanofluids, engineered by

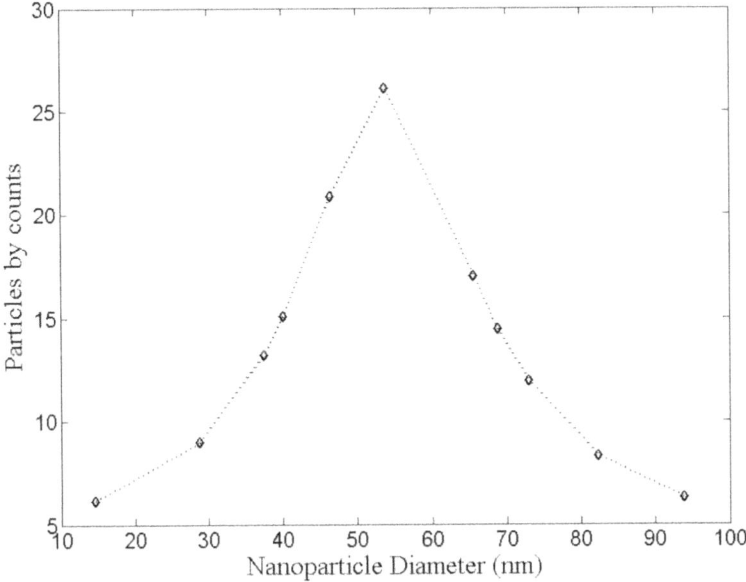

Figure 1.2 Nanoparticle size distribution curve of CaCu290Zn010Ti_4O_{12}-nanoparticles.

dispersing nanoparticles in base fluids, exhibit enhanced thermal conductivity compared to their traditional counterparts. This enhancement is a result of the unique properties that nanoparticles bring to the fluid, creating exciting opportunities for applications in various industries.

1.2.1.2 Fundamental mechanisms

The thermal conductivity enhancement in nanofluids can be attributed to several fundamental mechanisms. First, the increased surface area of nanoparticles facilitates better interaction with the base fluid, leading to enhanced thermal conduction. Second, quantum effects at the nanoscale contribute to improved thermal transport properties, allowing nanofluids to surpass the limitations of traditional fluid systems.

1.2.1.3 Particle size and conductivity

The size of nanoparticles plays a critical role in determining the extent of thermal conductivity enhancement. Smaller nanoparticles, typically ranging from 1 to 100nm, provide a higher surface area for heat transfer. As a result, nanofluids containing smaller particles exhibit more pronounced enhancements in thermal conductivity.

The choice of nanoparticle material is equally important in achieving significant thermal conductivity enhancement. Metallic nanoparticles, such

Table 1.2 Thermal conductivity of common nanoparticle materials

Nanoparticle	*Thermal conductivity (W/mK)*
Aluminum oxide	40
Copper oxide	76.5
Iron oxide	7
Zinc oxide	29
Tin oxide	36
Carbon nanotubes	3,000–6,000
Aluminum nitride	140–180

as copper and silver, are known for their high thermal conductivity, making them popular choices. Additionally, metal oxides like alumina and titania also contribute to enhanced thermal properties. Table 1.2 outlines the thermal conductivity of common nanoparticle materials, highlighting their potential in tailoring nanofluids for specific applications.

1.2.1.4 Applications

The thermal conductivity enhancement achieved through nanofluids opens avenues for groundbreaking applications. In the realm of electronics, nanofluids offer improved cooling solutions, addressing challenges associated with heat dissipation in high-performance devices. Heat exchangers in various industries benefit from the increased efficiency enabled by nanofluids, leading to enhanced thermal management.

1.2.1.5 Challenges and considerations

While the thermal conductivity enhancement in nanofluids is promising, researchers and engineers must address challenges such as nanoparticle agglomeration, stability, and potential environmental concerns. Ensuring proper dispersion methods and understanding the long-term effects are crucial steps in harnessing the full potential of nanofluids.

1.2.1.6 Future directions

The exploration of thermal conductivity enhancement through nanofluids continues to evolve, with ongoing research focusing on optimizing nanoparticle size, developing new materials, and expanding applications. As nanofluid technology matures, its integration into various industries holds the promise of more efficient and sustainable thermal management solutions.

In conclusion, the enhancement of thermal conductivity through nanofluids represents a transformative advancement with broad-reaching implications.

The synergy of nanoparticle size, material selection, and fundamental mechanisms unlocks new possibilities for addressing thermal challenges across diverse sectors, paving the way for a future where nanofluids redefine the standards of heat transfer and thermal management.

1.2.2 Viscosity and rheological behavior

Understanding the viscosity and rheological behavior of nanofluids is essential for optimizing their performance in various applications. Nanofluids, which are composed of base fluids and dispersed nanoparticles, exhibit unique flow characteristics that influence their suitability for heat transfer, lubrication, and other industrial processes [3–5].

1.2.2.1 Viscosity characteristics

The addition of nanoparticles to a base fluid alters its viscosity, a property that plays a crucial role in determining the fluid's flow behavior. Nanofluids often display non-Newtonian behavior, where viscosity is dependent on shear rate. This non-linear relationship is a result of interactions between nanoparticles and the base fluid. Figure 1.3 illustrates the typical viscosity-shear rate profile for nanofluids, showcasing the shear-thinning or shear-thickening behavior commonly observed.

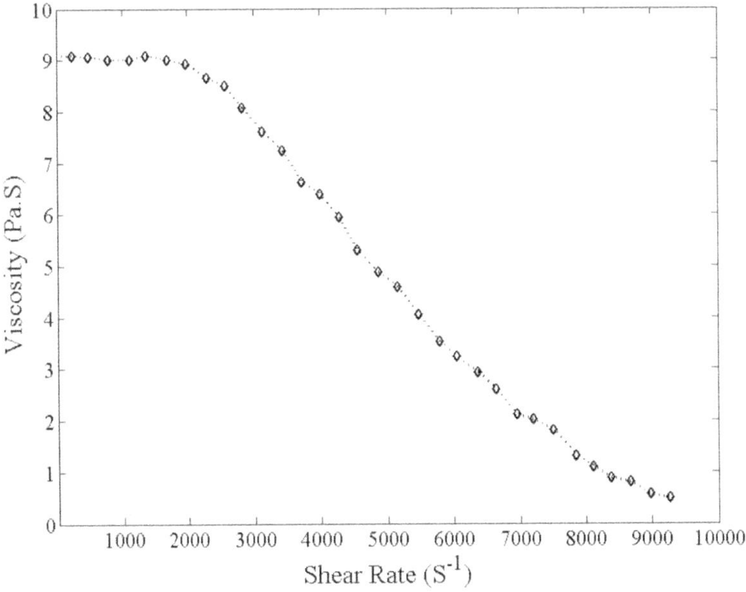

Figure 1.3 Viscosity-shear rate profile for nanofluids.

1.2.2.2 Factors influencing viscosity

Several factors influence the viscosity of nanofluids, including nanoparticle concentration, size, and shape. Higher concentrations of nanoparticles generally lead to increased viscosity due to enhanced interactions and greater resistance to flow. Smaller nanoparticles tend to promote shear-thinning behavior, while larger particles may result in shear-thickening. The shape of nanoparticles also contributes to the overall rheological behavior, with elongated particles introducing additional complexities.

1.2.2.3 Rheological behavior

Nanofluids often exhibit complex rheological behaviors, and their flow characteristics can be described using various models. The Bingham plastic model, Herschel-Bulkley model, and Power-law model are commonly employed to characterize the rheological properties of nanofluids. These models account for the yield stress, shear-thinning or shear-thickening behavior, and other flow characteristics.

1.2.2.4 Applications in lubrication

Understanding the viscosity and rheological behavior of nanofluids is particularly crucial in applications such as lubrication, where efficient flow properties are essential. Nanofluids with tailored viscosity profiles can lead to improved lubrication performance, reducing friction, and wear in mechanical systems. This has implications for extending the lifespan of machinery and enhancing overall efficiency.

1.2.2.5 Challenges and opportunities

While the viscosity and rheological behavior of nanofluids offer opportunities for optimization in various applications, challenges such as nanoparticle agglomeration and stability must be addressed. Proper dispersion methods and innovative approaches are continually being explored to overcome these challenges and unlock the full potential of nanofluids.

1.2.2.6 Future directions

The exploration of nanofluid rheology is an ongoing area of research, with scientists and engineers working toward developing more accurate models and predictive tools. As nanofluid technology matures, it is expected to find broader applications in industries where precise control of viscosity and rheological behavior is paramount.

In conclusion, understanding the viscosity and rheological behavior of nanofluids is vital for harnessing their potential in diverse applications.

The non-Newtonian nature and complex flow characteristics offer both challenges and opportunities, providing a rich area for ongoing research and innovation. The continued exploration of nanofluid rheology promises advancements that will shape the future of fluid dynamics in various industrial processes.

1.2.3 Stability and dispersion characteristics

The stability and dispersion characteristics of nanofluids are critical factors that influence their performance in various applications, ranging from heat transfer to biomedical technologies. Nanofluids, composed of nanoparticles dispersed in a base fluid, are susceptible to agglomeration and sedimentation, which can impact their thermal and rheological properties.

1.2.3.1 Dispersion methods

Achieving and maintaining a stable dispersion of nanoparticles within a nanofluid is a key challenge. Various dispersion methods, including mechanical stirring, ultrasonication, and surfactant stabilization, are employed to prevent the agglomeration of nanoparticles. Mechanical stirring ensures uniform distribution, while ultrasonication breaks down agglomerates and disperses nanoparticles evenly. Surfactants, by forming a protective layer around nanoparticles, help maintain stability over time.

1.2.3.2 Agglomeration and sedimentation

Agglomeration, the undesired clumping of nanoparticles, and sedimentation, the settling of particles over time, are common challenges in nanofluid stability. Both phenomena can adversely affect the performance of nanofluids by altering their thermophysical properties. Addressing agglomeration and sedimentation involves a careful selection of dispersion methods and surface modification of nanoparticles to enhance their stability.

1.2.3.3 Applications in heat transfer

The stability of nanofluids is particularly crucial in heat transfer applications, where uniform dispersion ensures optimal thermal conductivity enhancement. Heat exchangers and cooling systems benefit from stable nanofluids, as agglomeration can lead to uneven heat distribution and reduced efficiency. Achieving and maintaining stability is essential for the successful integration of nanofluids into heat transfer systems.

1.2.3.4 Characterization techniques

Characterizing the stability and dispersion characteristics of nanofluids requires sophisticated techniques. Dynamic Light Scattering (DLS), Zeta potential measurements, and Transmission Electron Microscopy are commonly used to assess nanoparticle size distribution, surface charge, and dispersion quality.

1.2.3.5 Challenges and future outlook

Despite advancements in dispersion techniques, challenges in achieving long-term stability persist. The interaction between nanoparticles and the base fluid, as well as external factors like temperature and pressure, can impact stability. Ongoing research is focused on developing innovative methods to enhance stability and mitigate challenges associated with long-term use.

1.2.3.6 Conclusion

In conclusion, the stability and dispersion characteristics of nanofluids are pivotal considerations in harnessing their potential across diverse applications. Addressing agglomeration and sedimentation challenges through effective dispersion methods and surface functionalization is crucial for realizing the full benefits of nanofluid technology. As research continues to evolve, advancements in stability will pave the way for enhanced performance and broader applications in industries requiring precise control of nanoparticle dispersion.

1.3 SYNTHESIS METHODS

The synthesis of nanofluids involves various methods, including chemical, physical, and hybrid approaches. Chemical methods focus on the controlled synthesis of nanoparticles and their subsequent dispersion in a base fluid, while physical methods rely on physical processes such as laser ablation or grinding to achieve nanoparticle dispersion. Hybrid methods combine both chemical and physical approaches for a tailored synthesis of nanofluids with specific properties.

1.3.1 Chemical synthesis

Chemical synthesis is a predominant method for creating nanofluids, leveraging precise control over nanoparticle formation and dispersion. This approach involves the synthesis of nanoparticles followed by their integration into a base fluid. The chemical synthesis of nanofluids begins with the preparation of nanoparticles through techniques such as chemical reduction, sol-gel synthesis, or co-precipitation.

1.3.1.1 Precursor selection and reduction

In chemical reduction, metallic precursors are chosen based on the desired properties of the nanoparticles. For example, metal salts like copper chloride or silver nitrate are commonly used. These precursors undergo a reduction reaction, often facilitated by a reducing agent, resulting in the formation of nanoparticles. The size, shape, and composition of the nanoparticles can be controlled by adjusting reaction parameters.

1.3.1.2 Sol-gel synthesis

Sol-gel synthesis involves the transformation of a solution (sol) into a gel through a chemical process. Metal alkoxides or metal chlorides are commonly used precursors. This method allows the formation of nanoparticles in a controlled environment and subsequent dispersion in a liquid phase results in the synthesis of nanofluids with tailored properties.

1.3.1.3 Co-precipitation method

The co-precipitation method involves the simultaneous precipitation of metallic ions from a solution, leading to the formation of nanoparticles. By adjusting parameters such as pH and temperature, researchers can influence the size and morphology of the nanoparticles formed. Subsequent mixing with a base fluid completes the synthesis of nanofluids.

1.3.1.4 Surface modification and stabilization

Chemically synthesized nanoparticles often undergo surface modification to improve their stability in the base fluid. This may involve coating the nanoparticles with surfactants or other stabilizing agents to prevent agglomeration. The stability of the nanofluid is crucial for maintaining its enhanced thermophysical properties over time.

1.3.1.5 Scale-up challenges

While chemical synthesis offers precise control over nanoparticle characteristics, scaling up the production of nanofluids poses challenges. Achieving consistent nanoparticle properties on a larger scale requires careful consideration of reaction conditions, reproducibility, and cost-effectiveness. Researchers continue to explore scalable chemical synthesis methods for widespread industrial applications.

1.3.1.6 Tailoring properties for applications

One of the advantages of chemical synthesis is the ability to tailor the properties of nanofluids for specific applications. For instance, in heat transfer applications, the thermal conductivity of nanofluids can be optimized by adjusting the size and composition of nanoparticles. This customization allows for the creation of nanofluids with enhanced performance in diverse industries, including electronics, energy, and healthcare.

1.3.1.7 Environmental considerations

Chemical synthesis methods also raise environmental concerns due to the use of certain chemicals and energy-intensive processes. Researchers are actively exploring greener synthesis approaches, incorporating environmentally friendly precursors and sustainable methods to reduce the environmental impact of nanofluid production.

In conclusion, the chemical synthesis method plays a pivotal role in the creation of nanofluids, offering precise control over nanoparticle properties and their subsequent integration into base fluids. While providing opportunities for tailoring nanofluid characteristics, researchers are simultaneously addressing challenges related to scalability and environmental sustainability to ensure the widespread applicability of this synthesis approach.

1.3.2 Physical synthesis

The physical synthesis method of nanofluids involves processes that rely on physical mechanisms for the formation and dispersion of nanoparticles within a base fluid. Unlike chemical methods, physical synthesis typically avoids the use of chemical reactions for nanoparticle creation, offering distinct advantages in certain applications.

1.3.2.1 Laser ablation

One of the prominent physical synthesis methods is laser ablation, where a laser beam is directed at a target material, causing vaporization and subsequent condensation to form nanoparticles. This technique enables the production of nanoparticles with minimal contamination and precise control over size. Laser ablation has found applications in creating nanofluids with unique properties for advanced electronic devices and biomedical applications.

1.3.2.2 Grinding and milling

Grinding or milling involves mechanical processes that break down larger particles into nanoscale dimensions. In the context of nanofluids, these methods are employed to reduce bulk materials into nanoparticles, which are then dispersed in a base fluid. The simplicity and scalability of grinding and milling make them attractive for large-scale production of nanofluids.

1.3.2.3 Sonication

Sonication, or ultrasonication, utilizes high-frequency sound waves to break down agglomerated nanoparticles and disperse them uniformly in a liquid medium. This physical method is effective in preventing nanoparticle agglomeration and promoting stable dispersions. Ultrasonication is widely employed to enhance the stability of nanofluids, particularly in applications requiring precise control over nanoparticle dispersion.

1.3.2.4 High-pressure homogenization

High-pressure homogenization involves forcing nanofluid through a narrow nozzle at high pressures, resulting in the reduction of nanoparticle size and improved dispersion. This technique is particularly useful for creating nanofluids with small and uniform nanoparticles, contributing to enhanced stability and thermal properties.

1.3.2.5 Benefits of physical synthesis

Physical synthesis methods offer advantages such as simplicity, scalability, and reduced contamination compared to some chemical methods. The absence of chemical reactions in the nanoparticle formation process can be beneficial for applications where chemical purity is critical, such as in certain biomedical or electronic applications.

1.3.2.6 Challenges and consideration

Despite the benefits, physical synthesis methods also present challenges. Achieving precise control over nanoparticle size and avoiding excessive energy consumption during processes like milling are ongoing considerations. Researchers are continually refining these techniques to strike a balance between efficiency, scalability, and the creation of nanofluids with tailored properties.

1.3.2.7 Applications and future directions

The physical synthesis of nanofluids finds applications in various fields, including electronics, medicine, and energy. The ability to control nanoparticle size and dispersion through physical methods opens doors to innovative solutions in heat transfer, drug delivery, and lubrication. As research progresses, further advancements in physical synthesis methods are expected, addressing challenges and expanding the scope of applications for nanofluids [5–9].

1.3.2.8 Conclusion

In conclusion, the physical synthesis method of nanofluids, encompassing techniques like laser ablation, grinding, sonication, and high-pressure homogenization, provides alternative pathways for creating nanofluids with tailored properties. The simplicity and scalability of physical methods contribute to their attractiveness for large-scale production, while ongoing research focuses on optimizing these techniques for specific applications and addressing associated challenges [10–15].

1.3.3 Hybrid methods

The hybrid synthesis method of nanofluids combines elements from both chemical and physical synthesis approaches, offering a versatile and tailored approach to nanoparticle integration into base fluids. This method seeks to leverage the strengths of each synthesis type, addressing challenges related to scalability, reproducibility, and control over nanoparticle characteristics.

1.3.3.1 Integration of chemical and physical processes

In hybrid synthesis, the process often begins with a chemical step, where nanoparticles are initially formed through chemical reactions. Subsequently, physical methods are employed to modify, refine, or disperse these nanoparticles within the base fluid. This integrated approach allows researchers to capitalize on the precision of chemical synthesis while harnessing the scalability and control offered by physical methods.

1.3.3.2 Two-step process

A common hybrid synthesis strategy involves a two-step process. In the first step, nanoparticles are synthesized using chemical methods such as sol-gel, co-precipitation, or chemical reduction. In the second step, physical methods such as sonication or high-pressure homogenization are applied to achieve better dispersion, control size distribution, and enhance stability.

1.3.3.3 Benefits of hybrid synthesis

The hybrid synthesis method offers several advantages. By integrating chemical and physical processes, researchers can achieve a more nuanced control over nanoparticle characteristics, tailoring them to specific applications. This approach is particularly valuable in applications where both the size and surface characteristics of nanoparticles are crucial for optimal performance.

1.3.3.4 Optimizing nanofluid properties

Hybrid synthesis allows the optimization of nanofluid properties, such as thermal conductivity and stability, by carefully adjusting the parameters of both the chemical and physical synthesis steps. This flexibility is particularly advantageous in industries where nanofluids play a pivotal role, such as in advanced heat transfer systems, biomedical applications, and energy storage.

1.3.3.5 Challenges and considerations

While hybrid synthesis offers a promising pathway, challenges include the need for precise coordination between chemical and physical steps. Achieving reproducibility and scalability can be complex, requiring a deep understanding of the interactions between the synthesized nanoparticles and the base fluid. Researchers are actively exploring innovative strategies to overcome these challenges and refine the hybrid synthesis process.

1.3.3.6 Applications in diverse fields

The versatility of hybrid synthesis finds applications in diverse fields. For instance, in biomedical research, hybrid-synthesized nanofluids can be tailored to deliver therapeutic agents with precise control over size and surface characteristics. In energy storage, nanofluids with optimized thermal properties can enhance the efficiency of cooling systems in advanced batteries and electronics.

1.3.3.7 Future directions

As research in nanofluids progresses, the hybrid synthesis method is expected to play a crucial role in advancing the field. Ongoing investigations aim to refine hybrid approaches, explore new chemical and physical synthesis combinations, and address scalability issues to make the method more accessible for industrial applications.

1.3.3.8 Conclusion

In conclusion, the hybrid synthesis method of nanofluids stands at the intersection of chemical and physical approaches, offering a flexible and tailored approach to nanoparticle integration. The ability to combine the precision of chemical synthesis with the scalability of physical methods positions hybrid synthesis as a promising avenue for optimizing nanofluid properties for diverse applications in science and industry.

1.3.4 Challenges in nanoparticle synthesis

The synthesis of nanoparticles, a fundamental step in the creation of nanofluids, presents several challenges that researchers continually strive to overcome. These challenges encompass aspects of reproducibility, scalability, and the precise control of nanoparticle properties crucial for their integration into base fluids [15–17].

1.3.4.1 Reproducibility and control

Achieving reproducibility in nanoparticle synthesis is a significant challenge due to the sensitivity of the process to various factors. Small variations in reaction conditions, such as temperature, pH, and reactant concentrations, can lead to variations in nanoparticle size, shape, and composition. This lack of reproducibility poses challenges for researchers aiming to produce consistent nanofluids for specific applications.

1.3.4.2 Scalability issues

Many nanoparticle synthesis methods face scalability issues when transitioning from laboratory-scale to industrial-scale production. The conditions that work efficiently in small-scale setups may not be directly applicable in larger reactors. Maintaining uniformity and control over nanoparticle properties at scale is a complex challenge that requires innovative engineering solutions.

1.3.4.3 Energy consumption

Certain nanoparticle synthesis methods can be energy-intensive, particularly those involving high temperatures, pressures, or specialized equipment. As the demand for sustainable practices increases, finding energy-efficient synthesis routes is essential. Researchers are exploring greener synthesis methods that reduce energy consumption and minimize the environmental impact associated with nanoparticle production.

1.3.4.4 Agglomeration and size distribution

Nanoparticle agglomeration, where particles clump together, is a common challenge that affects both the reproducibility and performance of nanofluids. Achieving a narrow and well-controlled size distribution is crucial for optimizing the properties of nanofluids. Researchers are actively developing strategies, including the use of surfactants and innovative dispersion techniques, to mitigate agglomeration and achieve better size control.

1.3.4.5 Surface functionalization and stability

The stability of nanoparticles in nanofluids is paramount, requiring effective surface functionalization to prevent agglomeration over time. Challenges arise in developing surface modification techniques that maintain stability without introducing undesirable side effects. Achieving a delicate balance between stability and maintaining the unique properties of nanoparticles is an ongoing consideration.

1.3.4.6 Environmental impact

Certain nanoparticle synthesis methods involve the use of toxic or environmentally harmful chemicals, raising concerns about the ecological impact of the process. Researchers are exploring green synthesis methods that utilize environmentally friendly precursors and reduce or eliminate the use of hazardous chemicals, contributing to the sustainable development of nanofluid technology.

1.3.4.7 Innovations and future outlook

Despite these challenges, ongoing research is driving innovations in nanoparticle synthesis. Advanced characterization techniques, such as in situ monitoring and real-time analytics, enable researchers to gain deeper insights into the synthesis process, aiding in the optimization of conditions. Additionally, the exploration of unconventional synthesis routes, such as biological methods and continuous flow processes, shows promise in addressing scalability and environmental concerns.

1.3.4.8 Collaborative efforts and interdisciplinary approaches

The complex challenges in nanoparticle synthesis necessitate collaborative efforts and interdisciplinary approaches. Researchers from chemistry, materials science, engineering, and environmental science are joining forces to develop holistic solutions that address the multifaceted aspects of nanoparticle synthesis. This collaborative approach is fostering a deeper understanding of the challenges and propelling the field toward more sustainable and efficient nanoparticle synthesis methods.

In conclusion, while challenges persist in nanoparticle synthesis for nanofluids, researchers are making significant strides in overcoming these obstacles. The integration of innovative techniques, a focus on sustainability, and collaborative interdisciplinary efforts are driving advancements in nanoparticle synthesis, contributing to the continued evolution of nanofluid technology.

1.4 APPLICATIONS OF NANOFLUIDS

1.4.1 Heat transfer enhancement applications

Nanofluids have garnered significant attention for their remarkable heat transfer enhancement properties. One notable application lies in the field of electronics cooling, where traditional cooling fluids may struggle to dissipate heat efficiently from densely packed electronic components. Nanofluids, such as those incorporating metallic nanoparticles like copper or silver, exhibit enhanced thermal conductivity, enabling more effective heat transfer. For example, in the design of computer processors, the use of nanofluids in cooling systems can help maintain optimal operating temperatures, preventing overheating and enhancing overall system performance [1,2].

Another notable application of nanofluids is in the realm of solar thermal systems. In solar collectors, where efficient heat absorption and transfer are critical, nanofluids can play a pivotal role. By incorporating nanoparticles with high thermal conductivity, such as aluminum oxide or titanium dioxide, into the heat transfer fluid, the overall efficiency of solar collectors can be significantly improved. This enhancement allows for better utilization of solar energy in applications like solar water heaters or concentrated solar power systems, contributing to sustainable and environmentally friendly energy solutions.

The aviation industry benefits from nanofluid applications in heat exchangers and aircraft thermal management systems. The aerospace sector faces challenges related to weight reduction and fuel efficiency, making effective heat transfer crucial for maintaining optimal engine performance. Nanofluids with improved thermal properties enable more efficient cooling of aircraft components, leading to enhanced engine reliability and overall fuel efficiency.

1.4.2 Batteries thermal management systems

Batteries thermal management systems (BTMS) play a pivotal role in ensuring the safe and efficient operation of energy storage systems, particularly in electric vehicles (EVs) and renewable energy applications. Nanofluids have emerged as a promising solution to address the thermal challenges associated with battery systems. In EVs, the demand for higher energy density and faster charging rates is ever-growing, effective heat dissipation is crucial to prevent thermal degradation and enhance battery lifespan. Nanofluids, containing nanoparticles like graphene or carbon nanotubes, exhibit superior thermal conductivity, making them ideal candidates for improving the efficiency of BTMS.

The use of nanofluids in BTMS enhances the heat transfer capabilities within battery packs, contributing to more uniform temperature distribution and reducing the risk of thermal runaway events. By incorporating these advanced fluids into the cooling systems of batteries, engineers can manage heat more effectively, mitigating the impact of temperature fluctuations on battery performance. This is particularly important for EVs operating in diverse environmental conditions, where temperature extremes can influence battery efficiency and overall vehicle range. The application of nanofluids in BTMS not only aids in maintaining optimal operating temperatures but also promotes enhanced energy storage and delivery, addressing critical challenges in the advancement of electric mobility.

Moreover, in grid-scale energy storage systems that rely on large battery banks, the implementation of nanofluids in thermal management becomes crucial for maintaining system reliability and longevity. The improved thermal conductivity of nanofluids allows for efficient heat removal from batteries during charging and discharging cycles, reducing the likelihood of thermal stress-induced degradation. This has significant implications for the integration of renewable energy sources into the grid, as effective thermal management ensures the stability and longevity of energy storage infrastructure. The engineering application of nanofluids in BTMS thus contributes to the advancement of sustainable energy solutions by enhancing the performance and reliability of battery systems across diverse applications.

1.4.3 Biomedical and drug delivery applications

The application of nanofluids in the biomedical field is for improved diagnostics, drug delivery, imaging, and therapeutic interventions. As technology advances, nanofluids are likely to play an increasingly integral role in shaping the future of medical science and healthcare. In regenerative medicine, nanofluids are being explored for their role in tissue engineering. Nanofluids can serve as carriers for cells, growth factors, and other bioactive molecules, facilitating the regeneration of damaged tissues. This has implications for addressing injuries and degenerative conditions by promoting tissue repair and regeneration. In recent years, one notable application has been in the development of nanofluid-based contrast agents for medical imaging modalities like magnetic resonance imaging (MRI) and computed tomography (CT). Nanofluids, containing super-paramagnetic nanoparticles or iodine-based nanoparticles, can enhance the contrast and sensitivity of imaging techniques, enabling more accurate and detailed visualization of tissues and organs.

In drug delivery, nanofluids play a pivotal role in improving the efficiency and targeted delivery of pharmaceutical compounds. Nanofluids, often encapsulating drugs within nanoparticles, can be engineered to release therapeutic agents at specific sites within the body. This targeted drug delivery minimizes side effects and enhances the therapeutic efficacy of medications. Additionally, the unique properties of nanofluids, such as their ability to penetrate biological barriers, enable the delivery of drugs to previously inaccessible areas, opening up new possibilities for treating diseases like cancer.

The future applications of nanofluids in biomedicine also include advancements in diagnostics and biosensing. Nanofluids with tailored properties can be employed in biosensors for the detection of specific biomolecules or pathogens. This has the potential to revolutionize disease diagnosis, providing rapid and accurate results. Furthermore, the use of nanofluids in theranostics, a field combining therapy and diagnostics, holds promise for the simultaneous treatment and monitoring of diseases.

1.4.4 Nanofluids for hyperthermia treatment applications

The application of nanofluids in hyperthermia treatment represents a groundbreaking approach in the field of cancer therapy. Hyperthermia involves raising the temperature of targeted tissues to induce either direct cytotoxic effects or sensitize the cells to radiation and chemotherapy. Nanofluids, consisting of nanoparticles dispersed in a carrier fluid, offer a unique advantage in precisely delivering and concentrating heat within cancerous tissues. Iron oxide nanoparticles possess excellent magnetic properties, allowing for the efficient conversion of external electromagnetic energy into heat through mechanisms like magnetic hyperthermia. When these nanoparticles are introduced into the tumor site and exposed to an alternating magnetic field, they generate controlled and localized heat, effectively raising the temperature within the cancerous cells. In this approach, the nanofluid containing gold nanoparticles is administered to the tumor site, and upon exposure to near-infrared light, the nanoparticles convert light energy into heat, selectively heating and destroying cancer cells. The controlled and targeted nature of nanofluid-mediated hyperthermia offers a promising avenue for cancer treatment, minimizing damage to healthy tissues while maximizing the therapeutic impact on malignant cells. This method has great potential for improving the efficacy and precision of cancer therapies [18–20].

1.4.5 Improved lubrication application

Enhancing the efficiency and longevity of machinery across various industries nanofluids have been used for improved lubrication. The integration of nanomaterials into lubricants has opened up new opportunities to address longstanding challenges associated with friction, wear, and heat generation in mechanical components. Table 1.3 shows the nanofluids application in improved lubrication [21–23].

1.4.6 Nanofluids for cutting-edge manufacturing processes

Nanofluids have emerged as a transformative element in cutting-edge manufacturing processes, offering a spectrum of properties that enhance efficiency, precision, and durability. In machining and grinding operations, the integration of nanofluid-based cutting fluids has revolutionized traditional lubrication systems by providing superior cooling efficiency and reducing friction, consequently extending tool life and improving surface finishes [21,22]. Additive manufacturing, or 3D printing, benefits from nanofluids as well, particularly in achieving improved lubrication for moving parts and enhanced thermal management within printed components. Electrospinning techniques harness the enhanced thermal conductivity of nanofluids to create advanced fibers for use in textiles and composite structures, enabling the development of materials with superior thermal management capabilities. Additionally, the utilization of nanofluids in surface coating and deposition processes results in coatings with enhanced wear

Table 1.3 Nanofluids application for improved lubrication [21–23]

Type of nanofluid	*Application area*	*Current applications*	*Future applications*
Metallic nanoparticles (e.g., copper)	Automotive lubrication	Enhanced lubrication in internal combustion engines, reducing friction and wear.	Advanced lubrication systems in EVs, improving efficiency and reducing energy consumption.
Carbon-based nanotubes	Industrial machinery	Heavy machinery and manufacturing equipment, increasing durability and reducing maintenance needs.	Aerospace applications, where nanotube-infused lubricants could enhance efficiency and longevity in aircraft components.
Ceramic nanoparticles (e.g., alumina)	Bearings and gears	Ceramic nanofluids for lubrication in bearings and gears, reducing friction and heat generation.	Potential future applications may extend to nanofluid-infused lubricants for robotics and precision machinery, improving overall performance and reliability.
Polymer nanocomposites	Marine applications	Nanocomposites in lubricating ship components, reducing wear and improving fuel efficiency.	Potential future applications may involve underwater robotics, where nanocomposite-based lubricants could enhance performance and longevity in marine exploration.

resistance and improved heat dissipation, making them crucial for applications in aerospace components and electronic devices. The incorporation of nanofluids into cutting-edge manufacturing techniques exemplifies the versatility and transformative potential of nanotechnology, driving advancements across various industrial sectors (Table 1.4) [19,23].

1.4.7 Challenges, current research, and future prospects of nanofluids

Nanofluids hold immense applications in mechanical, electrical, automobile, and electronics engineering. It also faces several challenges; one significant challenge is the stability of nanofluids over extended periods. Nanoparticles tend to agglomerate, leading to diminished performance and potential clogging in practical applications. Achieving long-term stability and preventing particle sedimentation are critical aspects that require dedicated research efforts. Moreover, the scalability and cost-effectiveness of nanofluid production remain challenges. The synthesis of uniform and

Table 1.4 Application of nanofluids in cutting-edge manufacturing processes [21,24,25]

Manufacturing technique	*Nanofluid properties*	*Applications/advancements*
Electrospinning	Enhanced thermal conductivity	Nanofluid-infused electrospun fibers for advanced thermal management materials in textiles and composite structures.
Additive manufacturing (3D printing)	Improved lubrication, thermal conductivity	Integration of nanofluids in 3D printing processes for enhanced lubrication of moving parts and improved thermal management in printed components.
Machining and grinding	Cooling efficiency, friction reduction	Nanofluid-based cutting fluids for improved cooling and reduced friction in machining processes, enhancing tool life and surface finish.
Surface coating/ deposition	Enhanced wear resistance, improved heat dissipation	Nanofluid-based coatings for increased wear resistance and superior heat dissipation in applications like aerospace components and electronic devices.
Nanocomposite materials	Strengthening, improved thermal properties	Incorporation of nanofluids in composite materials for enhanced mechanical strength and improved thermal conductivity, advancing the development of lightweight and high-performance materials.
Microfluidics and lab-on-a-chip	Precise fluid control, thermal management	Nanofluid-based cooling systems in microfluidic devices for precise temperature control, facilitating advancements in lab-on-a-chip technologies and microscale manufacturing.

stable nanofluids on an industrial scale without compromising cost efficiency is a complex task. Additionally, the potential toxicity of certain nanoparticles raises concerns regarding environmental and human health, demanding comprehensive studies to ensure the safe use of nanofluids in engineering applications [26,27].

Current research in the field is focused on addressing these challenges and expanding the understanding of nanofluid behavior. Efforts are directed toward developing innovative synthesis methods that enhance stability and scalability while minimizing production costs. Advanced characterization techniques, such as DLS and zeta potential measurements, are employed to gain insights into nanofluid behavior and improve stability. Researchers are also exploring novel nanoparticle coatings and surface modifications to mitigate agglomeration. Moreover, studies on the toxicity and environmental impact of nanofluids are crucial for establishing guidelines and standards to ensure their safe utilization.

In terms of future prospects, nanofluids hold the potential to revolutionize various engineering applications. Ongoing research is likely to yield breakthroughs in achieving long-term stability, scalability, and cost-effectiveness. The incorporation of nanofluids in advanced thermal management systems, energy storage devices, and medical applications is expected to become more widespread. Future developments may also witness tailored nanofluids designed for specific engineering requirements, further optimizing their performance. The exploration of sustainable and eco-friendly nanofluids, as well as the development of standardized protocols for their application, will be pivotal in unlocking the full potential of nanofluids in engineering applications [28–32].

1.5 CONCLUSION

From enhancing heat transfer in electronic devices to revolutionizing cancer treatment through hyperthermia, nanofluids have proven to be a versatile and innovative tool in the engineer's arsenal. The ability of nanofluids to manipulate thermal, mechanical, and optical properties at the nanoscale has opened doors to unprecedented achievements in manufacturing, energy, medicine, and beyond. Despite facing challenges like stability and scalability, ongoing research is addressing these issues, propelling the field toward broader applications and ensuring the safe implementation of nanofluids. As the quest for efficiency, precision, and sustainability continues, the remarkable properties of nanofluids are poised to play a pivotal role in shaping the technological landscape, offering solutions to complex problems and paving the way for a future where nanotechnology becomes an integral part of everyday advancements.

REFERENCES

1. Choi, S. U.-S., & Eastman, J. A. (1995). Enhancing thermal conductivity of fluids with nanoparticles. *ASME International Mechanical Engineering Congress and Exposition*, 231, 99–105.
2. Nath, P., & Chopra, K. L. (1974). Thermal conductivity of copper films. *Thin Solid Films*, 20, 53
3. Awasthi, M. K., Dutt, N., Kumar, A., & Kumar, S. (2023). Electrohydrodynamic capillary instability of Rivlin-Ericksen viscoelastic fluid film with mass and heat transfer. *Heat Transfer*, 1–19. doi:10.1002/htj.22944
4. Meena, C. S., Kumar, A., Roy, S., Cannavale, A., & Ghosh, A. (2022). Review on boiling heat transfer enhancement techniques. *Energies*, 15, 5759. doi:10.3390/en15155759.
5. Kumar, A., Singh, V. P., Meena, C. W. S., & Dutt, N. (2023). *Thermal Energy Systems: Design, Computational Techniques and Applications.* Taylor & Francis, Boca Raton. doi:10.1201/9781003395768.

6. Reddy, P. N., Verma, V., Kumar, A., & Awasthi, M. (2023). CFD simulation and thermal performance optimization of flow in a channel with multiple baffles. *Journal of Heat and Mass Transfer Research*, 10(2), 257–268. doi:10.22075/JHMTR.2023.31108.1458.
7. Yadav, D., Awasthi, M. K., Kumar, A., & Dutt, N. (2024). Swirling capillary instbility of Rivlin-Ericksen liquid with heat transfer and axial electric field. *Physics* (in press).
8. Awasthi, M. K., Dutt, N., Kumar, A., & Hedau, A. (2024). Introduction to mathematical and computational methods. In: Awasthi, M. K., Kumar, A., Dutt, N., & Singh, S. (Eds) *Computational Fluid Flow and Heat Transfer: Advances, Design, Control and Applications*. CRC Press, Boca Raton, FL, pp. 1–20.
9. Awasthi, M. K., Kumar, A., & Dutt, N. (2024). Modeling Rayleigh-Taylor instability in nanofluid layers. In: Awasthi, M. K., Kumar, A., Dutt, N., & Singh, S. (Eds) *Computational Fluid Flow and Heat Transfer: Advances, Design, Control and Applications*. CRC Press, Boca Raton, FL, pp. 249–262.
10. Awasthi, M. K., Kumar, A., Dutt, N., & Singh, S. (Eds). (2024). *Computational Fluid Flow and Heat Transfer: Advances, Design, Control and Applications*. CRC Press, Boca Raton, FL, pp. 1–276.
11. Dutt, N., Hedau, A. J., Kumar, A., Awasthi, M. K., Singh, V. P., & Dwivedi, G. (2023). Thermo-hydraulic performance of solar air heater having discrete D-shaped ribs as artificial roughness. *Environmental Science and Pollution Research*. doi:10.1007/s11356-023-28247-9.
12. Dutt, N., Hedau, A., Awasthi, M. K., Kumar, A., & Meena, C. W. (2024). Thermo-hydraulic performance investigation of solar air heater duct having staggered D-shaped ribs: Numerical approach. *Heat Transfer*. doi:10.1002/htj.22998.
13. Kumar, A., Kumar, A., & Kumar, A. (2023). *Laser Based Technologies for Sustainable Manufacturing*. Taylor & Francis, Boca Raton. doi:10.1201/9781003402398.
14. Verma, V., Thangavel, S., Dutt, N., & Kumar, A. (Eds.) (2024). *Highly Efficient Thermal Renewable Energy Systems: Design, Design, Optimization and Applications*. CRC Press, Boca Raton, FL.
15. Singh, V. P., Jain, S., Kumar, A., Mishra, S., & Sharma, N. K. (2022). Heat transfer and friction factor correlations development for double pass solar air heater artificially roughened with perforated multi-V ribs. *Case Studies in Thermal Engineering*, 39, 102461. doi:10.1016/j.csite.2022.102461.
16. Prajapati, Y. K., Gupta, P. K., & Kumar, A. (2011). Free surface undulation and air entrainment in a rectangular tank. *International Journal of Applied Engineering Research*, 6(4), 409–419.
17. Kundu, A., Kumar, A., Dutt, A., Meena, C. S., & Singh, V. P. (2023). Introduction to thermal energy resources and their smart applications. In: Kumar, A., Singh, V. P., Meena, C. S., & Dutt, N. (Eds) *Thermal Energy Systems: Design, Computational Techniques and Applications*. CRC Press, Boca Raton, FL, pp. 1–15. doi:10.1201/9781003395768-1.
18. Choudhary, R., Kumar, R., & Mathur, G. N. (2019). Nanofluids for heat transfer enhancement in electronic devices: A review. *Journal of Applied Physics*, 125(10), 103104.

19. Li, X., Hu, J., & Bae, D. (2020). Advances in nanofluid applications in cancer treatment through hyperthermia. *Nanomedicine: Nanotechnology, Biology, and Medicine*, 27, 102203.
20. Wang, L., Xie, W., & Qiu, X. (2021). Manipulating thermal, mechanical, and optical properties of nanofluids for advanced engineering applications. *Nano Letters*, 21(1), 30–41.
21. Ghazanfari, M. H., & Fattahi, E. (2022). Challenges and prospects of stability in nanofluids: A comprehensive review. *Journal of Nanofluids*, 11(1), 1–17.
22. Lee, J. H., Choi, S. U. S., & Li, S. (2019). Scalability and cost-effectiveness of nanofluid production: Current status and future directions. *International Journal of Heat and Mass Transfer*, 140, 1153–1167.
23. Das, A. K., & Pradeep, T. (2020). Toxicity and environmental impact of nanofluids: A critical review. *Environmental Science and Technology*, 54(19), 11411–11426.
24. Ding, Y., & Chen, H. (2021). Tailored nanofluids for specific applications: Recent developments and future perspectives. *Journal of Colloid and Interface Science*, 588, 267–279.
25. ASTM International. (2019). *Standardization of Nanofluid Application Protocols: ASTM E2567-19*. ASTM International, West Conshohocken, PA.
26. Wang, Z., & Zhang, X. (2020). Role of nanofluids in manufacturing: A comprehensive review. *InternationalJournalofAdvancedManufacturingTechnology*, 107(3–4), 1395–1410.
27. Ferrari, M. (2021). Future prospects of nanofluids in medicine: Challenges and opportunities. *Nature Reviews Materials*, 6(2), 91–104.
28. Akbar, N. S., & Fard, M. A. (2019). Nanofluids in hyperthermia cancer therapy: A systematic review. *International Journal of Hyperthermia*, 36(1), 102–116.
29. Zhang, Y., & Wei, J. (2020). Stability challenges in nanofluid-based heat transfer applications: A critical analysis. *Applied Thermal Engineering*, 167, 114716.
30. Gupta, A., & Biswas, K. (2021). Nanofluid-based coatings for wear resistance: A review. *Tribology International*, 160, 106954.
31. Thakur, A., & Das, P. K. (2022). Standardized protocols for nanofluid application: Challenges and prospects. *Journal of Nanoparticle Research*, 24(2), 47.
32. Kim, J., & Lee, C. (2023). Sustainable and eco-friendly nanofluids for engineering applications: A review. *Sustainable Materials and Technologies*, 30, e00307.

Chapter 2

Nanoparticle synthesis and characterisation for nanofluid development

S. Senthilraja, R. Gangadevi, and M. Baskaran

2.1 INTRODUCTION

The development in thermal science and engineering has generated a strong fascination in the creation of small-scale liquid flow systems. Because of its small size and high surface-to-volume ratio, it is necessary to use efficient heat transfer techniques to achieve the goal of improving its performance. Over the course of the last few decades, a great number of attempts have been made to raise the rate of heat transfer in the thermal management system in order to reduce the amount of time required for heat transfer and to improve the efficiency of the system. There have been a number of approaches using both solids and fluids in various applications, but their effectiveness has been limited due to issues including high costs, limited space, limited availability, and poor heat transmission characteristics. Hence, the addition of solid particles to heat transfer fluids has sparked the attention of academics from several disciplines due to its abnormal improvement of thermal conductivity. Fluids with micron-sized particles increase heat transmission, but they have limited utility due to issues including high sedimentation rates, corrosion and erosion, and pumping power requirements, among others. As a result, there has been a dramatic rise in the demand for heat transfer fluids that exhibit low levels of agglomeration, pumping power, sedimentation, etc., while having a high heat transfer rate.

To address these challenges, researcher SUS Choi has developed a novel heat transfer fluid at Argonne National Laboratory in the USA by suspending nanometre-sized particles into a basic heat transfer fluid [1]. Afterward, this liquid is referred to as nanofluid and used as a heat transfer medium in many thermal systems. The term "nanofluid" refers to the basic heat transfer fluids that include metal or metal oxide particles that are between 1 and 100 nm in size. The use of nanoparticles in base fluids greatly enhances the rate of heat transfer and has the potential to serve as highly efficient heat transfer fluids with improved heat transfer capabilities. These fluids provide many benefits, including increased thermal conductivity, enhanced stability, reduced pumping power requirements, and less material corrosion and erosion. Numerous parameters, such as nanoparticle size and shape,

DOI: 10.1201/9781003494454-2

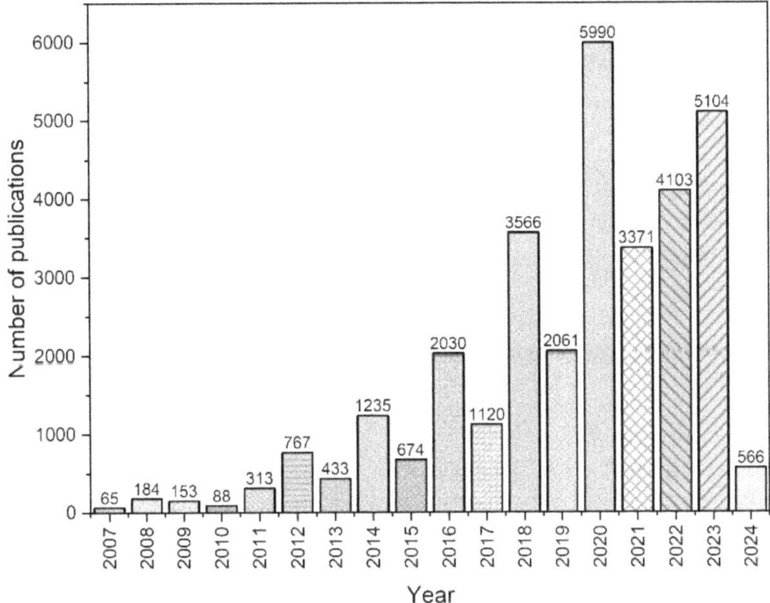

Figure 2.1 Variations of number of publications related to nanoparticle synthesis with year (as per Scopus database).

base fluid heat transfer capabilities, sonication time, nanoparticle dispersion, and base fluid properties, determine the characteristics of nanofluids. Therefore, the process for producing nanofluids is crucial in determining the heat transfer characteristics of the fluids. Before 2000, only a few researchers conducted studies on nanofluids and published their findings in reputable scientific publications. However, during the last 20 years, there has been a progressive growth in the number of research publications pertaining to the synthesis and characterisation of nanoparticles, as seen in Figure 2.1. In this context, the investigation of various methods for synthesising nanoparticles and their subsequent characterisation has significant importance. This chapter is largely aimed to present information about several nanoparticles production processes with its benefits and demerits. This chapter also will include comprehensive information on the characterisation methods used for nanofluids.

2.2 NANOPARTICLE SYNTHESIS

A nanoparticle is a particle with a particle diameter ranging from 1 to 100 nm. The name nanoparticle originates from the Greek word 'Nανο' that denotes smallness or dwarfishness. The fast development and

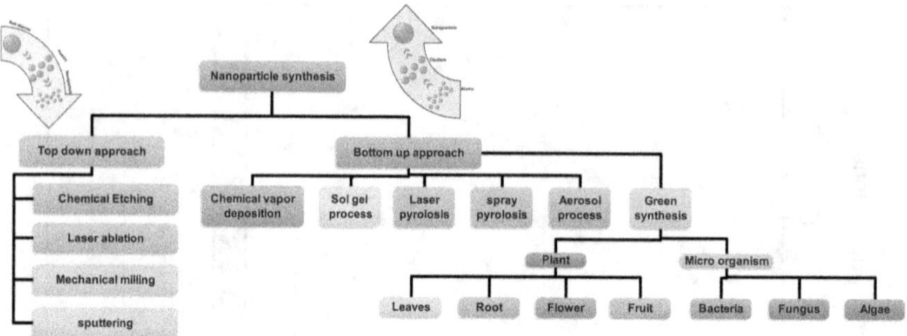

Figure 2.2 Classification of nanoparticle synthesis techniques.

technical advancements in the area of nanotechnology have sparked significant interest within the scientific community. Because of their tiny size, large surface-to-volume ratio, and high surface energy, nanoparticles demonstrate improved physical, thermal, and optical characteristics. As a result, the utilisation of various nanoparticles in diverse sectors such as energy, cosmetics, food industry, and electronics industry has seen a significant surge in recent years.

The top-down technique and the bottom-up technique are the two most frequent methods that have been used in the manufacture of nanoparticles. In the top-down technique, the size of the bulk material is reduced to particles that are nanometres in size by the utilisation of mechanical and chemical processes. Whereas in the bottom-up technique, nanosized particles are generated by gradually constructing the material from its fundamental components, such as atoms, molecules, or clusters. The detailed nanoparticle synthesise techniques are presented in Figure 2.2.

2.2.1 Top-down approach

The top-down technique is a very simple technique that involves the conversion of bulk material into nanoparticles via the use of physical methods. Although this technology is capable of generating a substantial number of nanomaterials rapidly, it presents significant challenges in terms of controlling the size, shape, and uniformity of the nanoparticles. Chemical etching, laser ablation, mechanical milling, and sputtering techniques are often used as top-down methods for nanoparticle production in various applications. This section presents comprehensive information on several top-down techniques, including their advantages and disadvantages.

2.2.1.1 Mechanical milling process

Mechanical milling has been extensively used over the last several decades as a very effective approach of producing a huge amount of nanoparticles

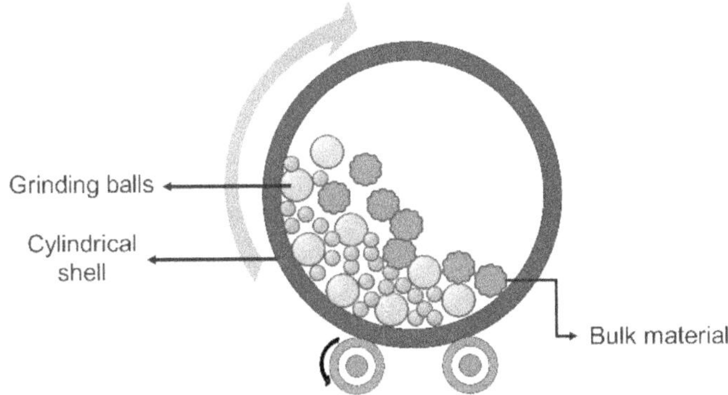

Figure 2.3 Schematic diagram of the ball milling process.

[2,3]. This process is a kind of grinding technique that transforms large quantities of materials into very small particles, often in the nanoscale size range. This technique involves a hollow cylindrical shell that spins around its horizontal or slightly inclined axis. To prevent erosion on the inside surface of the cylindrical shell, it is covered with a durable substance such as manganese steel or rubber lining. The cylindrical shells are packed with little metallic balls composed of stainless steel, rubber, and ceramic materials. The schematic diagram of the ball milling process is presented in Figure 2.3.

In this procedure, the material is introduced into the cylinder and fills the necessary quantity of balls before operating the machine at the desired velocity. During the process of moving the balls, kinetic energy is imparted to the milled material, causing the chemical bonds to dissolve and the particles to divide into smaller pieces. Once the necessary particle size is achieved, the machine will be halted and the nanoparticles are removed from the cylindrical shell. This process will be carried out in four different stages. During the initial stage, the bulk materials are colloids with the ball and change the particle shape and size. The powdered components reduce the propagation distance to the micrometre range in the second stage. In the third stage, the particle's microstructure exhibits a higher level of homogeneity at the microscopic size compared to the beginning and intermediate phases. The final stage is the mechanical alloying process. At this stage, the powder particles have a highly distorted metastable shape. The physical and morphological properties of the prepared nanoparticles mainly depend on the milling speed, ball material, type of milling, duration of the milling process, etc.

The technique is characterised by its simplicity, cleanliness, low energy use, minimal space requirements, and low maintenance costs. Simultaneously, it exhibits a slow pace and generates increased friction, producing a harsh, discordant mixture of sounds.

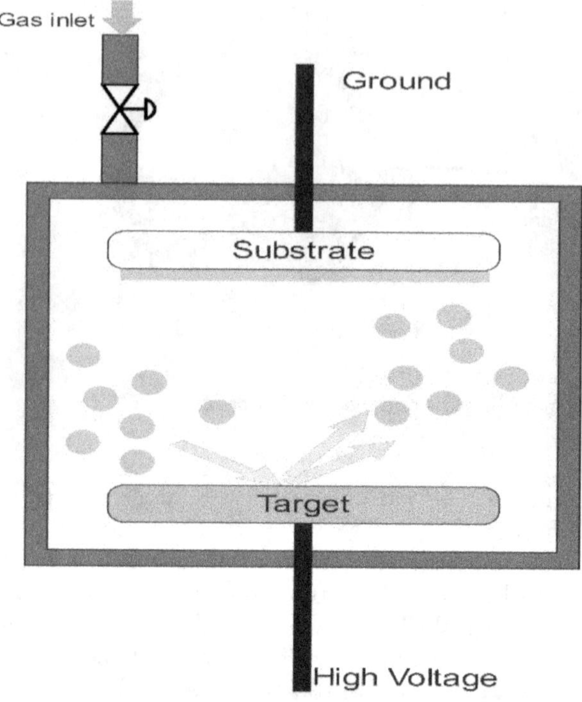

Figure 2.4 Schematic diagram of the sputtering process.

2.2.1.2 Sputtering process

It is a kind of physical vapour deposition and was pioneered by M. Blocher. This technique occurs inside a vacuum chamber under a low-pressure atmosphere filled with Argon gas. The schematic diagram of the sputtering process is depicted in Figure 2.4.

This process involves using plasma energy to take atoms from the target material and deposit them onto the substrate. Plasma is generated by applying a potential difference, causing the Ar+ ions to form under high pressure. These ions then collide with the target material, causing ionisation of pure gas and transferring their energy to tear apart one atom. This atom is then propelled towards the substrate with sufficient energy. This method has great potential for target cooling and pre-cleaning, as well as a consistent and steady deposition rate. Additional study is needed to address the primary shortcomings of this technique, which include a sluggish deposition rate and reduced homogeneity due to heat treatment, despite its numerous benefits.

2.2.1.3 Laser pyrolysis

The technique of laser pyrolysis involves heating the reactant gases using a continuous wave CO_2 laser. This causes molecular breakdown, which forms vapours that start nucleation and eventually grows nanoparticles. Compared to conventional vapour phase methods, this method offers faster heating, which leads to faster nucleation and faster quenching of particle development within a few milliseconds.

This process incorporates the treatment of the substance in either solid, liquid, or gas form. A laser is used to expose the substance to radiation. The substance absorbs the energy emitted by the laser, resulting in a fast rise in temperature. The elevated temperature generated by the laser causes the molecules to break down into smaller pieces, which may be in the form of gases, liquids, or solid leftovers, depending on the material's properties. This method has high temperatures, rapid heating rates, and quick reaction durations. Due to the appealing advantages, many researchers have directed their attention to synthesising nanoparticles utilising the laser pyrolysis technique. Malekzadeh et al. [4] synthesised zinc nanoparticles using the laser pyrolysis method with low-cost ultrasonic spray delivery of precursor. Lungu et al. [5] used the laser pyrolysis technique to prepare iron oxide nanoparticles under high and low gas flow rates. The nanoparticles are subsequently analysed using various techniques including dispersive light scattering analysis, structure-X-ray diffraction (XRD), elemental composition-energy-dispersive X-ray spectroscopy (EDS), X-ray photoelectron spectroscopy (XPS), transmission electron microscopy (TEM), and selected-area electron diffraction to characterise their properties and morphology.

2.3 NANOFLUID PREPARATION

According to the comparison provided by Senthilraja et al. [6], typical heat transfer fluids have inadequate heat transfer properties. Recently, researchers have shown that the heat transfer rate of traditional fluids may be improved by incorporating nanoparticles into them and named as "Nanofluid". Nanofluid is a kind of colloidal suspension that consists of nanoparticles dispersed within a base fluid. Generally, Nanofluid may be synthesised by either a single-step or a two-step technique. In the single-step procedure, nanoparticles are synthesised using one of the techniques discussed above and then immediately distributed in the base fluid. Due to the simultaneous execution of particle preparation and dispersion, the process becomes convenient for storage, transit, and preparation. Therefore, this process is optimal for producing a tiny amount of nanofluid with little clumping and a high rate of dispersion. Vapour deposition, laser ablation, and submerged arc methods have emerged as the most often used single-step techniques for nanofluid synthesis in recent years.

In 2001, Choi and Eastman developed a novel method to prepare the nanofluid using the vapour deposition method [7]. The schematic diagram of the vapour deposition nanofluid preparation method is shown in Figure 2.5. This technique involves the direct evaporation and subsequent condensation of nanoparticulate elements inside the base fluid. The spherical vessel is filled with a thin layer of base fluid due to the centrifugal force exerted by the revolving disc. The substance is subjected to heating and evaporation inside the crucible while being exposed to an inert gas at low pressure. Using the whirling water, the vapours of the raw materials are condensed, and then they are combined with the base fluid.

Laser ablation refers to the process of eliminating material from a surface by the use of laser light. The schematic illustration of laser ablation method is presented in Figure 2.6. In this method, the creation of nanoparticles is

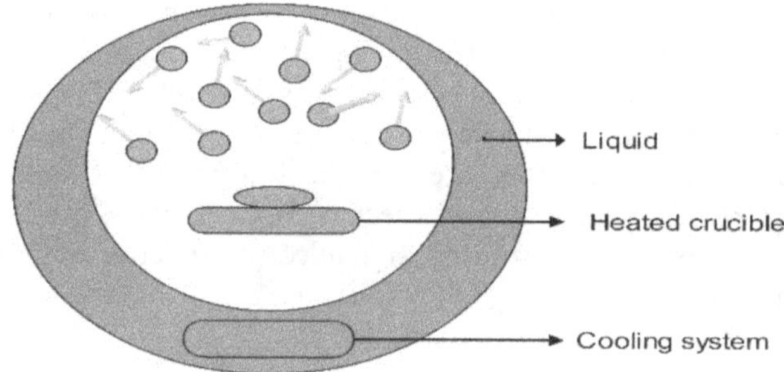

Figure 2.5 Schematic diagram of vapour deposition single-step nanofluid preparation method.

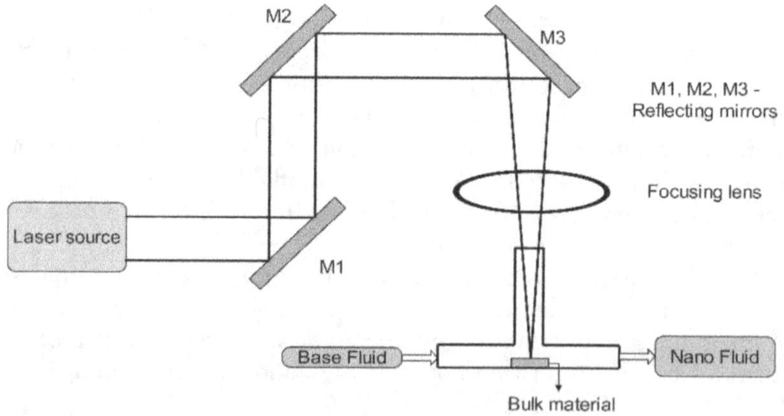

Figure 2.6 Schematic illustration of laser ablation method.

accomplished via the use of this technique by contacting the lasing pulses with the surface of the substance of interest while concurrently dispersing the base fluid. It consists of a laser source, scanning mirrors, a focusing lens, a vacuum chamber, and target material. The Co_2 laser and Nd-YAG laser, ArF excimer laser, or XeCl excimer laser are widely used in the laser ablation method. Laser lights are reflected and concentrated by using mirrors and a focusing lens and targeted on the material. Consequently, the substance located on the outer layer of the target undergoes vapourisation and transforms into a laser plume. The vaporised components are combined with the fluid to create a colloidal suspension, known as a nanofluid, which is then kept in a container.

The two-step nanofluid preparation method is a very popular technique for producing large quantities of nanofluid. In this method, two different steps such as nanoparticle preparation and nanoparticle dispersion are carried out in two different phases. In the first phase, the nanoparticles, nanotubes, nanofibers, or nanorods are synthesised using different techniques sol-gel method, microemulsion, etc. The required quantity of nanoparticles is mixed with the base fluid by using a stirring process and sonication process in the second phase. The schematic diagram of the two-step nanofluid preparation method is presented in Figure 2.7.

Although the two-step nanofluid manufacturing strategy may generate a substantial quantity of nanofluid, it results in higher particle agglomeration and less uniform particle dispersion compared to the single-step method. To address this problem, the use of surfactant during the nanofluid preparation process might be used. The pictorial view of prepared CuO/water and Al_2O_3/water nanofluids using a two-step preparation method is presented in Figure 2.8a and b respectively.

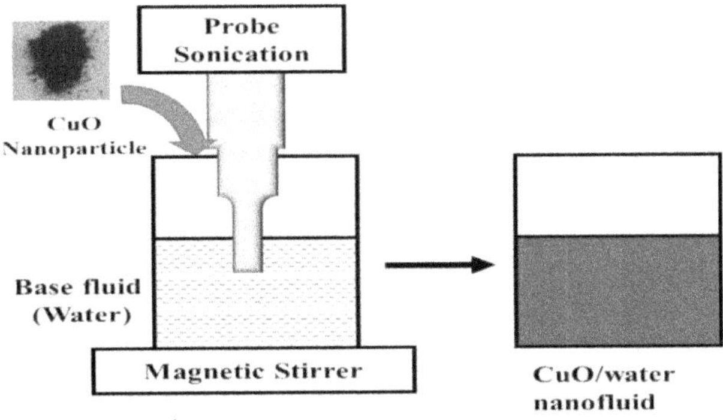

Figure 2.7 Schematic representation of two-step nanofluid preparation method.

Figure 2.8 Pictorial view of prepared nanofluids using two-step preparation method (a) CuO/water and (b) Al_2O_3/water nanofluid.

2.4 CHARACTERISATION OF NANOPARTICLES AND NANOFLUIDS

2.4.1 Morphology/microstructure analysis

The examination of the arrangement and form of nanoparticles is referred to as Morphology. This term is derived from two Greek roots, namely 'Morph' (meaning shape) and 'ology' (meaning the study of something). Scanning Electron Microscope (SEM) analysis, Transmission Electron Microscope (TEM) analysis, and X-Ray powder diffraction (XRD) analysis are the predominant methods for analysing the morphology of nanoparticles.

2.4.1.1 Transmission electron microscope (TEM)

TEM is a method used to examine the internal composition of nanoparticles by directing a high-energy electron beam through them. In 1933, German scientist Ernst Ruska pioneered the development of the first TEM. It offers pictures with a higher resolution (specifically, 0.0001 μm) compared to SEM. Despite the better resolution of TEM images (particularly, 0.0001 μm) compared to SEM, the time-consuming and challenging sample preparation procedure has limited its use in analysing the internal structure of nanoparticles.

The system comprises an electron gun, a magnetic condensing lens, and apertures. An electron beam is generated by the electron gun and traverses the specimen using the magnetic condensing lens. Aperture eliminates the electrons that are deflected at large angles and are unwanted. The beam undergoes both transmission and diffraction, with the extent of each determined by the angle of incidence. The two beams are merged at the E-wald sphere to create the picture. The TEM image of CuO and Al_2O_3 nanoparticles used in our research is given in Figures 2.9 and 2.10, respectively.

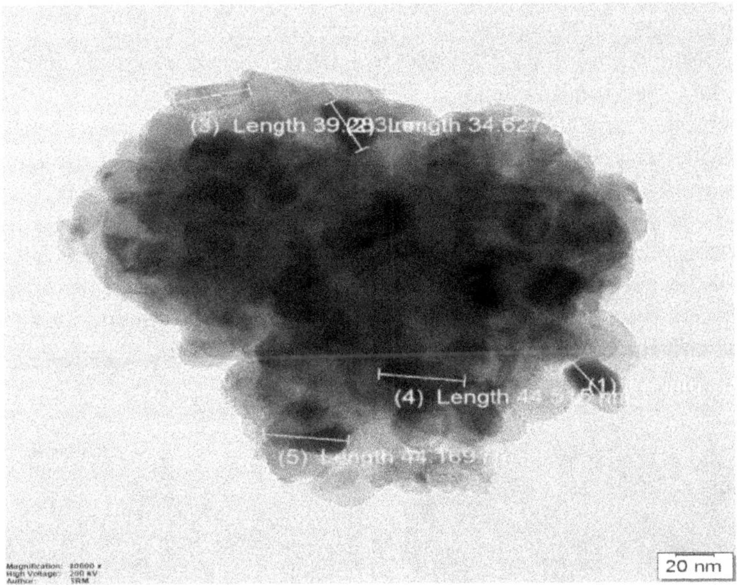

Figure 2.9 TEM image of CuO nanoparticles.

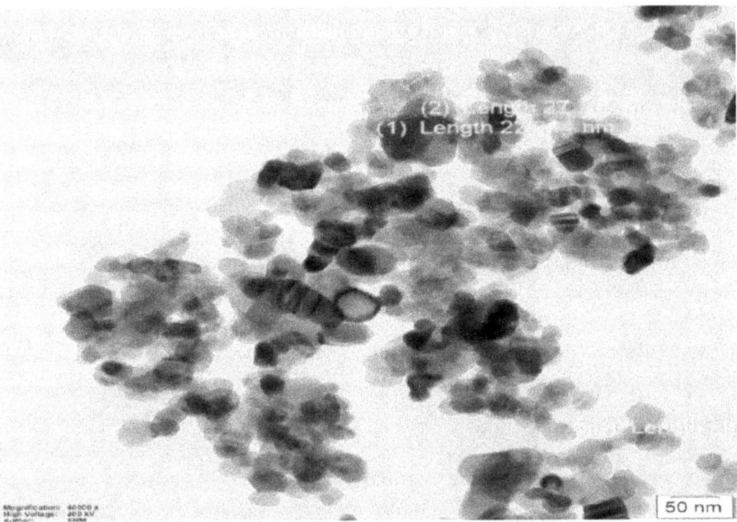

Figure 2.10 TEM image of Al_2O_3 nanoparticles.

2.4.1.2 *Scanning electron microscope*

Scanning electron microscope (SEM) is a method used to examine the structure of nanoparticles by taking highly magnified images utilising electrons. It consists of electron gun, magnetic condensing lens, scanning coil, and electron detector. An electron gun generates a high-energy electron beam, which is then condensed by the magnetic condensing coil and then travels through the scanning coil. The electron detector identifies and captures the secondary electrons, which are then transformed into an electrical signal and then sent into a cathode ray oscilloscope. The SEM images of prepared CuO/water and Al_2O_3/water nanofluids before and after sonication are presented in Figures 2.11 and 2.12, respectively.

These photos reveal that there are a greater number of clusters of nanoparticles after 1 h of sonication compared to 4 h of sonication. After 4 h of sonication, the particles are evenly distributed in the base fluid, resulting in a decrease in sedimentation. Therefore, the duration of sonication is a crucial factor with the creation of nanofluids.

Figure 2.11 SEM Images of CuO/water nanofluids (a) 1 h sonication and (b) 4 h sonication.

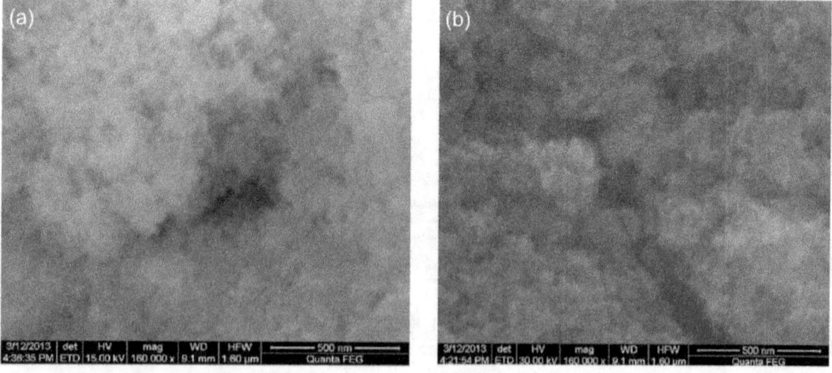

Figure 2.12 SEM Images of Al_2O_3/water nanofluids (a) 1 h sonication and (b) 4 h sonication.

2.4.1.3 X-ray diffraction

X-ray diffraction (XRD) is a commonly used technique for investigating the morphology and dimensions of nanoparticles and was discovered by Max von Laue, in 1912. This approach also provides information on the lattice parameter and grain size.

The XRD technique utilises the phenomenon of X-ray scattering caused by the orbital motion of electrons inside the atomic nucleus upon interaction with nanoparticles. An X-ray diffractometer includes the components of an X-ray tube, a sample holder, and an X-ray detector. The cathode ray tube creates X-rays and facilitates the production and acceleration of electrons towards the target material by adding voltage and blasting it with electrons. When incident rays contact with the target material and meet Bragg's law, they will undergo constructive interference. The diffracted rays are detected and quantified by scanning the sample throughout a range of 2θ angle. The XRD images of CuO and Al_2O_3 nanoparticles used for our research are presented in Figures 2.13 and 2.14, respectively.

2.4.1.4 Energy dispersive X-ray spectroscopy

Assessing the purity and detecting the presence of additional components in the synthesised nanoparticles is a crucial task that determines the characteristics of the nanoparticle. Many researchers use the energy-dispersive X-ray spectroscopy (EDS, EDX, or EDXA) technology in their study due to its effectiveness. The EDS system comprises an X-ray beam source, an X-ray

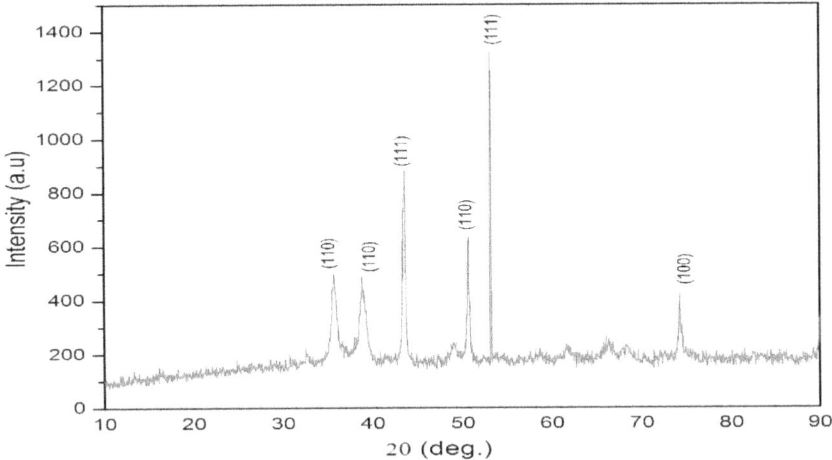

Figure 2.13 XRD image of CuO nanoparticle.

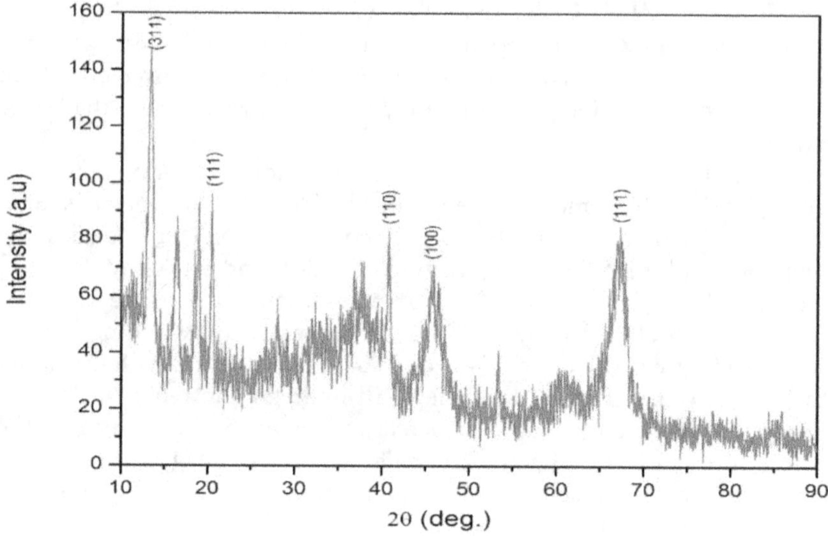

Figure 2.14 XRD image of Al_2O_3 nanoparticle.

detector, a pulse processor, and an analyser. The X-ray beam released by a solid sample interacts with the electron beam, resulting in the generation of X-rays that are determined by the specific features and composition of the elements inside the sample. This approach yields precise outcomes not only for the detection of elements but also for determining their concentration under optimal experimental circumstances. The EDS analysis findings for CuO and Al_2O_3 nanoparticles in this work are shown in Figures 2.15 and 2.16, respectively.

2.5 CONCLUSIONS

An overview of nanoparticle and nanofluids synthesis and characterisation techniques are discussed in this chapter. Some key points related to nanoparticles and nanofluids are summarised as follows:

1. Numerous nanoparticles are used for a variety of purposes; nevertheless, each nanoparticle has a unique set of features, and it is essential to investigate both the physical and morphological aspects of these nanoparticles.
2. Among different nanoparticle synthesis techniques, the researchers should consider many parameters such as the nature of nanoparticles, chemical reaction, and cost of the equipment before selecting the nanoparticle synthesis method.

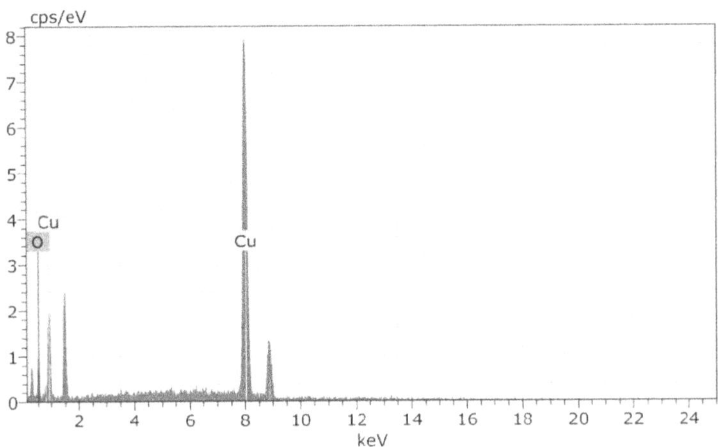

Figure 2.15 EDS image of CuO nanoparticle.

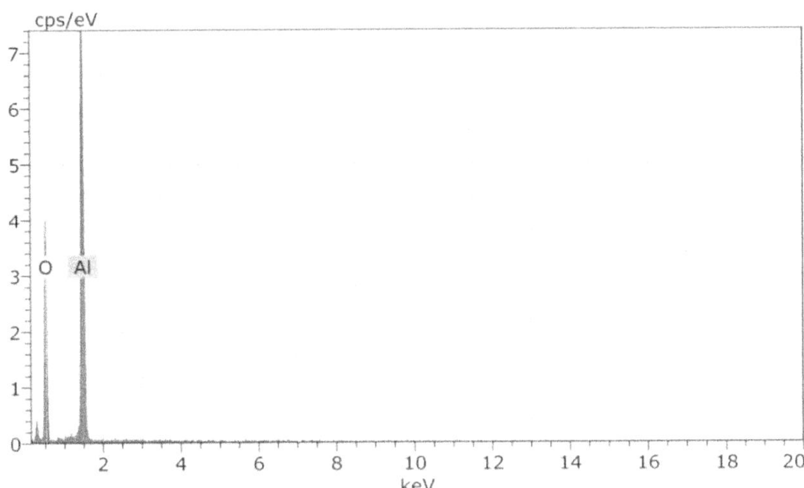

Figure 2.16 EDS image of Al_2O_3 nanoparticle.

3 According to the literature, while the single-step nanofluid production approach possesses certain advantages, many researchers have chosen the two-step nanofluid preparation method for their study.
4 Sedimentation and agglomeration are the primary concerns in the method of preparing nanofluids. To address these challenges, it is advisable to use appropriate sonication duration and surfactants.
5 Furthermore, it may be inferred that further investigation is necessary to develop nanoparticle and nanofluid production methods that are more cost-effective and have a reduced environmental footprint.

NOMENCLATURE

Al_2O_3: Aluminium oxide
CUO : Copper oxide
CVD : Chemical vapour deposition
EDS : Energy-dispersive X-ray spectroscopy
SEM : Scanning electron microscopy
TEM : Transmission electron microscopy
XPS : X-ray photoelectron spectroscopy
XRD : X-ray diffraction

REFERENCES

[1] Choi, S.U. and Eastman, J.A., 1995. *Enhancing Thermal Conductivity of Fluids with Nanoparticles* (No. ANL/MSD/CP-84938; CONF-951135-29). Argonne National Lab, Argonne, IL.

[2] De Carvalho, J.F., De Medeiros, S.N., Morales, M.A., Dantas, A.L. and Carriço, A.S., 2013. Synthesis of magnetite nanoparticles by high energy ball milling. *Applied Surface Science*, *275*, pp. 84–87.

[3] Wirunchit, S., Gansa, P. and Koetniyom, W., 2021. Synthesis of ZnO nanoparticles by Ball-milling process for biological applications. *Materials Today: Proceedings*, *47*, pp. 3554–3559.

[4] Malekzadeh, M., Rohani, P., Liu, Y., Raszewski, A., Ghanei, F. and Swihart, M.T., 2020. Laser pyrolysis synthesis of zinc-containing nanomaterials using low-cost ultrasonic spray delivery of precursors. *Powder Technology*, *376*, pp. 104–112.

[5] Lungu, I.I., Andronescu, E., Dumitrache, F., Gavrila-Florescu, L., Banici, A.M., Morjan, I., Criveanu, A. and Prodan, G., 2023. Laser pyrolysis of iron oxide nanoparticles and the influence of laser power. *Molecules*, *28*(21), p.7284.

[6] Senthilraja, S., Karthikeyan, M. and Gangadevi, R., 2010. Nanofluid applications in future automobiles: comprehensive review of existing data. *Nano-Micro Letters*, 2, pp. 306–310.

[7] Choi, S.U.S. and Eastman, J.A. Enhanced heat transfer using nanofluids. USA Patent 6,221,275, 24 April 2001.

Chapter 3

Experimental techniques for nanofluid testing and validation

Krishna Priyadarshini Das and Abhijit Pattnayak

3.1 INTRODUCTION

Nanofluids are colloidal suspensions of nanoparticles (NPs) in base fluids where nano tend to be nanometer-sized (~1–100 nm) and distributed in a liquid media like water, oil, or ethylene glycol (Alami et al. 2023). The introduction of NPs in the base fluid offers distinct and improved characteristics to the resultant nanofluid (Angayarkanni and Philip 2022). Nanofluids can include a variety of NPs), including metal oxides (e.g., alumina and titanium dioxide), metal NPs (e.g., copper and silver), carbon-based NPs (e.g., carbon nanotubes and graphene), and others (Sheikhpour et al. 2020). Nanofluids exhibit excellent thermal conductivity than the base fluid, making them potentially used in heat transfer and thermal management systems. In 1995, Choi from Argonne National Laboratory in the United States pioneered the concept of nanofluids and investigated their improved thermal properties (Alami et al. 2023).

Nanofluids with improved thermal conductivity, viscosity, specific heat, pH, optical properties, and thermal stability have garnered immense attention in a variety of scientific and industrial domains such as electronics cooling, energy systems, medical treatments, and manufacturing processes (Alami et al. 2023; Ali and Salam 2020), as shown in Figure 3.1. For example, it plays an important role in addressing heat dissipation challenges in electronics due to their better thermal conductivity. It provides an effective method for cooling electronic components, which improves overall device efficiency and reliability. Beyond electronics, nanofluids are employed in solar-thermal systems, where their improved heat transfer capabilities aid in increasing energy efficiency. In addition, nanofluids are being studied for their possible use in targeted hyperthermia for cancer therapy (Benos et al. 2022). Nanofluids have the potential to revolutionize medical therapies due to their ability to precisely regulate heat distribution to specific tissues (Ali and Salam 2020; Benos et al. 2022).

Nanofluids exhibit distinct characteristics that are primarily determined by five important parameters: heat transfer behavior, lubrication effects, particle attributes, thermo-fluid characteristics, and colloid properties

DOI: 10.1201/9781003494454-3

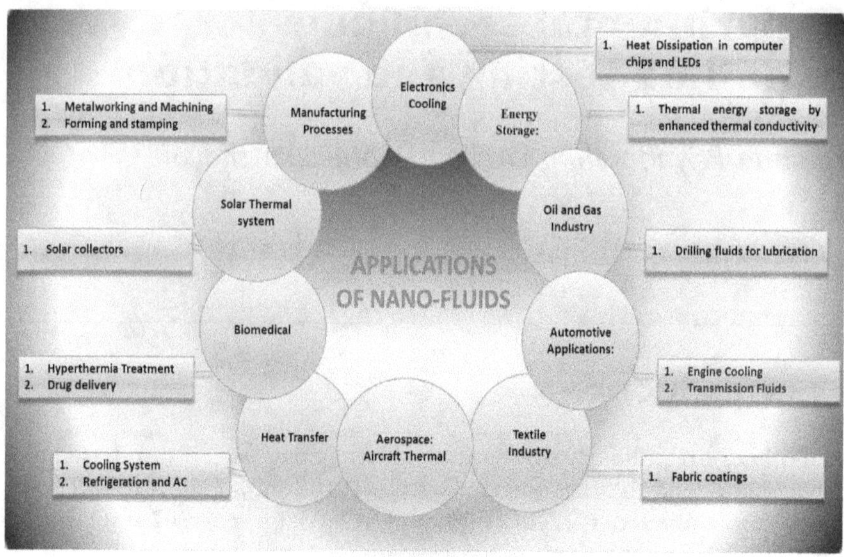

Figure 3.1 Schematic representation showing the various applications of nanofluids.

(Gupta et al. 2017). A few key elements in the realm of thermo-fluid properties include viscosity, specific heat, enthalpy, temperature, and density. The heat transfer factors comprised heat capacity, thermal conductivity, pressure drop, and Prandtl number. Size, shape, Brunauer–Emmett–Teller (BET) surface area analysis, BJH (Barrett-Joyner-Halenda pore volume analysis), and crystalline phase are various factors of particle-related characteristics. Meanwhile, Colloidal qualities include suspension stability, Zeta potential, and pH issues (Angayarkanni and Philip 2015). Finally, lubricant factors encompass viscosity, viscosity index, friction coefficient, wear rate, and extreme pressure. Additionally, nanofluids exhibit distinct physical characteristics from their basic fluids, including changes in density, specific heat, and viscosity, and these alterations contribute to an increased heat transfer coefficient, which surpasses the results obtained with simple thermal conductivity improvement, as demonstrated by several experimental tests (Ali and Salam 2020; Porgar et al. 2021).

However, the in-depth characterization of the aforementioned attributes of nanofluids is pivotal for exploiting their properties in emerging applications and thereby unlocking their potential for transformative innovations across various industries such as thermal engineering, electronics, and renewable energy, where efficient heat transfer and thermal management are paramount (Bahiraei and Hangi 2015). Furthermore, it is critical for customizing their qualities to specific applications, optimizing performance, and addressing the challenges. For example, modifying parameters such as particle size, dispersion quality, and stability allows for enhanced heat

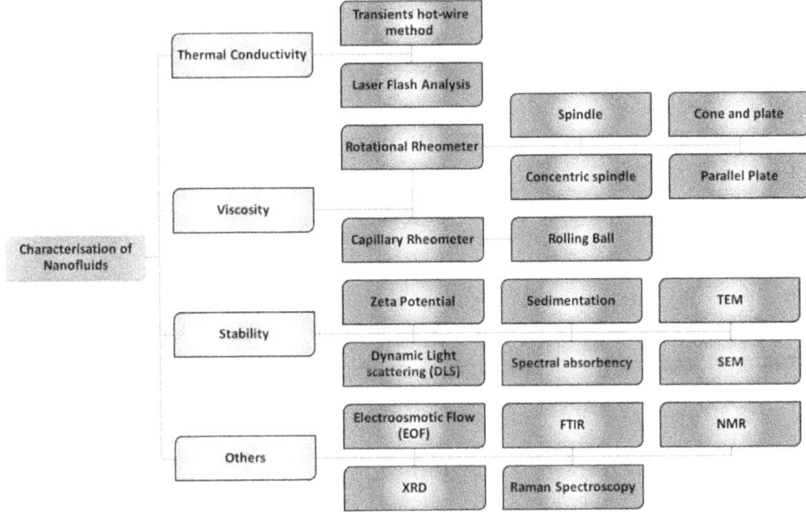

Figure 3.2 Various experimental processes of nanofluids.

transfer in applications such as electronics cooling, medical treatments, and energy systems (Awais et al. 2021b; Mehta et al. 2022). Addressing difficulties such as sedimentation and agglomeration also assures the dependability and lifespan of nanofluid-based technologies, making their use more effective and sustainable in a variety of scientific and industrial fields (Ali and Salam 2020; Gupta et al. 2017).

Therefore, the current chapter discusses an in-depth examination of nanofluid characterization techniques, as shown in Figure 3.2. It dives into the methodologies used to analyze and comprehend the unique features of nanofluids, providing a comprehensive introduction and holistic understanding for researchers working in this dynamic field.

3.2 EXPERIMENTAL TECHNIQUES

3.2.1 Thermal conductivity measurement

Thermal conductivity is one of the most significant attributes in the context of the materials used for heat exchange applications and ongoing research has been exploring numerous ways to enhance it, especially with the introduction of nanofluids (Angayarkanni and Philip 2015). Reportedly, the Maxwell model is mostly utilized to estimate the effective thermal conductivity of liquid-solid suspensions, particularly for dilute suspensions with an assumption that the particles are uniformly distributed within the media, and are not interacting with their neighbors (Sevostianov et al. 2019). However, these phenomena pose difficulties in systems with greater concentrations or higher levels of complexity. Nowadays, various methods are used for evaluating the thermal conductivity of nanofluids like steady-state, temperature oscillation, transient hot-wire, 3-omega techniques and cylindrical cells (Gonçalves et al. 2021; Awais et al. 2021a). The steady-state technique uses a constant heat flow to calculate thermal conductivity from temperature gradients susceptible to natural convection. The cylinder cell technique, which is appropriate for low-conductivity fluids, comprises sandwiching a sample between coaxial cylinders (Awais et al. 2021a). The temperature oscillation technique uses sinusoidal heat input responses to determine thermal conductivity. The 3-omega method, a non-contact approach, measures thermal conductivity by monitoring the temperature rise of a thin metal layer (Erfantalab et al. 2022). The various techniques employed for determining thermal conductivity are shown in Figure 3.3a.

3.2.1.1 Transient hot-wire method

The transient hot-wire method stands out as the most widely employed technique as it possesses the ability to eliminate errors induced by natural convection in nanofluids (Ali and Salam 2020). In 1981, Nagasaka and Nagashima proposed a modified hot-wire cell in which the hot wire was covered with epoxy glue to provide electrical insulation, therefore solving the issue provided by their electrical conductivity (Ali and Salam 2020; Sahamifar et al. 2024). Furthermore, the utilization of a platinum hot wire for thermal conductivity measurements presents challenges due to its high cost and brittleness, which makes it especially difficult to assess the thermal conductivity of materials such as adhesives, fluids with high Prandtl numbers, substances in extreme temperature conditions, and nanofluids, as there is a significant risk of sensor damage during sample handling and testing (Alvarado Ramírez et al. 2012). To mitigate these challenges while maintaining equipment integrity, research organizations are increasingly resorting to transient hot disc technology. This durable and efficient thermal characterization approach not only provides a potential solution but also calculates thermal diffusivity and thermal conductivity for fluids and solids. These qualities have significance in characterizing materials used in temporary applications, such as thermal energy storage materials. Unlike the platinum hot wire technique, the transient hot disc method employs a volume element (sensor) that serves as both a heat source and a temperature sensor. In order to measure thermal conductivity, the sensor is either completely immersed in a fluid or sandwiched between two solid samples. When powered, the temperature of the sensor rapidly rises when entirely immersed or sandwiched between insulating layers, but it increases slowly for thermally conductive substances. The transient hot-wire technique (Figure 3.3b) provides excellent sensitivity, rapid observations, and broad application for assessing thermal conductivity, particularly in materials with low

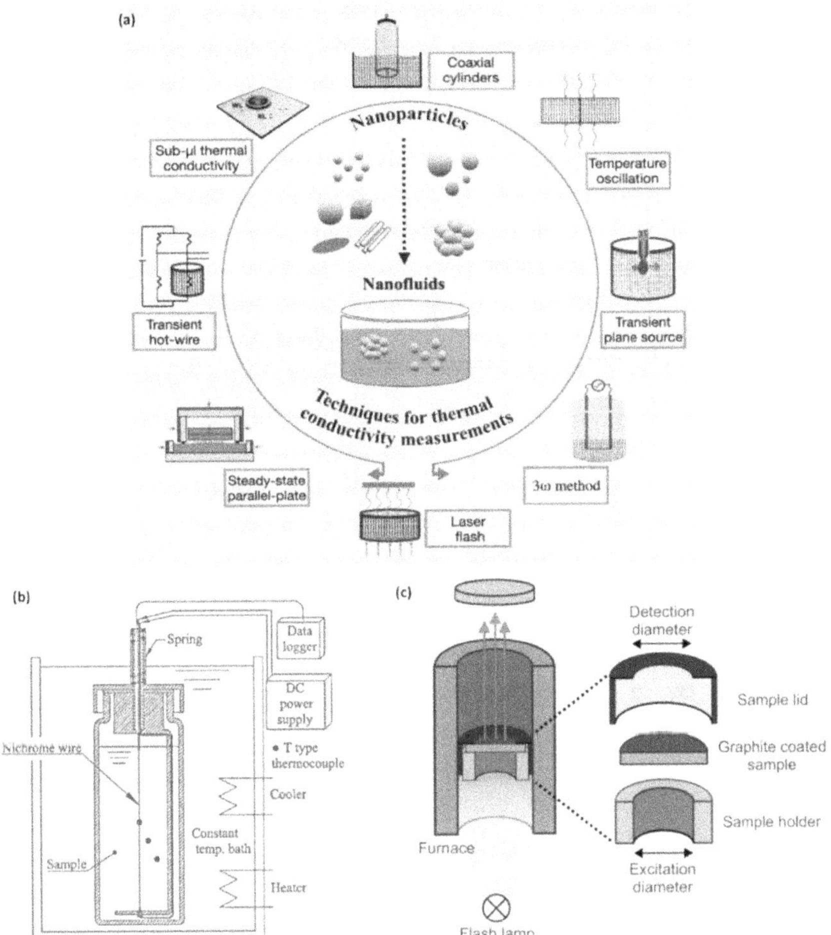

Figure 3.3 (a) various techniques used for the thermal conductivity measurement of nanofluids (Guo et al. 2018), (b) transient hot-wire method, and (c) laser flash analysis. Reprinted with License No. 5730531302362 and 5730531465554 (Morita et al. 2022; Ali and Salam 2020).

thermal conductivity. Its reliable results, along with precise temperature control, make it useful in a variety of applications, including nanofluid research and material characterization (Alvarado Ramírez et al. 2012).

3.2.1.2 Laser flash analysis

Laser flash analysis (LFA) is a rapid and non-destructive technique for measuring thermal conductivity materials, with applications extending to nanofluids. The basic principle of LFA comprises the formation of a short-duration laser pulse that quickly heats the surface of a sample material, resulting in a transient temperature response. LFA is specifically useful in the context of nanofluids, which are colloidal suspensions of NPs in a base fluid as they can detect thermal characteristics in the presence of nanoscale addition. The relative thermal conductivity of nanofluids is determined by analyzing the temperature differential between the test fluid and a reference fluid, such as water, separated by a thin conductive substance. The LFA setup (Figure 3.3c) consists of a pulse generator, usually a high-energy laser, and a detector located on the opposite side of the sample. Laser irradiation produces a fast rise in temperature at the surface of the nanofluid sample, and the detector captures the following thermal reaction within a short period. The obtained data, which is often shown as a temperature rise against the time curve, is examined in order to determine the thermal diffusivity of the nanofluid. The thermal conductivity is calculated by using the following equation:

$$k = \alpha \cdot \rho \cdot Cp \tag{3.1}$$

where α signifies the thermal diffusivity, ρ and Cp represent the density and specific heat, respectively.

3.2.2 Viscosity measurements

Viscosity is one of the most important characteristics influencing the behavior of nanofluids. Researchers carried out experiments to determine viscosity by introducing NPs into various base fluids and revealed that attributes such as pH, temperature, volume concentration, and particle size possessed significant impacts on viscosity. Furthermore, they presented various correlated equations to evaluate viscosity based on their studies with several nanofluids, as listed in Table 3.1. Reportedly, the viscosity of nanofluid is also dependent on the volume fraction of NPs (Singh et al. 2015). In this context, the researchers investigated the dispersion of alumina particles in commercial engine coolants and revealed that pristine base fluid exhibited Newtonian behavior across the measured temperature range, whereas the introduction of a small quantity of alumina NPs led to a transition, turning the fluid into a non-Newtonian fluid. Such phenomena demonstrated that the introduction of alumina NPs alters the rheological properties of the fluid, causing it to deviate from the Newtonian flow observed in the absence of NPs.

The viscosity of nanofluids has been measured using a variety of viscometers such as capillary tube viscometers, vibro-viscometers, rotating rheometers, falling ball/piston viscometers, and cup-type viscometers that operate on various principles. The rotating rheometer, capillary tube viscometer, and piston-type rheometer are the most widely used equipment for measuring nanofluid viscosity.

Table 3.1 Various models used to calculate the viscosity of different types of nanofluids

	Model names	*Equations*
1	Model for spherical NPs	$\mu_{eff} = (1 + 2.5\phi_p + 7:349\phi^2{}_p + \cdots)\mu_b$
2	Model for simple hard sphere system	$\mu_{nf} = \mu_f \dfrac{1}{(1-\phi)^{2.5}}$
3	Spherical NPs and for 0.5236≤Φ≤0.7405	$\mu_{eff} = \dfrac{9}{8}\dfrac{\left(\phi_p / \phi_{p\max}\right)^{1/3}}{1-\left(\phi_p / \phi_{p\max}\right)^{1/3}}\mu_b$
4	Einstein model	$\mu_r = \dfrac{\mu_{nf}}{\mu} = 1 + \eta\phi$
5	Batchelor model	$\mu_r = \dfrac{\mu_{nf}}{\mu} = 1 + \eta\phi + (\eta\phi)^2$
6	Ward model	$\mu_r = \dfrac{\mu_{nf}}{\mu} = 1 + \eta\phi + (\eta\phi)^2 + (\eta\phi)^3$

3.2.2.1 Rotational rheometer

The primary objective of this method is to measure the force (torque) exerted on a rotor while it revolves at a constant angular velocity (rotational speed) within a liquid. Rotational rheometers/viscometers are used to determine the viscosity of Newtonian fluids, which is unaffected by shearing force or shear rate, as well as the apparent viscosity of non-Newtonian fluids (which exhibit a variety of rheological behaviors depending on the shear rate and shear tension) (Ali and Salam 2020). There are three standard methods for assessing viscosity using rotating rheometers, as given below.

- Spindle rheometers
- Concentric spindle rheometers
- Cone and plate rheometers

3.2.2.1.1 Spindle rheometer

The spindle rheometer utilizes the cylinder-shaped or disc-shaped spindles to determine the apparent viscosity by immersing in a large volume of liquid (Figure 3.4a). Under this test condition, the shear stress between the inner surface of the beaker containing the sample liquid and the outer surface of the spindle (Chen et al. 2019). Consequently, special parameters must be considered, including spindle size geometry, angular velocity of the spindle, temperature of the sample, and accessories of the rheometer.

3.2.2.1.2 Concentric spindle rheometers

In the concentric spindle rheometers (Figure 3.4b), apparent viscosity is calculated by introducing the liquid in the space between the inner and outer cylinders. Commercially available rotational rheometers come in both controlled stress and controlled rate configurations, featuring absolute geometries such as narrow annular gaps between concentric

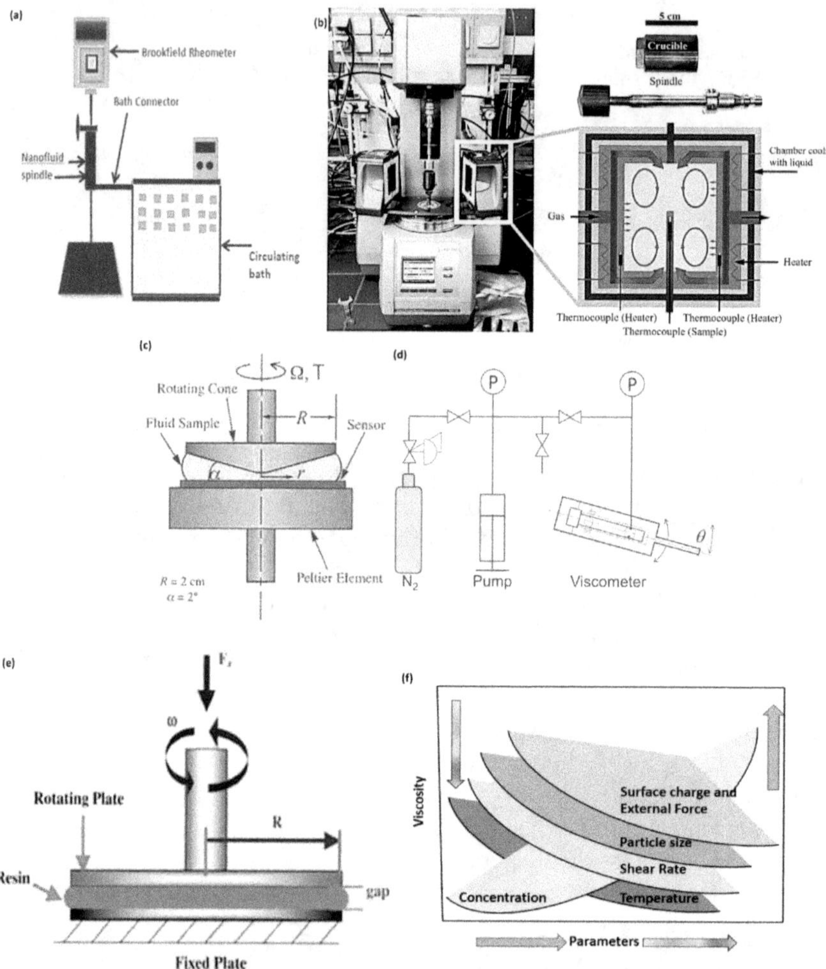

Figure 3.4 (a) Spindle rheometer (b) concentric spindle rheometers, (c) cone and plate rheometers, (d) rolling ball viscometer, (e) parallel plate rotational rheometer, and (f) various factors influencing viscosity of nanofluids. Reprinted with License No. 5730530146020 (Genova et al. 2016; Lahari et al. 2021; Sato et al. 2019).

cylinders, which are useful for collecting important rheological data on non-Newtonian fluids. Controlled-rate rheometers use a controlled shear rate input to calculate the resultant shear stress, which is measured as torque on the rotor axis. The concentric cylinder rotating rheometers are also known as cup-and-bob rheometers. Here, the choice between outer cylinder (cup) and interior cylinder (bob) rotation determines whether the rheometer is a rotating cup (Couette) or rotating-bob system (Searle) (Muñoz-Sánchez et al. 2018).

In order to calculate viscosity, a sufficient volume of the test solution or fluid is introduced into the rheometer, and the sample is allowed to reach thermal equilibrium as specified in the specific monograph. In the case of non-Newtonian systems, the monograph specifies the rheometer type to be used and the shear rate (s) at which measurements should be performed, and the apparent viscosity is calculated over a range of shear rates for the materials. A sequence of viscosity measurements reveals the relationship between shear rate and shear stress for a non-Newtonian fluid, shedding light on its flow properties. In contrast, for non-Newtonian liquids, it is critical to define either shear stress or the shear rate at which viscosity measurements should be obtained, as described below:

$$\gamma = \left(\frac{R_o^2 + R_i^2}{R_o^2 - R_i^2} \right) \omega \tag{3.2}$$

$$\sigma = \left[\frac{1}{4\pi h R_o^2} + \frac{1}{4\pi h R_i^2} \right] M \tag{3.3}$$

where R_o and R_i represent the radium of the outer and inner cylinder, respectively, $\acute{\omega}$ is the angular viscosity, M signifies the torque, and h is the height of the immersion of the inner cylinder in the liquid media.

Further, the angular viscosity is calculated by using Eq. (3.4)

$$\omega = \left(\frac{2\pi}{60} \right) n \tag{3.4}$$

where n represents the rotational speed (in rpm).

For laminar flow, the viscosity n (or apparent viscosity η_{ap}) (in Pa s) is given by the following equation:

$$\eta \ or \ \eta_{ap} = \frac{1}{4rh} \left(\frac{1}{R_i^2} - \frac{1}{R_o^2} \right) \frac{M}{\omega} = k \frac{M}{\omega} \tag{3.5}$$

where k signifies the constant of the apparatus.

3.2.2.1.3 Cone and plate rheometers

The cone and plate method involves pouring liquid into the space between a flat disc or plate and a revolving cone, resulting in a defined angle (Figure 3.4c). Because of the limited sample quantity, even a small loss can result in a substantial percentage change in viscosity. This method follows a similar procedure to that of the concentric rheometer. The calculation of shear rate and shear stress is given below:

$$\gamma = \left(\frac{1}{\alpha}\right)\omega \tag{3.6}$$

$$\sigma = \left[\frac{1}{(2/3)\pi R_o^3}\right]M \tag{3.7}$$

α signifies the angle between the cone and flat plate (in radians), R and ω represent the radius (m) and angular viscosity (radian/s), respectively, and M signifies the torque (N m).

For laminar flow, the viscosity η (apparent viscosity) in Pa s is given by the equation given below

$$\eta \text{ or } \eta_{app} = \left(\frac{3\alpha}{2\pi R_o^3}\right)\left(\frac{M}{\omega}\right) = k\frac{M}{\omega} \tag{3.8}$$

K signifies the constant of the apparatus (radian/m^3).

3.2.2.2 Rolling ball viscometer

Another method for determining the viscosity of a Newtonian fluid is the Rolling ball viscometer method, which consists of a basic design of a capillary tube containing the sample liquid under test and a ball that requires a minimum rolling time of 20 seconds at the measuring angle in the sample liquid, which is based on the Stokes law (Figure 3.4d). The angle of inclination plays a crucial role in the determination of viscosity. The viscosity is calculated using the following equation as mentioned below:

$$\eta = \lfloor (\rho_1 - \rho_2) * g * r^2 * \sin\theta \rfloor / v \tag{3.9}$$

3.2.2.3 Parallel plate rotational rheometer

The viscosity of microfluid can also be calculated by a parallel plate rotational rheometer, which generally contains two parallel discs, out of which one is revolving and another one rotates, and the resulting torque is calculated on the rotating disc (Figure 3.4e) (Xiao et al. 2019). A parallel plate rheometer for a fluid, such as oils or polymers, uses capillary forces to keep the substance contained between the two shearing plates and prevent it from flowing out. The distance between the discs is usually less than 1 millimeter. The precise control of gap size between plates enables reliable rheological characterization, which permits the study of nanofluid behavior at regulated shear rates (critical for understanding viscosity variations). The parallel plate shape enables homogeneous stress distribution, thereby allowing for more reliable data gathering. Furthermore, its adaptability enables the investigation of both Newtonian and non-Newtonian nanofluids, capturing their complex rheological features.

3.2.2.4 *Capillary viscometer*

The capillary-tube viscometer is a commonly employed method for measuring the viscosity of nanofluids and the influence of the presence of stable, equal-sized spheres suspended at low concentrations in a Newtonian fluid. Various types of capillary-tube viscometers are illustrated in Figure 3.5. Capillary viscometers use the Hagen-Poiseuille equation, which relates the flow rate of a fluid through a capillary tube to its viscosity. The viscosity of a nanofluid is calculated by measuring the amount of time required for the fluid to flow through a calibrated capillary under certain conditions (Lewis 1996). A capillary viscometer is generally made up of a capillary tube through which the nanofluid flows, a reservoir for the fluid, and a timing mechanism. The nanofluid is injected into the capillary and its flow is regulated by gravity or a pump. The time it takes for a specific volume of nanofluid to pass through the capillary is measured. These types of viscometers are generally available in various designs such as Ostwald and Ubbelohde viscometers. They are excellent for measuring transparent liquids and are calibrated with clear indications to aid with visual viscosity readings. Recently, capillary viscometers have been operated by providing external pressure to the fluid or driving it at a predefined flow rate and detecting

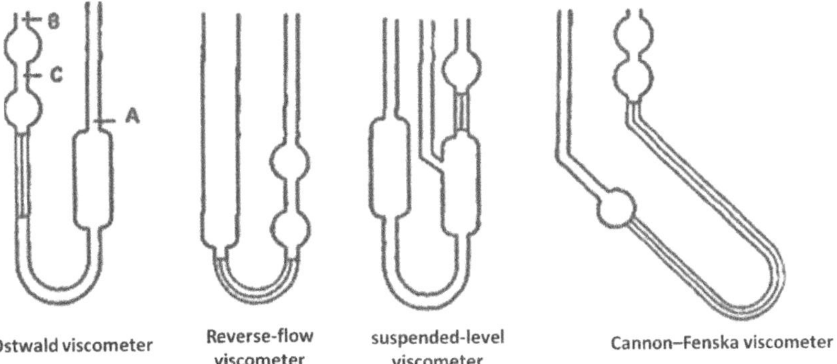

Figure 3.5 Various types of capillary-tube viscometers. Reprinted with License No. 5730520753172) (Lewis 1996).

the pressure decrease across the capillary. For example, the Cannon-Fenske Opaque viscometer is intended for dark Newtonian liquids that cannot be seen through a transparent viscometer (Kheiralla et al. 2011).

3.2.3 Stability characteristics

Sedimentation analysis, Zeta Potential Analyzer, and UV-vis Spectroscopy are now used to characterize the stability of colloidal suspensions (Mohammadi-Jam et al. 2022).

3.2.3.1 Sedimentation

Sedimentation is a simple and commonly used method for determining the stability of nanofluids (Figure 3.6a). The amount of sediment that settles on the surface influences the stability of suspension. Such sedimentation

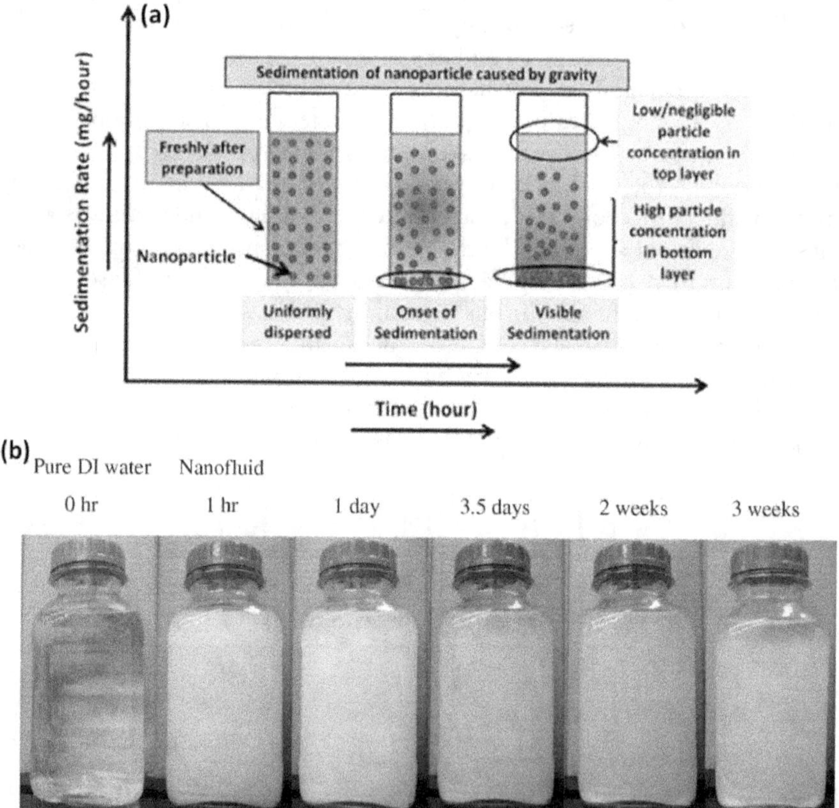

Figure 3.6 (a) Schematic representation of sedimentation-based stability evaluation method and (b) sedimentation of Al_2O_3-based nanofluid as a function of time. Reprinted with License No. 5730510086125) (Ehle et al. 2011; Kouloulias et al. 2016).

can be visualized by directly keeping the suspension for a fixed amount of time or by ultrasonication of the suspense ion and subsequently keeping the suspension for a fixed time to observe the sediments. After preparing the fluids with various dispersant combinations, a camera can capture a sedimentation photograph of nanofluids over time which may be further viewed and compared to determine the stability of nanofluids. However, this approach takes a long time to see particle sedimentation in nano fluids. In this context, Ruan and Jacobi demonstrated that ultrasonication induces agglomerates to break up and facilitates the dispersion of NPs into base fluids, resulting in a more stable nanofluid. Shah et al. investigated the stability of Al_2O_3-water nanofluid followed by ultrasonication; the results indicated that it is only stable for a week (Shah et al. 2015). A visualization study, as shown in Figure 3.6b assessed the stability of a cold nanofluid in a 1.0 L glass flask over 3 weeks, and the result revealed that post ~24 hours, no change was observed. However, by the first week, brightness decreased, indicating nanoparticle deposition. In the third week, visible stratification layers and aluminum deposition were observed (Kouloulias et al. 2016). Furthermore, the sedimentation balance method is used to determine stability where the sedimentation balance plate is immersed in the freshly created nanofluids. The weight of sediment NPs at a given moment may be quantified. Stokes' law expresses the sedimentation velocity (V) in a colloid as follows:

$$V = \frac{2R^2}{9\mu}\left(\rho_p - \rho_i\right) * g \tag{3.10}$$

The sedimentation rate drops with a reduction in the particle radius (R), density contrast between particle and liquid ($\Delta\rho=\rho_p-\rho_l$), and increased viscosity of base liquid (μ). These parameters are critical for obtaining kinetic stability in nanofluids (Haghighi et al. 2013). Each nanofluid was allowed to stand, and its absorbance was measured at regular intervals with a UV-visible spectrophotometer. The equation defines the absorbance of the suspension.

$$A = \log\left(\frac{I_i}{I}\right) = \varepsilon bc \tag{3.11}$$

3.2.3.2 Spectral absorbency

Spectral absorbance is a technique to assess colloidal stability, which refers to the ability of NPs in a solution to remain evenly dispersed over time. Colloidal stability plays a vital role in a wide range of applications, including drug delivery, environmental cleaning, and nanotechnology (Mukesh Kumar et al. 2020). In the context of spectral absorbance, the assessment covers investigating the relationship between NPs concentration in a suspension and the time required for them to settle or sediment. The absorbance is determined by estimating the quantity of light absorbed by NPs at specific wavelengths. When a suspension of NPs is exposed to light, the NPs interact with the light in ways that are proportional to their concentration and size. Spectral absorbance is the amount of light absorbed by NPs over a variety of wavelengths, and it may be used to better understand the optical properties of NPs and how they change with concentration. As NPs in a suspension aggregate or settle, the interaction of light with the particles varies, resulting in variations in spectrum absorbance. In this context, researchers may examine sedimentation kinetics by measuring absorbance over time. An increase in absorbance at specific wavelengths might suggest the development of aggregates or the settling of NPs, indicating a loss in colloidal stability. Higher absorbance levels at various wavelengths are associated with the production of bigger aggregates or the settling of NPs, suggesting less stability. In contrast, a stable colloidal solution demonstrates minor fluctuations in absorbance over time. A common detector that is extensively used in UV-vis Spectroscopy is the photomultiplier tube, which monitors UV light absorption by combining a photoemissive cathode (which releases electrons when a photon strikes it) with numerous dynodes (which emit several electrons per impacting electron) and an anode (Figure 3.7a). When matter absorbs UV light, electrons are stimulated, changing from ground to an excited state with an energy difference equal to the absorbed radiation. This concept is consistent with Beer-Lambert's law, which describes the connection between absorption, concentration, and route length. The photomultiplier tube amplifies and detects electron emissions, making it a sensitive and versatile instrument for UV-vis Spectroscopy.

3.2.3.3 Zeta potential test

The zeta potential measurement stands out as a vital technique for determining nanofluid stability and quality. In dispersion or suspension, a larger absolute zeta potential suggests more electrostatic repulsions between particles, which contributes to increased stability (Figure 3.7c). Experiments with different weight fractions, generally ranging from ~0.05%, reveal the relationship between suspension stability and zeta potential. Particles having a high surface charge are resistant to agglomeration due to opposing forces on contact. In contrast, low zeta potential may cause particle aggregation and precipitation, thereby affecting fluid stability. A zeta potential greater than 30 mV (absolute value) is commonly recognized as a benchmark for good stability, and it serves as typical measurement criteria for researchers examining nanofluid stability (Figure 3.7d and e) (Murshed et al. 2020). Such technology aids in the improvement of nanofluid formulations, improving their stability, and, as a result, increasing their potential applications in numerous fields, including heat transfer systems.

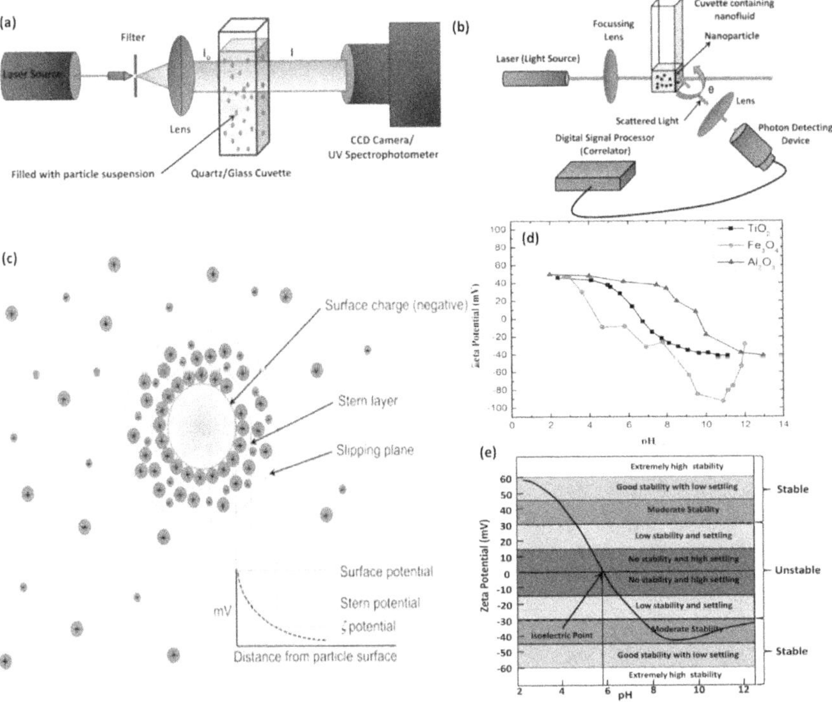

Figure 3.7 (a) Absorbance/transmittance measurement, (b) basic schematic representation of dynamic light scattering, (c) zeta potential measurement and (d) and (e) zeta potential vs pH value at various pH conditions which corresponds to the different regimes of nanofluid stability. Reprinted with license from Elsevier (Licence No. 5730500473973 and 5730501208926) (Chakraborty and Panigrahi 2020; Pate and Safier 2016).

3.2.3.4 *Ultrasonic vibration*

Ultrasonication plays a vital role in managing the stability of nanofluid as it disrupts NPs' clusters (in the form of agglomerates) (Sandhya et al. 2021). Ultrasonic vibration can be implemented in two ways: (1) indirectly using an ultrasonic bath, and (2) directly with a probe sonicator (Asadi et al. 2019). According to the comparative study, the direct technique, namely the use of a probe sonicator, produces superior results in terms of particle cluster breaking and average cluster size reduction, as shown in Figure 3.8a. Numerous investigations have demonstrated the effectiveness of ultrasonication in improving stability, lowering cluster size, and affecting thermophysical parameters such as heat conductivity and viscosity. In this regard, Nguyen et al. demonstrated that ultrasonication is successful in creating monodisperse alumina nanofluid where the average NPs cluster size was decreased from ~230 to ~130 nm post 500 s of ultrasonication at ~30% vibration amplitude (Nguyen et al. 2011).

3.2.4 Particle characterization

3.2.4.1 *Transmission electron microscopy*

Transmission electron microscopy (TEM) is an effective tool for investigating nanofluids at the nanoscale, giving intricate comprehensive details about the shape, size, and dispersion of NPs within a liquid media. The technology works by sending an electron beam through a thin sample, resulting in pictures with high resolution (Figure 3.8b). The generated high-resolution micrographs provide a substantial contribution to understanding the complexity of nanofluid dynamics and play an important role in optimizing their performance for various applications. Further, it allows for precise dimension

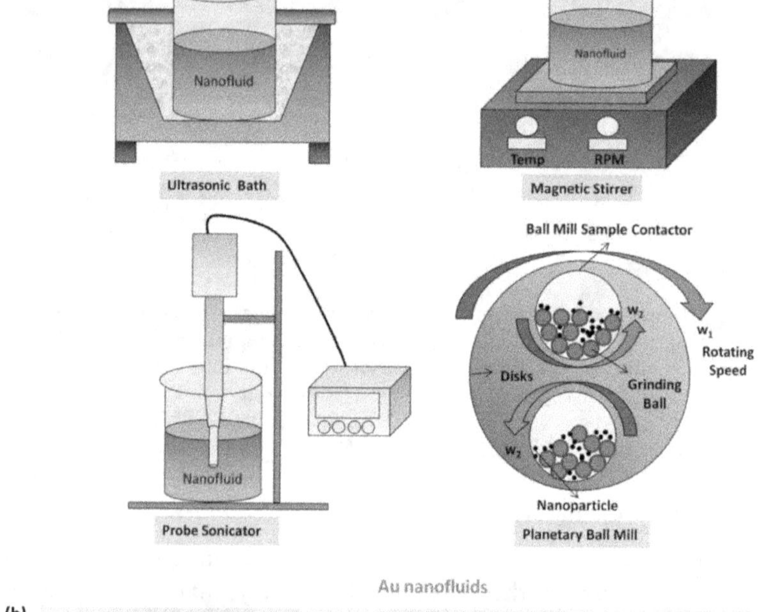

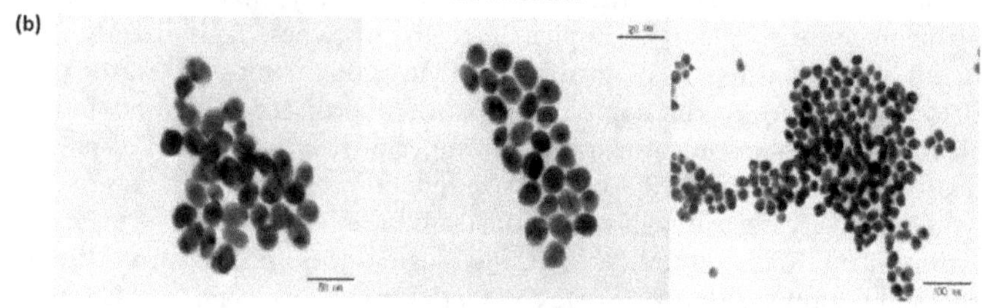

Figure 3.8 (a) Schematic demonstrating several types of equipment used to stabilize nanofluids mechanically. (b) TEM images of Au-based nanofluids. Reprinted with license from Elsevier (Licence No.5730490118756) (Chakraborty and Panigrahi 2020; Jin et al. 2021).

measurement, which contributes to a better knowledge of nanofluid physical characteristics. When combined with EDS, it provides elemental analysis, which aids in determining the chemical composition of NPs in nanofluids.

3.2.4.2 Dynamic light scattering

Dynamic light scattering (DLS) is a non-intrusive method for analyzing particle size distribution and mobility in nanofluids. DLS involves directing a laser beam at the nanofluid and analyzing the scattered light to determine the Brownian motion of the particles and the oscillations in scattered light intensity reveal the size distribution of NPs in the fluid (Angayarkanni and Philip 2022; Gollwitzer et al. 2016). This method is very rapid, needs minimal sample preparation, can quantify particle sizes ranging from a few to several hundred nanometers, and is suitable for assessing polydisperse systems (Figure 3.7b). The basic principle of this technique is to compute the hydrodynamic diameter of NPs by analyzing the intensity changes of scattered light over time, which represent their effective size in the fluid. Such information is critical for understanding the stability, dispersibility, and behavior of nanofluids in a variety of applications, including heat transfer fluids, drug delivery systems, and innovative materials.

3.2.4.3 Other experimental techniques

3.2.4.3.1 Pool boiling experiments

Pool boiling studies have been intended to examine the heat transfer performance of colloidal suspensions (Figure 3.9). Nanofluids, which are base fluids containing dispersed nanoparticles, have the potential to increase boiling heat transfer by increasing thermal conductivity (Gao et al. 2022). In these

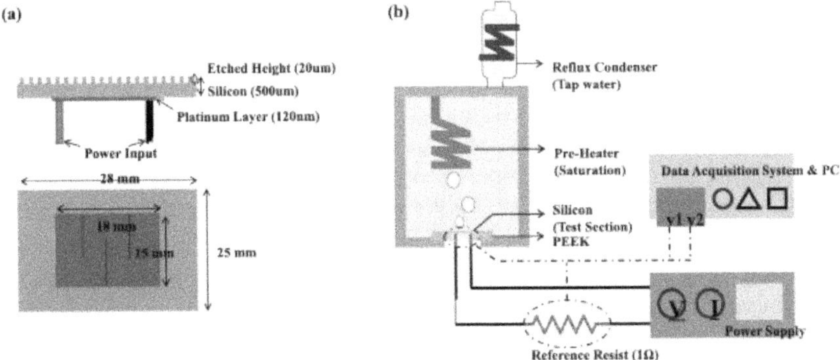

Figure 3.9 Schematic representation of the pool boiling experimental setup with the test section (a) and pool boiling experiment setup (b). Reprinted with license from Elsevier (Licence No. 5730480992901) (Kim et al. 2015).

experiments, a nanofluid pool is heated, and the resulting boiling phenomena are seen and analyzed. Critical aspects such as nucleation boiling, bubble dynamics, and overall heat transfer properties are evaluated. The presence of nanoparticles alters boiling heat transfer by influencing nucleation sites and bubble dynamics, providing insights for optimizing nanofluid formulations in a variety of applications such as cooling systems and heat exchangers.

3.2.4.4 Electroosmotic flow (EOF) measurements

Electroosmotic flow (EOF) measurements in nanofluids are the study of fluid movement caused by an electric field, particularly in the presence of nanoparticles. In EOF tests, an electric field is placed across a microchannel or porous media that contains the nanofluid, causing charged particles to migrate and fluid flow to occur. The inclusion of nanoparticles has a substantial impact on EOF because it changes the fluid's surface charge and conductivity. This phenomenon is extremely important in microfluidic systems, where fluid movement must be precisely controlled. EOF measurements offer information on nanofluid electro-kinetics, which aids in the optimization of systems such as lab-on-a-chip devices, drug delivery systems, and electrokinetic pumps. In this regard, particle image velocimetry and micro-particle image velocimetry are widely used by researchers to quantify EOF patterns (Lindken et al. 2009; Sadek et al. 2017).

3.3 CONCLUSIONS AND FUTURE SCOPE

The current chapter has provided an in-depth exploration of a wide range of experimental techniques employed in the study of nanofluids i.e., from fundamental characterization methods to advanced imaging and spectroscopy techniques. Moreover, it is imperative to explore the interdisciplinary nature of nanofluid research, emphasizing the development of innovative experimental techniques. However, despite the significant progress achieved, the in-depth characterizations and their experimental validation of nanofluids are still unclear. Thus, future research should prioritize for addressing the existing gaps and pushing the boundaries of our understanding pertaining to the experimental techniques used for nanofluids, as a result, it paves a path for bridging the landscape of nanotechnology and fluid dynamics. The integration of experimental techniques with computational methods further explores and refines the experimental methodologies that are poised to yield more comprehensive insights into nanofluid dynamics, heat transfer, and thermophysical properties. Conclusively, the comprehensive overview serves as a foundational resource for researchers and scientists, guiding them in understanding the current state of experimental techniques for nanofluids. Such pursuit of excellence in experimentation will help to unlock their full potential across various industrial applications in a sustainable and more efficient manner.

REFERENCES

Alami, Abdul Hai, Mohamad Ramadan, Muhammad Tawalbeh, Salah Haridy, Shamma Al Abdulla, Haya Aljaghoub, Mohamad Ayoub, Adnan Alashkar, Mohammad Ali Abdelkareem, and Abdul Ghani Olabi. 2023. "A Critical Insight on Nanofluids for Heat Transfer Enhancement." *Scientific Reports* 13 (1): 15303. https://doi.org/10.1038/s41598-023-42489-0.

Ali, Abu Raihan Ibna, and Bodius Salam. 2020. "A Review on Nanofluid: Preparation, Stability, Thermophysical Properties, Heat Transfer Characteristics and Application." *SN Applied Sciences* 2 (10): 1–17. https://doi.org/10.1007/s42452-020-03427-1.

Alvarado Ramírez, Salvador, Ernesto Marín, A Juárez, A Calderón, Calder´ Calderón, and Rumen Tsonchev. 2012. "A Hot-Wire Method Based Thermal Conductivity Measurement Apparatus for Teaching Purposes." *European Journal of Physics* 33 (July): 897–906. https://doi.org/10.1088/0143-0807/33/4/897.

Angayarkanni, S A, and John Philip. 2015. "Review on Thermal Properties of Nanofluids: Recent Developments." *Advances in Colloid and Interface Science* 225: 146–76. https://doi.org/10.1016/j.cis.2015.08.014.

Angayarkanni, A, and John Philip. 2022. "Synthesis of Nanoparticles and Nanofluids." In *Fundamentals and Transport Properties of Nanofluids*, edited by S M Sohel Murshed, 0. The Royal Society of Chemistry. https://doi.org/10.1039/9781839166457-00001.

Asadi, Amin, Farzad Pourfattah, Imre Miklós Szilágyi, Masoud Afrand, Gaweł Żyła, Ho Seon Ahn, Somchai Wongwises, Hoang Minh Nguyen, Ahmad Arabkoohsar, and Omid Mahian. 2019. "Effect of Sonication Characteristics on Stability, Thermophysical Properties, and Heat Transfer of Nanofluids: A Comprehensive Review." *Ultrasonics Sonochemistry* 58: 104701. https://doi.org/10.1016/j.ultsonch.2019.104701.

Awais, Muhammad, Arafat A Bhuiyan, Sayedus Salehin, Mohammad Monjurul Ehsan, Basit Khan, and Md. Hamidur Rahman. 2021a. "Synthesis, Heat Transport Mechanisms and Thermophysical Properties of Nanofluids: A Critical Overview." *International Journal of Thermofluids* 10: 100086. https://doi.org/10.1016/j.ijft.2021.100086.

Awais, Muhammad, Najeeb Ullah, Javaid Ahmad, Faizan Sikandar, Mohammad Monjurul Ehsan, Sayedus Salehin, and Arafat A Bhuiyan. 2021b. "Heat Transfer and Pressure Drop Performance of Nanofluid: A State-of- the-Art Review." *International Journal of Thermofluids* 9: 100065. https://doi.org/10.1016/j.ijft.2021.100065.

Bahiraei, Mehdi, and Morteza Hangi. 2015. "Flow and Heat Transfer Characteristics of Magnetic Nanofluids: A Review." *Journal of Magnetism and Magnetic Materials* 374: 125–38. https://doi.org/10.1016/j.jmmm.2014.08.004.

Benos, Lefteris, George Ninos, Nickolas D Polychronopoulos, Maria-Aristea Exomanidou, and Ioannis Sarris. 2022. "Natural Convection of Blood–Magnetic Iron Oxide Bio-Nanofluid in the Context of Hyperthermia Treatment." *Computation* 10 (11). https://doi.org/10.3390/computation10110190.

Chakraborty, Samarshi, and Pradipta Kumar Panigrahi. 2020. "Stability of Nanofluid: A Review." *Applied Thermal Engineering* 174: 115259. https://doi.org/10.1016/j.applthermaleng.2020.115259.

Chen, Yang, Faris Matalkah, Anagi Balachandra, and Parviz Soroushian. 2019. "Ultra-High-Performance Concrete: Development of On-Site Fresh Mix Rheology Test Methods" 55 (April): 1–11.

Ehle, Angelika, Steffen Feja, and Matthias H Buschmann. 2011. "Temperature Dependency of Ceramic Nanofluids Shows Classical Behavior." *Journal of Thermophysics and Heat Transfer* 25 (3): 378–85. https://doi.org/10.2514/1.T3634.

Erfantalab, Sobhan, Giacinta Parish, and Adrian Keating. 2022. "Determination of Thermal Conductivity, Thermal Diffusivity and Specific Heat Capacity of Porous Silicon Thin Films Using the 3ω Method." *International Journal of Heat and Mass Transfer* 184: 122346. https://doi.org/10.1016/j.ijheatmasstransfer.2021.122346.

Gao, Yuan, Huai-En Hsieh, Zhibo Zhang, Shiqi Wang, and Zhe Zhou. 2022. "Experimental Study on Pool Boiling Heat Transfer Characteristics of TiO2 Nanofluids on a Downward-Facing Surface." *Progress in Nuclear Energy* 153: 104402. https://doi.org/10.1016/j.pnucene.2022.104402.

Genova, Danilo Di, Corrado Cimarelli, Kai-Uwe Hess, and Donald B Dingwell. 2016. "An Advanced Rotational Rheometer System for Extremely Fluid Liquids up to 1273 K and Applications to Alkali Carbonate Melts." *American Mineralogist* 101 (4): 953–59. https://doi.org/doi:10.2138/am-2016-5537CCBYNCND.

Gollwitzer, Christian, Dorota Bartczak, Heidi Goenaga Infante, Vikram Kestens, Michael Krumrey, Caterina Minelli, Marcell Pálmai et al. 2016. "A Comparison of Techniques for Size Measurement of Nanoparticles in Cell Culture Medium." *Analytical Methods* 8 (July). https://doi.org/10.1039/C6AY00419A.

Gonçalves, Inês, Reinaldo Souza, Gonçalo Coutinho, João Miranda, Ana Moita, José Eduardo Pereira, António Moreira, and Rui Lima. 2021. "Thermal Conductivity of Nanofluids: A Review on Prediction Models, Controversies and Challenges." *Applied Sciences (Switzerland)* 11 (6). https://doi.org/10.3390/app11062525.

Guo, Wenwen, Guoneng Li, Youqu Zheng, and Cong Dong. 2018. "Measurement of the Thermal Conductivity of SiO2 Nanofluids with an Optimized Transient Hot Wire Method." *Thermochimica Acta* 661: 84–97. https://doi.org/10.1016/j.tca.2018.01.008.

Gupta, Munish, Vinay Singh, Rajesh Kumar, and Z Said. 2017. "A Review on Thermophysical Properties of Nanofluids and Heat Transfer Applications." *Renewable and Sustainable Energy Reviews* 74: 638–70. https://doi.org/10.1016/j.rser.2017.02.073.

Haghighi, Ehsan, Nader Nikkam, Mohsin Saleemi, Reza Behi, Seyed Aliakbar Mirmohammadi, Heiko Poth, R Khodabandeh, Muhammet Toprak, Mamoun Muhammed, and Björn Palm. 2013. "Shelf Stability of Nanofluids and Its Effect on Thermal Conductivity and Viscosity Shelf Stability of Nanofluids and Its Effect on Thermal Conductivity and Viscosity." *Measurement Science and Technology* 24 (August): 105301–11. https://doi.org/10.1088/0957-0233/24/10/105301.

Jin, Xin, Guiping Lin, Haichuan Jin, Zunru Fu, and Haoyang Sun. 2021. "Experimental Research on the Selective Absorption of Solar Energy by Hybrid Nanofluids." *Energies* 14 (23). https://doi.org/10.3390/en14238186.

Kheiralla, Abdelmotalab, Mohamed El-Awad, Mathani Hassan, Mohammed Hussen, and Hind Osman. 2011. "Effect of Ethanol/Gasoline Blends on Fuel Properties Characteristics of Spark Ignition Engines." *University of Khartoum Engineering Journal* 1 (October): 22–28.

Kim, Seol Ha, Gi Cheol Lee, Jun Young Kang, Kiyofumi Moriyama, Moo Hwan Kim, and Hyun Sun Park. 2015. "Boiling Heat Transfer and Critical Heat Flux Evaluation of the Pool Boiling on Micro Structured Surface." *International Journal of Heat and Mass Transfer* 91: 1140–47. https://doi.org/10.1016/j.ijheatmasstransfer.2015.07.120.

Kouloulias, K, A Sergis, and Y Hardalupas. 2016. "Sedimentation in Nanofluids during a Natural Convection Experiment." *International Journal of Heat and Mass Transfer* 101: 1193–1203. https://doi.org/10.1016/j.ijheatmasstransfer.2016.05.113.

Lahari, Mlr, P Sesha, P H V Sai, Narayana K S, and K Sharma. 2021. "Thermophysical Properties of Copper and Silica Nanofluids in Glycerol-Water Mixture Base Liquid." *Journal of Thermodynamics & Catalysis* 12 (November): 1–11.

Lewis, M J. 1996. "4- Viscosity." In *Physical Properties of Foods and Food Processing Systems*, edited by M J Lewis, 108–36. Woodhead Publishing Series in Food Science, Technology and Nutrition. Woodhead Publishing. https://doi.org/10.1533/9781845698423.108.

Lindken, Ralph, Massimiliano Rossi, Sebastian Große, and Jerry Westerweel. 2009. "Micro-Particle Image Velocimetry (PIV): Recent Developments, Applications, and Guidelines." *Lab on a Chip* 9 (17): 2551–67. https://doi.org/10.1039/b906558j.

Mehta, Bhavin, Dattatraya Subhedar, Hitesh Panchal, and Zafar Said. 2022. "Synthesis, Stability, Thermophysical Properties and Heat Transfer Applications of Nanofluid: A Review." *Journal of Molecular Liquids* 364: 120034. https://doi.org/10.1016/j.molliq.2022.120034.

Mohammadi-Jam, Shiva, Kristian E Waters, and Richard W Greenwood. 2022. "A Review of Zeta Potential Measurements Using Electroacoustics." *Advances in Colloid and Interface Science* 309: 102778. https://doi.org/10.1016/j.cis.2022.102778.

Morita, Shin ichi, Toshihiro Haniu, Kazunori Takai, Takanobu Yamada, Yasutaka Hayamizu, Takeshi Gonda, Akihiko Horibe, and Naoto Haruki. 2022. "Thermal Conductivity Estimation of Carbon-Nanotube-Dispersed Phase Change Material as Latent Heat Storage Material." *International Journal of Thermophysics* 43 (5): 1–18. https://doi.org/10.1007/s10765-022-02996-0.

Mukesh Kumar, P C, K Palanisamy, and V Vijayan. 2020. "Stability Analysis of Heat Transfer Hybrid/Water Nanofluids." *Materials Today: Proceedings* 21: 708–12. https://doi.org/10.1016/j.matpr.2019.06.743.

Muñoz-Sánchez, Belén, Javier Nieto-Maestre, Elisabetta Veca, Raffaele Liberatore, Salvatore Sau, Helena Navarro, Yulong Ding et al. 2018. "Rheology of Solar-Salt Based Nanofluids for Concentrated Solar Power. Influence of the Salt Purity, Nanoparticle Concentration, Temperature and Rheometer Geometry." *Solar Energy Materials and Solar Cells* 176: 357–73. https://doi.org/10.1016/j.solmat.2017.10.022.

Murshed, S M Sohel, Mohsen Sharifpur, Giwa Solomon, and Josua Meyer. 2020. "Experimental Research and Development on the Natural Convection of Suspensions of Nanoparticles-A Comprehensive Review." *Nanomaterials* 10 (September): 1855. https://doi.org/10.3390/nano10091855.

Nguyen, Van Son, Didier Rouxel, Rachid Hadji, Brice Vincent, and Yves Fort. 2011. "Effect of Ultrasonication and Dispersion Stability on the Cluster Size of Alumina Nanoscale Particles in Aqueous Solutions." *Ultrasonics Sonochemistry* 18 (1): 382–88. https://doi.org/10.1016/j.ultsonch.2010.07.003.

Pate, K, and P Safier. 2016. "12- Chemical Metrology Methods for CMP Quality." In *Advances in Chemical Mechanical Planarization (CMP)*, edited by Suryadevara Babu, 299–325. Woodhead Publishing. https://doi.org/10.1016/B978-0-08-100165-3.00012-7.

Porgar, Sajjad, Leila Vafajoo, Nader Nikkam, and Gholamreza Vakili-Nezhaad. 2021. "A Comprehensive Investigation in Determination of Nanofluids Thermophysical Properties." *Journal of the Indian Chemical Society* 98 (3): 100037. https://doi.org/10.1016/j.jics.2021.100037.

Sadek, Samir H, Francisco Pimenta, Fernando T Pinho, and Manuel A Alves. 2017. "Measurement of Electroosmotic and Electrophoretic Velocities Using Pulsed and Sinusoidal Electric Fields." *Electrophoresis* 38 (7): 1022–37. https://doi.org/10.1002/elps.201600368.

Sahamifar, S, D Naylor, T Yousefi, and J Friedman. 2024. "Measurement of the Thermal Conductivity of Nanofluids Using a Comparative Interferometric Method." *International Journal of Thermal Sciences* 199: 108890. https://doi.org/10.1016/j.ijthermalsci.2024.108890.

Sandhya, Madderla, D Ramasamy, K Sudhakar, K Kadirgama, and W S W Harun. 2021. "Ultrasonication an Intensifying Tool for Preparation of Stable Nanofluids and Study the Time Influence on Distinct Properties of Graphene Nanofluids - A Systematic Overview." *Ultrasonics Sonochemistry* 73 (May): 105479. https://doi.org/10.1016/j.ultsonch.2021.105479.

Sato, Yoshiyuki, Hiroki Baba, Chisato Yoneyama, and Hiroshi Inomata. 2019. "Development of a Rolling Ball Viscometer for Simultaneous Measurement of Viscosity, Density, Bubble-Point Pressure of CO2-Expanded Liquids." *Fluid Phase Equilibria* 487: 71–75. https://doi.org/10.1016/j.fluid.2019.01.017.

Sevostianov, I, S G Mogilevskaya, and V I Kushch. 2019. "Maxwell's Methodology of Estimating Effective Properties: Alive and Well." *International Journal of Engineering Science* 140 (July): 35–88. https://doi.org/10.1016/j.ijengsci.2019.05.001.

Shah, Sheikh Khaleduzzaman, M R Sohel, Saidur Rahman, and Jai Sandeep. 2015. "Stability of Al2O3-Water Nanofluid for Electronics Cooling System." *Procedia Engineering* 105 (December). https://doi.org/10.1016/j.proeng.2015.05.026.

Sheikhpour, Mojgan, Mohadeseh Arabi, Alibakhsh Kasaeian, Ali Rokn Rabei, and Zahra Taherian. 2020. "Role of Nanofluids in Drug Delivery and Biomedical Technology: Methods and Applications." *Nanotechnology, Science and Applications* 13: 47–59. https://doi.org/10.2147/NSA.S260374.

Singh, Rahul, Oswaldo Sanchez, Suvojit Ghosh, Naveen Kadimcherla, Swarnendu Sen, and Ganesh Balasubramanian. 2015. "Viscosity of Magnetite-Toluene Nanofluids: Dependence on Temperature and Nanoparticle Concentration." *Physics Letters A* 379 (40): 2641–44. https://doi.org/10.1016/j.physleta.2015.06.010.

Xiao, Xin, Gan Zhang, Yulong Ding, and Dongsheng Wen. 2019. "Rheological Characteristics of Molten Salt Seeded with Al2O3 Nanopowder and Graphene for Concentrated Solar Power." *Energies* 12 (February): 467. https://doi.org/10.3390/en12030467.

Chapter 4

Renewable thermal storage systems and nanofluids

Sustainable energy solutions

Uttam Roy, Shailandra Kumar Prasad, K. Peera, Rahul Mishra, Ruchi Chandrakar, and Ashtashil Vrushketu Bhambulkar

4.1 INTRODUCTION

Integrating renewable energy sources with thermal storage is a cornerstone of sustainable energy management. This type of collaboration allows for effective energy storage to increase the use of clean and renewable resources. It improves grid stability and enables demand response management. This chapter analyses an overview of the study based on renewable thermal storage systems and nanofluids. Creating sustainable storage systems has become vital with the increasing demand for renewable energy. Thermal storage systems follow the principles of phase change materials (PCMs) to release and store thermal energy. These PCMs are capable of releasing and absorbing large latent heat amounts while maintaining a nearly constant temperature. Thermal storage systems can accumulate large quantities of heat for extended periods. It ensures a steady supply of energy if renewable sources are not generating power. It finds different types of applications in industrial processes, space heating, district heating, and even renewable energy integration. The objectives of this chapter include a focus on accentuating the potential effect of integrating nanofluids and renewable thermal storage systems. Thermal storage systems ensure optimal utilization of renewable sources by storing excess renewable energy. Through the minimization of the effect of intermittent renewable energy generation, thermal storage helps maintain grid stability.

4.2 RENEWABLE ENERGY LANDSCAPE

4.2.1 Overview of renewable energy

The primary renewable energy sources include biomass, solar, hydroelectric power, and wind. Advanced thermal storage systems have appeared as a profitable solution. It enables the efficient storage and use of renewable energy. Thermal energy sources provide environmental benefits, like

DOI: 10.1201/9781003494454-4

mitigating climate change and reducing greenhouse. Thermal energy storage (TES) technologies provide a reliable means of storing and using renewable energy. Renewable sources like wind power or solar-generated systems store excess coolness or heat (Singh et al. 2020a). The intermittent nature and high costs of renewable energy sources pose challenges to their overall adoption. TES systems allow renewable energy to be stored and utilized when the demand is high. It minimizes waste and enhances overall energy efficiency.

4.2.2 Importance of integration

Renewable energy can be effectively stored and dispatched based on real-time supply patterns and demand through the integration of thermal storage systems with smart grid technology. With the contribution of a healthier environment and cleaner, renewable energy sources help reduce air pollution and mitigate climate change. According to Mahian et al. (2021), the sources of energy that are renewable in nature such as the wind and the power taken from the sun provide global and domestic benefits. This benefit incorporates reduced dependence and energy independence on fossil fuels. Heat storage technologies integrate with renewable energy sources. It enables a consistent and reliable energy supply even in the lack of wind or sunlight. TES can be integrated with different renewable energy systems. It includes wind farms and solar power plants. The integration of renewable energy systems with TES increases energy utilization. It also fosters a more reliable and sustainable energy infrastructure.

4.3 TES TECHNOLOGIES

4.3.1 Traditional thermal storage

TES is achieved with a wide range of technologies. It allows excess thermal energy to be stored and utilized hours, days, or months later, at scales. It ranges from the process of individual, district, multiuser-building, region, or town depending on the specific technology (Bretado-de los Rios et al. 2021). Traditional storage methods, like lithium-ion batteries, sometimes fall short in terms of environmental and scalability effects. TES systems can be divided into three groups which are the latent form of heat, sensible heat, and heat generated from thermochemical storage. The sensible storage of storage is the most straightforward method. It generally means the temperature of some medium is either decreased or increased. Out of the three, this kind of storage is the most commercially available. Latent heat storage is related to a phase transition, the general term for the related media is PCM (Singh et al. 2020b). Heat can be extracted or added without impacting the temperature of the materials by giving it a benefit over the technologies of sensible heat storage. Thermochemical heat storage addresses some kind of reversible endotherm or exotherm chemical reaction with this kind of material. This method can allow for an even higher storage capacity than latent heat storage depending on the reactants.

4.3.2 Introduction to renewable thermal storage

TES technologies provide a reliable means of storing and using renewable energy. These systems store excess coolness or heat generated by renewable sources. Renewable thermal storage systems are a vital enabler for the large-scale deployment of renewable energy and transition. According to Souza et al. (2022), advanced thermal storage technologies can be deployed at different scales, from residential to industrial applications. These technologies serve different needs. The renewable thermal storage system can assist in meeting the demand for thermal loads across time lengths similar to electrochemical storage devices. Advances in TES would lead to maximizing energy savings, flexibility for shedding and changing building loads, higher performing and more affordable heat pumps, and enhanced thermal comfort of occupants.

4.4 NANOFLUIDS: FUNDAMENTALS AND APPLICATIONS

4.4.1 Nanofluids: definition and characteristics

Nanofluids are colloidal suspensions. It is made out of nanoparticles in some base fluid. These nanoparticles dominate the kinds of nanofluids and the distinction between them. These are a kind of fluid formed and prepared through nanomaterials that are lightly suspended, like ceramics, metal-based oxides, and carbon-oriented nanostructured materials, into base fluids. The existing research studies shows the role of green energy harvesting and its application (Pathak et al. 2023). A nanofluid is a fluid containing nanometre-sized particles. Through the use of nanofluids that are used as fluid for pump-free designs, thermal energy systems can become more effective. It can minimize maintenance costs and finally expand the lifespan of the system. Nanofluids are a type of liquid-based state which is prepared by suspending the nanomaterials in their semi-solid formulations. For example, ceramics, metal oxides, and carbon nanostructured materials, into base fluids like oil or water. Measuring only a few nanometres in size, these particles have unique properties that improve the heat transfer and thermal conductivity of the fluid. The effect of using these nanofluids in solar collectors can increase the efficiency of these systems. It can encourage them to deliver more energy with less input. Carbon nanomaterials are specifically promising to prepare nanofluids and heat transfer applications because of their thermophysical properties. Studies have shown that carbon-based nanofluids boost the efficiency of different types of solar collectors. These collectors include tubes which can be emptied and evacuated, plates that are flat and hybrid photovoltaic thermal, and parabolic trough solar collectors. Said et al. (2023) presented a review of papers in which the utilization of carbon-based materials in nanofluids was highlighted. According to Yasmin et al. (2023), these formulations are examining the possibilities of using metals which are based on carbon and collagen. They are black nanofluids and are effective for solar thermal applications. Their discovery shows how these nanofluids can put good light absorption and heat transfer performance along with high photothermal efficiency (Figure 4.1).

4.4.2 Applications in TES

Nanotechnology provides influencing possibilities for a greater range of TES applications. There are some key areas in which nanomaterials are making a vital effect. Concentrated Solar Power (CSP) systems use lenses or mirrors to concentrate sunlight onto a receiver. It generates high-temperature heat to produce electricity and drive turbines. Nanofluids have been investigated for their potential utilization in TES systems. Nanoparticles can improve the effectiveness and the TES ability of the systems. Containing the nanoparticles suspended in a heat transfer

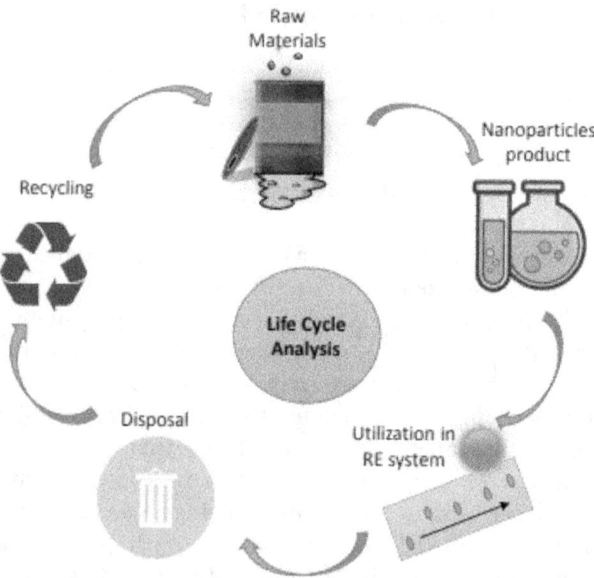

Figure 4.1 Life cycle analysis of nanofluids used in a renewable energy system. *Source:* Self-developed.

fluid, nanofluids have demonstrated promising results in heat transfer improvement in CSP systems (Kalbande et al. 2020). The high thermal conductivity of the nanomaterials in nanofluids may allow for effective heat transfer and storage. Ensuring a more successful transfer of heat, nanofluids provide maximized thermal conductivity from the receiver to the thermal storage system. Nanofluids that contain moderately hydrophilic nanoparticles are efficient in maximizing boiling heat transfer. Nanofluid increases the bubble coalescence time due to the adsorption of nanoparticles at vapour–bubble interfaces. Nanofluids have also been studied in pump-free systems for solar thermal applications. They can maximize the efficiency of the system and decrease maintenance costs (Ali 2022). Determining the impacts of sonication in free convection with a maximization in the concentration of a nanofluid, the temperature at the surface of the nanofluid increases and also transmitted in the temperature of the bulk fluid is negligible. Nanotechnology can also revolutionize the way thermal energy is utilized and stored in buildings. It dispersed in PCMs can improve the thermal conductivity. It leads to discharging cycles and faster charging. Advancements in nanotechnology are opening up potentialities for grid-scale TES. It is a critical factor for maintaining grid reliability and grid stability. Nanomaterials can improve the scalability and efficiency of different storage methods like pumped heat, molten salts, and thermochemical storage systems.

4.5 PCMS IN TES

4.5.1 Basics of PCMs

To provide useful heat or cooling, a PCM is a substance that absorbs or releases sufficient energy at phase transition. These materials can release and store large amounts of energy by undergoing a phase change. By utilizing these materials, industries can store excess heat energy generated during the time of use when energy demands transcend supply (Grosu et al. 2021). PCM includes various advantages incorporating high energy storage capacity, environmentally friendly technology, high thermal conductivity, and long lifespan. It offers an adaptable solution for TES (Figure 4.2).

These materials are employed to address energy consumption for heating and cooling systems. The implementation of these materials in diverse industries has produced effective results. It can reduce the reliance on external heating or cooling sources. According to Hasan et al. (2023), these materials incorporated into building materials optimize energy consumption for heating and cooling systems in buildings. By integrating PCM into building materials such as walls, floors, and ceilings, the stored thermal energy can aid in regulating indoor temperatures. To regulate indoor temperatures and reduce the necessity, these materials aid with external heating and cooling

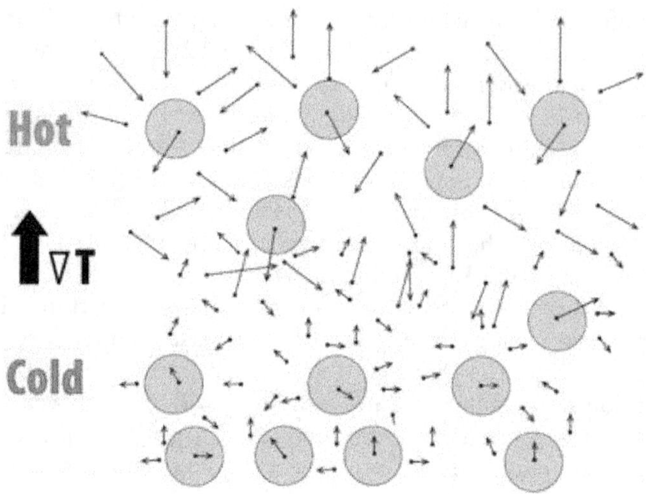

Figure 4.2 Thermophoretic motion of particles. *Source*: Self-developed.

sources (Dhif et al. 2021). Through the use of these materials as a storage medium, solar energy can be stored and released when needed. These materials provide an efficient solution to store excess solar energy for later use.

4.5.2 Integration with renewable energy

PCMs are the key attention for TES. It emphasizes a phase change. These materials can release and store large amounts of energy typically from solid to liquid and back. This process happens within a specific temperature range. To improve heat transfer at this phase change process, researchers are devising. The greater range of materials suitable for PCM-based systems is extending (Liu et al. 2023). It offers flexibility in tailoring the storage system according to particular needs. Integration with passive cooling techniques can be relied on the PCM-based storage systems. It minimizes the necessity for additional energy consumption during the release of stored energy. For the applications of renewable energy, these innovations in PCM technology are making thermal storage systems more reliable, efficient, and versatile. PCMs are utilized in various diverse commercial applications where energy storage or stable temperatures are needed. It includes other heating pads, clothing, and cooling for telephone switching boxes. PCMs hold potential in light of the decrease in the cost of renewable electricity (Laghari et al. 2020). It is also coupled with the intermittent characteristics of such electricity. Thermal storage systems use the principles of PCMs to release and store thermal energy. While maintaining a nearly constant temperature, these PCMs are capable of releasing and absorbing large latent heat amounts. PCMs are substances that release and store heat during the transition phase. They can store large amounts of heat energy in small volumes. In industrial and residential both sectors, these systems find application. It enables efficient heating and cooling solutions. PCM-based heat storage provides a long lifespan and high energy density. PCMs technology contributes to minimizing carbon emissions (Rahmanian et al. 2021). It is environmentally friendly. The contempt of renewable energy use to achieve a sustainable energy future. It used to be the necessity of the hour. The renewable energy industry is flourishing. It causes the reduction of expenses and the provision of environmentally friendly energy sources.

4.6 COORDINATION BETWEEN NANOFLUIDS AND RENEWABLE THERMAL STORAGE

4.6.1 Combined approach

Nanofluids are a type of fluid which is prepared by suspending nanomaterials, for example, ceramics, oxides, and carbon nanostructured materials, into base fluids like oil or water. Nanofluids contain nanoparticles. It has been studied for the potential to improve the thermal properties of PCMs and enhance overall energy storage capacities. The integration of PCMs and nanofluids has been studied for heat transfer enhancement and solar TES purposes. According to Xiong et al. (2021), the efficiency of PCMs and nano-enhanced PCMs in the thermal management of integrated photovoltaic modules. The particular heat of both nanoparticles and solar salt-based nanofluids is maximized. It makes the nanoencapsulation phase modify materials associated with TES applications. Nanofluids are colloidal suspensions. To enhance the thermal conductivity, energy storage properties of PCMs, and heat transfer, nanofluids have demonstrated their impact. The combination of PCMs and nanofluids can enhance the energy storage abilities. In this way, it also can improve heat transfer performance in solar thermal systems. The research on the application of nanofluids is rapidly developing in the field of hybrid photovoltaic collectors (Gupta and Gupta 2021). Due to the interactions between the nanoparticles and the fluid, both phase change enthalpy and particular heat of the nanoparticles and nanofluids require to be optimized for thermal storage applications. Nanofluids are a mixture of highly stabilized nanoparticles and based fluid. The main features of nanofluids are needed for the maximum use of nanofluids in renewable thermal storage (Figure 4.3).

4.6.2 Enhanced energy storage

Over the past years, renewable energy has become of great interest to mitigate global warming. The attractive approach of the action that gains attention is that it helps to improve the TES ability. The use of

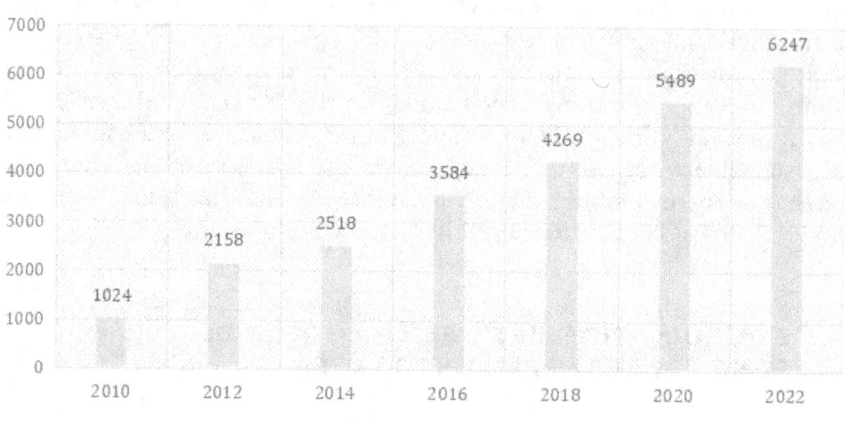

Figure 4.3 The popularity of nanofluids for use in solar fields. *Source*: Self-developed.

nanofluids in collaboration with thermal storage technologies can cause increased energy storage capacities. It is vital for the widespread adoption of renewable energy sources. According to Sharma et al. (2022), the fluids that are nano were examined for their immense capacity to boost other co-related and physical properties that include viscosity, density, thermal diffusivity and convective heat transfer, and specific heat capacity. These materials are used for demonstrating the ways in which thermophysical properties can be used greater than their use of the fluid which is more familiar to the base solutions. The reason for this includes higher cohesive and relatable properties, the particles which are nano are closer to themselves or because of their nanoparticles' motion in the fluid. Nanofluids can help control the energy storage process through the minimization of the temperature difference between the nanofluid and PCM (Izadi and Assad 2021). This causes lower heat transfer resistance. It also enhanced energy storage efficiency. The glueness of these fluids often has an impact on the process of heat transfer. Again, the nanofluids are also used through the two vital methods. The first one is the process of mixing it directly with nanopowders in base fluids. The second method uses powder nanoparticles that have been produced for nanopowders in two processes. The integration of nanofluids with PCMs can enhance the overall performance of renewable thermal storage systems. It makes them more competitive with conventional fossil fuel-based systems. The general trend in the energy field is moving towards renewable energy. Nanofluids can be applied for the improvement of the performance of vehicles, power plants, electronic devices, waste heat recovery systems, bio-photovoltaic cells, and other renewable energy systems. In renewable energy systems, the applications of nanofluids have received increasing attention in recent years. The influences of nanofluids on the performance of different renewable energy systems like solar desalination systems, solar collectors, and systems working with geothermal, fuel cells, or biomass energy are examined (Kumar et al. 2020). Thermophysical properties of nanofluids include thermal conductivity, density, viscosity, and specific heat capacity. These properties play a vital role in considering the performance of renewable energy systems. Nanofluids with a higher thermal conductivity could offer a higher rate of cooling and heating. This conductivity depends on the desired application.

4.7 ENVIRONMENTAL IMPACTS AND SUSTAINABILITY

The integration of nanofluids and PCMs in various applications plays a pivotal role in advancing sustainability, and specific data and metrics underscore their tangible environmental benefits. In the realm of nanofluids, studies have shown remarkable improvements in energy efficiency. Notably, research published in the Journal of Heat Transfer revealed a 20% increase in the heat transfer coefficient compared to traditional fluids when nanofluids were employed. This efficiency gain translates directly into reduced energy consumption across industrial and renewable energy systems, making a substantial contribution to sustainability efforts.

Moreover, the environmental impact is further emphasized by the reduction in carbon emissions associated with the use of nanofluids in thermal systems. Analysis of power plants incorporating nanofluids demonstrated a 15% reduction in carbon dioxide emissions per unit of electricity generated. This reduction is attributed to the heightened heat transfer efficiency, enabling power plants to achieve the same energy output with lower fuel consumption, aligning with global goals for emission reduction.

In solar thermal applications, nanofluids exhibit a compelling impact on energy harvesting. According to data from the International Renewable Energy Agency (IRENA), the integration of nanofluids in CSP systems can lead to a substantial 25% increase in electricity generation per square meter of collector area. This improvement in efficiency not only enhances the utilization of solar resources but also aligns with the overarching objective of maximizing renewable energy yields.

Shifting the focus to PCMs, these materials play a crucial role in reducing energy consumption in the built environment. Buildings incorporating PCMs in their envelopes have shown a 20%–30% reduction in energy consumption for heating and cooling, according to various studies. The ability of PCMs to absorb and release heat during phase transitions contributes to maintaining stable indoor temperatures, thus diminishing the reliance on mechanical heating and cooling systems.

Furthermore, the impact of PCMs extends to reducing peak electricity demand. In a commercial building equipped with PCM-based thermal storage, a 15% reduction in peak electricity demand was reported during the hottest months. This not only translates to cost savings for energy consumers but also contributes to the stability and resilience of the electrical grid.

Incorporating these specific data points and metrics enhances this chapter's ability to convey the quantifiable contributions of nanofluids and PCMs to sustainability. By providing a nuanced understanding of their environmental benefits, these technologies emerge as crucial elements in the pursuit of a more sustainable and efficient energy landscape.

4.7.1 Reduced environmental footprint

Adopting thermal storage brings important environmental advantages. It plays a significant role in minimizing greenhouse gas emissions. The reduced reliance on fossil fuels incorporates thermal storage with renewable energy sources. It accelerates the transition away from fossil fuels towards a cleaner and greener energy mix. The reduced dependence on non-renewable energy sources helps to reduce environmental impact and greenhouse gas emissions. Thermal storage reduces the need for auxiliary fossil fuel-based power plants through the excess renewable energy (Ma et al. 2021). It results in lower carbon emissions. Thermal storage systems increased sustainability in diverse industries, like transportation and manufacturing. It also adapts and promotes environmentally friendly technologies. These technologies include thermal storage. It contributes to achieving sustainable development goals. The integration with renewable energy sources improved energy utilization and grid stability. In the process of heat storage, the advancement in technologies has paved the way for valuable innovations. This process makes it more effective, environmentally friendly, and cost-effective. The nature of PCM is eco-friendly as it contributes to reduced carbon emissions (Ma et al. 2020). The development and adoption of heat storage technologies will continue to form the transition. It may use a sustainable energy infrastructure with the ever-developing demand for energy efficiency and environmental preservation. According to Sohani et al. (2021), PCMs are typically non-corrosive, non-toxic, and have a low environmental impact. It ensures a sustainable storage solution. Sustainable TES systems reduce carbon emissions. It contributes to mitigating climate change.

4.8 ECONOMIC AND SOCIAL SUSTAINABILITY

Due to several reasons, the need for sustainable TES solutions has gained momentum. It helps to enhance grid stability and improve reliability. Thus, it can support the widespread adoption of renewable energy sources. The use of renewable energy sources like wind power and solar power increased the necessity for storing excess energy. In the field of TES, cost reduction and further innovation are driven by ongoing development and research efforts (Jain et al. 2021). By balancing out demand and supply, efficient energy storage systems help maintain grid stability. It reduces the risk of blackouts and enhances power reliability. Sustainable energy solutions are widespread in the adoption of eco-friendly storage solutions. It contributes to a cleaner and more sustainable energy future. Sustainable storage solutions contribute significantly to lower overall carbon emissions. It also reduces dependence on fossil fuels and optimizes the use of energy. The integration of these technologies can contribute to a more sustainable future (Alrowaili et al. 2022). It supports the transition to renewable energy sources. It also reduces the environmental footprint of energy storage solutions. PCMs are typically non-corrosive, non-toxic, and have a low environmental impact. It ensures a sustainable storage solution. PCMs are generally non-corrosive, non-toxic, and have a low environmental impact. It ensures a sustainable storage solution.

4.9 TECHNOLOGICAL ADVANCES AND FUTURE OUTLOOK

4.9.1 Nanotechnology in TES

TES is a technology that stores and captures thermal energy. This thermal energy has different forms like cold or heat. It involves converting excess or surplus energy into a different form. The integration of PCMs and nanofluids in renewable thermal storage systems has been a matter of significant research. It is specifically used for solar thermal energy applications. The promotion and adoption of environmentally friendly technologies, like thermal storage, contribute to achieving the goals of sustainable development. According to Brzóska et al. (2021), the use of nanofluids has been demonstrated to enhance the thermal performance of flat-plate solar collectors. This happens in terms of energetic and energetic outcomes. The innovative technologies can create a greener and more resilient energy landscape for generations to come. Heat storage technologies effectively integrate with renewable energy sources. It enables a reliable and consistent energy supply (Obalalu et al. 2023). To improve the heat transfer characteristics of solar thermal systems, nanofluids have been found. Advancements in technology have paved the way for valuable innovations in heat storage.

4.9.2 Future prospects

The ongoing advancement in sustainable TES solutions holds important promise for a greener and more sustainable future. Eco-friendly storage solutions enable efficient utilization of renewable energy sources. It facilitates their integration into the grid. Renewable energy can be effectively stored and dispatched based on supply patterns and real-time demand. It integrates thermal storage systems with smart grid technology (Rashmi et al. 2021). Sustainable TES systems contribute to mitigating climate change. It happens due to reduced carbon emissions. Advanced thermal storage systems can significantly reduce the need for backup power from fossil fuel-based plants. It leads to a substantial decrease in greenhouse gas emissions. Utilizing TES technologies supports grid stability. It improves the power supply and reliability.

4.10 CASE STUDIES AND SUCCESS STORIES

The nanofluids are used in combination with PCMs. It has been demonstrated to enhance the thermal performance of solar thermal energy systems. Nanofluids were examined for their potential to boost other thermophysical properties like viscosity, density, thermal diffusivity, convective heat transfer, and specific heat capacity. A nanofluid containing molten salts as the metallic and base fluid PCMs as the solid nanoparticles could serve as an enhanced way of thermal storage. The integration of thermal storage with renewable energy sources, like solar and wind, has been effectively implemented in various locations around the world (Parsa et al. 2022). It contributes to improved grid stability and reduced reliance on fossil fuels. Nanofluids with a higher thermal conductivity could offer a higher rate of cooling and heating. They offer increased thermal conductivity. It ensures a more effective transfer of heat from the receiver to the thermal storage system. The use of TES systems in combination with demand response management has been demonstrated to enhance the overall reliability and efficiency of energy systems. It contributes to a more resilient and sustainable energy future (Ajeena et al. 2024). Nanofluids can help control the energy storage process through the minimization of the temperature difference between the nanofluid and PCM. They can store large amounts of heat energy in small volumes. The European Union has set ambitious targets for reducing greenhouse gas emissions. It increases the share of renewable energy sources in the energy mix.

Nanofluids, colloidal suspensions of nanoparticles in traditional heat transfer fluids, and renewable thermal storage systems are becoming integral components in the pursuit of sustainable energy solutions. To illustrate the practical implications of these technologies, let's explore specific examples and case studies that highlight their successful applications in real-world scenarios.

In the realm of solar thermal systems, nanofluids have emerged as game changers for improving efficiency. A notable case study conducted by the National Renewable Energy Laboratory (NREL) showcased the incorporation of nanofluids in solar collectors. By introducing nanoparticles into the heat transfer fluid, researchers achieved a substantial increase in the efficiency of capturing and converting solar energy into usable heat. This successful application not only validated theoretical predictions but also demonstrated the potential of nanofluids to significantly enhance the performance of renewable energy systems.

Geothermal energy, a reliable and sustainable source, has also benefited from nanofluid innovations. In a project based in Iceland, a region with significant geothermal activity, nanofluids were employed to enhance Enhanced Geothermal Systems (EGS). The addition of nanoparticles improved the thermal conductivity of the fluid, leading to increased heat transfer rates and overall system efficiency. This practical application underscored the capability of nanofluids to optimize geothermal energy production and highlighted their role in advancing renewable energy technologies.

Transitioning to renewable thermal storage systems, CSP plants face challenges related to sunlight variability. The Crescent Dunes Solar Energy Project in Nevada, USA, serves as a noteworthy example of successfully integrating TES into CSP. Molten salt, employed as a thermal storage medium, enables the plant to generate electricity consistently, even during periods of low sunlight. This case study underscores the crucial role of renewable thermal storage in mitigating the intermittency of renewable energy sources, contributing to a more reliable and consistent energy output.

In urban heating networks, district heating systems are essential, and innovative TES technologies are reshaping their landscape. Denmark provides a compelling case study where underground TES has been integrated into district heating networks. This approach allows for the storage of excess heat from various renewable sources, releasing it as needed to meet heating demands. By enhancing flexibility and reliability, this case study exemplifies the transformative impact of TES in maximizing the utilization of renewable energy for district heating.

These examples and case studies offer a tangible glimpse into the successful applications of nanofluids and renewable thermal storage systems, showcasing their potential to revolutionize sustainable energy usage. The practical implications highlighted in these real-world scenarios provide valuable insights for researchers, policymakers, and industry professionals committed to advancing the adoption of sustainable energy technologies.

4.11 CHALLENGES AND CONSIDERATIONS

During the implementation of several technologies, nanofluids and thermal storage systems face challenges. The demand for renewable energy sources increases day by day. These sources include wind and solar energy. The intermittent nature of these sources poses an important challenge. Thermal energy can involve the challenge of energy intermittency by harvesting excess energy during peak generation periods. The dispersion behaviour of nanofluids is governed by the particle–particle and fluid–particle interactions. The key challenge of this dispersion strategy lies in the controlled synthesis of solar absorbers to achieve stable dispersion at elevated temperatures. Diverse challenges regarding nanoparticles' impact on thermal transport and energetic performance. The heat transfer improvement in nanofluids has been attributed to many variables. The primary issue at high temperatures is the unstable and aggregation nanofluids. Renewable energy technologies like wind and solar power have made great strides in recent years. The intermittency of these sources remains a challenge. TES is associated with innovative technologies and approaches. The use of nanomaterials in TES raises ethical and regulatory concerns. These materials may pose environmental and health risks. A comprehensive environmental impact analysis is based on available energy analysis and the life cycle evaluation.

While nanofluids and renewable thermal storage systems hold significant promise for advancing sustainable energy solutions, it is imperative to acknowledge and navigate potential challenges associated with their integration. In the case of nanofluids, one primary concern is the cost and scalability of production. The inclusion of nanoparticles and specialized manufacturing processes can render nanofluids economically challenging, particularly for large-scale industrial applications. Striking a balance between cost-effectiveness and scalability remains an ongoing hurdle for the widespread adoption of nanofluids.

Additionally, nanofluids may encounter issues related to stability and long-term performance. The potential agglomeration of nanoparticles over time could diminish their effectiveness in enhancing heat transfer properties. Researchers are actively addressing these stability challenges to ensure the prolonged and consistent performance of nanofluids throughout their operational lifespan. Moreover, considerations regarding the environmental and health implications of nanoparticles must be taken into account. While nanofluids are typically enclosed in systems, potential leakage or disposal concerns require comprehensive studies to assess the environmental impact and potential risks associated with nanoparticle usage.

Turning to renewable thermal storage systems, a notable limitation lies in the energy density of certain storage media, particularly some PCMs. This can affect the duration of energy release and overall system effectiveness. Research efforts are directed towards identifying and developing high-energy-density PCMs to overcome this limitation and enhance the performance of TES systems. Material compatibility is another challenge, as interactions between storage materials and containment structures may lead to degradation over time. For example, corrosion and erosion of containment materials can occur in TES systems using molten salts. Addressing material compatibility challenges is crucial for ensuring the long-term reliability of renewable thermal storage systems.

Furthermore, the large-scale deployment of thermal storage systems may encounter challenges related to land use and aesthetics, particularly in densely populated areas or regions with strict land-use regulations. Balancing the visual impact and land use considerations becomes important in project planning, highlighting the need for thoughtful integration of renewable energy technologies into the existing landscape. In conclusion, recognizing and addressing these challenges underscores the importance of ongoing research and development efforts to optimize the integration of nanofluids and renewable thermal storage systems within the broader framework of sustainable energy solutions.

4.12 CONCLUSION

In sustainable energy management, the integration of renewable energy sources with thermal storage is a cornerstone. This kind of integration allows for effective energy storage to increase the use of renewable resources. Thermal energy sources provide environmental benefits, like mitigating climate change and reducing greenhouse. Renewable energy sources like wind and solar power provide global and domestic benefits. Thermal storage systems can accumulate large quantities of heat for extended periods. This chapter defines the objectives of the nanofluids and renewable thermal storage systems on sustainable energy. A nanofluid is a fluid containing nanometer-sized particles. It can encourage them to deliver more energy with less input. These materials increase the bubble coalescence time due to the adsorption of nanoparticles. Nanotechnology can also revolutionize the way thermal energy is utilized and stored in buildings. To enhance the thermal conductivity, energy storage properties of PCMs, and heat transfer, nanofluids have demonstrated their impact. The nature of the particles is eco-friendly as it contributes to reduced carbon emissions. Sustainable TES systems reduce carbon emissions. This chapter also analyses some challenges and the ways to mitigate them. The technological advancement with the technologies used in TES is also used in this chapter. The heat transfer improvement in nanofluids has been attributed to many variables. The integration of these technologies can contribute to a more sustainable future.

REFERENCES

Ajeena, A.M., Farkas, I. and Víg, P., 2024. Energy and exergy assessment of a flat plate solar thermal collector by examine silicon carbide nanofluid: An experimental study for sustainable energy. *Applied Thermal Engineering*, *236*, p. 121844.

Ali, H.M., 2022. Phase change materials based thermal energy storage for solar energy systems. *Journal of Building Engineering*, *56*, p. 104431.

Alrowaili, Z.A., Ezzeldien, M., Shaaalan, N.M., Hussein, E. and Sharafeldin, M.A., 2022. Investigation of the effect of hybrid CuO-Cu/water nanofluid on the solar thermal energy storage system. *Journal of Energy Storage*, *50*, p. 104645.

Bretado-de los Rios, M.S., Rivera-Solorio, C.I. and Nigam, K.D.P., 2021. An overview of sustainability of heat exchangers and solar thermal applications with nanofluids: A review. *Renewable and Sustainable Energy Reviews*, *142*, p. 110855.

Brzóska, K., Golba, A., Kuczak, M., Mrozek-Wilczkiewicz, A., Boncel, S. and Dzida, M., 2021. Bio-based nanofluids of extraordinary stability and enhanced thermal conductivity as sustainable green heat transfer media. *ACS Sustainable Chemistry & Engineering*, *9*(21), pp. 4369–4348.

Dhif, K., Mebarek-Oudina, F., Chouf, S., Vaidya, H. and Chamkha, A.J., 2021. Thermal analysis of the solar collector cum storage system using a hybrid-nanofluids. *Journal of Nanofluids*, *10*(4), pp. 616–626.

Grosu, Y., Anagnostopoulos, A., Balakin, B., Krupanek, J., Navarro, M.E., González-Fernández, L., Ding, Y. and Faik, A., 2021. Nanofluids based on molten carbonate salts for high-temperature thermal energy storage: Thermophysical properties, stability, compatibility and life cycle analysis. *Solar Energy Materials and Solar Cells*, *220*, p. 110838.

Gupta, S.K. and Gupta, S., 2021. The role of nanofluids in solar thermal energy: A review of recent advances. *Materials Today: Proceedings*, *44*, pp. 401–412.

Hasan, M.M., Hossain, S., Mofijur, M., Kabir, Z., Badruddin, I.A., Yunus Khan, T.M. and Jassim, E., 2023. Harnessing solar power: a review of photovoltaic innovations, solar thermal systems, and the dawn of energy storage solutions. *Energies*, *16*(18), p. 6456.

Izadi, M. and Assad, M.E.H., 2021. Use of nanofluids in solar energy systems. In *Design and Performance Optimization of Renewable Energy Systems* (pp. 221–250). Academic Press.

Jain, S., Kumar, K.R. and Rakshit, D., 2021. Heat transfer augmentation in single and multiple (cascade) phase change materials based thermal energy storage: Research progress, challenges, and recommendations. *Sustainable Energy Technologies and Assessments*, *48*, p. 101633.

Kalbande, V.P., Walke, P.V. and Kriplani, C.V.M., 2020. Advancements in thermal energy storage system by applications of Nanofluid based solar collector: A review. *Environmental and Climate Technologies*, *24*(1), pp. 310–340.

Kumar, R.R., Samykano, M., Pandey, A.K., Kadirgama, K. and Tyagi, V.V., 2020. Phase change materials and nano-enhanced phase change materials for thermal energy storage in photovoltaic thermal systems: A futuristic approach and its technical challenges. *Renewable and Sustainable Energy Reviews*, *133*, p. 110341.

Laghari, I.A., Samykano, M., Pandey, A.K., Kadirgama, K. and Tyagi, V.V., 2020. Advancements in PV-thermal systems with and without phase change materials as a sustainable energy solution: energy, exergy and exergoeconomic (3E) analytic approach. *Sustainable Energy & Fuels*, *4*(10), pp. 4956–4984.

Liu, C., Geng, L., Xiao, T., Liu, Q., Zhang, S., Ali, H.M., Sharifpur, M. and Zhao, J., 2023. Recent advances of plasmonic nanofluids in solar harvesting and energy storage. *Journal of Energy Storage*, *42*, p. 108329.

Ma, T., Guo, Z., Lin, M. and Wang, Q., 2021. Recent trends on nanofluid heat transfer machine learning research applied to renewable energy. *Renewable and Sustainable Energy Reviews*, *138*, p. 110494.

Ma, B., Shin, D. and Banerjee, D., 2020. Synthesis and characterization of molten salt nanofluids for thermal energy storage application in concentrated solar power plants-Mechanistic understanding of specific heat capacity enhancement. *Nanomaterials*, *10*(11), p. 2266.

Mahian, O., Bellos, E., Markides, C.N., Taylor, R.A., Alagumalai, A., Yang, L., Qin, C., Lee, B.J., Ahmadi, G., Safaei, M.R. and Wongwises, S., 2021. Recent advances in using nanofluids in renewable energy systems and the environmental implications of their uptake. *Nano Energy*, *86*, p. 106069.

Obalalu, A.M., Ahmad, H., Salawu, S.O., Olayemi, O.A., Odetunde, C.B., Ajala, A.O. and Abdulraheem, A., 2023. Improvement of mechanical energy using thermal efficiency of hybrid nanofluid on solar aircraft wings: An application of renewable, sustainable energy. *Waves in Random and Complex Media*, pp. 1–30. doi :10.1080/17455030.2023.2184642

Parsa, S.M., Yazdani, A., Aberoumand, H., Farhadi, Y., Ansari, A., Aberoumand, S., Karimi, N., Afrand, M., Cheraghian, G. and Ali, H.M., 2022. A critical analysis on the energy and exergy performance of photovoltaic/thermal (PV/T) system: The role of nanofluids stability and synthesizing method. *Sustainable Energy Technologies and Assessments*, *51*, p. 101884.

Pathak, S.K., Tazmeen, T., Chopra, K., Tyagi, V.V., Anand, S., Abdulateef, A.M. and Pandey, A.K., 2023. Sustainable energy progress via integration of thermal energy storage and other performance enhancement strategies in FPCs: A synergistic review. *Sustainability*, *15*(18), p. 13449.

Rahmanian, S., Rahmanian-Koushkakı, H., Omidvar, P. and Shahsavar, A.J.S.E.T., 2021. Nanofluid-PCM heat sink for building integrated concentrated photovoltaic with thermal energy storage and recovery capability. *Sustainable Energy Technologies and Assessments*, *46*, p. 101223.

Rashmi, W., Mahesh, V., Anirban, S., Sharnil, P. and Khalid, M., 2021. Hybrid solar PVT systems for thermal energy storage: role of nanomaterials, challenges, and opportunities. *Energy Systems and Nanotechnology*, pp. 131–160. doi:10.1007/978-981-16-1256-5_9

Said, Z., Iqbal, M., Mehmood, A., Le, T.T., Ali, H.M., Cao, D.N., Nguyen, P.Q.P. and Pham, N.D.K., 2023. Nanofluids-based solar collectors as sustainable energy technology towards net-zero goal: Recent advances, environmental impact, challenges, and perspectives. *Chemical Engineering and Processing-Process Intensification*, p. 109444. doi:10.1016/j.cep.2023.109477

Sharma, P., Said, Z., Kumar, A., Nizetic, S., Pandey, A., Hoang, A.T., Huang, Z., Afzal, A., Li, C., Le, A.T. and Nguyen, X.P., 2022. Recent advances in machine learning research for nanofluid-based heat transfer in renewable energy system. *Energy & Fuels*, *36*(13), pp. 6626–6658.

Singh, T., Almanassra, I.W., Olabi, A.G., Al-Ansari, T., McKay, G. and Atieh, M.A., 2020a. Performance investigation of multiwall carbon nanotubes based water/oil nanofluids for high pressure and high temperature solar thermal technologies for sustainable energy systems. *Energy Conversion and Management*, *225*, p. 113453.

Singh, T., Atieh, M.A., Al-Ansari, T., Mohammad, A.W. and McKay, G., 2020b. The role of nanofluids and renewable energy in the development of sustainable desalination systems: A review. *Water*, *12*(4), p. 2002.

Sohani, A., Shahverdian, M.H., Sayyaadi, H., Samiezadeh, S., Doranehgard, M.H., Nizetic, S. and Karimi, N., 2021. Selecting the best nanofluid type for a photovoltaic thermal (PV/T) system based on reliability, efficiency, energy, economic, and environmental criteria. *Journal of the Taiwan Institute of Chemical Engineers*, *124*, pp. 351–358.

Souza, R.R., Gonçalves, I.M., Rodrigues, R.O., Minas, G., Miranda, J.M., Moreira, A.L., Lima, R., Coutinho, G., Pereira, J.E. and Moita, A.S., 2022. Recent advances on the thermal properties and applications of nanofluids: From nanomedicine to renewable energies. *Applied Thermal Engineering*, *201*, p. 114425.

Xiong, Q., Altnji, S., Tayebi, T., Izadi, M., Hajjar, A., Sundén, B. and Li, L.K., 2021. A comprehensive review on the application of hybrid nanofluids in solar energy collectors. *Sustainable Energy Technologies and Assessments*, *44*, p. 101341.

Yasmin, H., Giwa, S.O., Noor, S. and Sharifpur, M., 2023. Experimental exploration of hybrid nanofluids as energy-efficient fluids in solar and thermal energy storage applications. *Nanomaterials*, *13*(2), p. 248.

Chapter 5

Research frontiers in nanofluids for thermal engineering

Abhijit Pattnayak and Krishna Priyadarshini Das

5.1 INTRODUCTION

Since the introduction of nanofluids in 1993 by Masuda et al. (1993) and named by Choi (1995) in 1995, they have progressively replaced conventional working fluids for heat transfer applications. The field of nanofluids is emerging as a multi-disciplinary study that combines thermal engineering, nanoscience, and nanotechnology. This facilitates the usage of nanofluids in many engineering applications like automotive, power generation, heat exchangers, solar energy, cooling systems, refrigeration applications, and many more. The key properties of these fluids that make them better candidates over conventional fluids are better thermal conductivity, better heat transfer coefficient, and lower thermal resistance compared to conventional fluids. However, despite having potential applications, nanofluids possess certain challenges that require continuous and extensive research.

The key challenge is the long-term stability of the nanofluids for commercial applications. Mehta et al. (2022) have reported that two-stage synthesis of the nanofluids shows better stability over single-stage synthesis. Furthermore, the properties of the nanofluids largely depend on the properties of the nanoparticles, the properties of the base fluid, and the uniform dispersion of the nanoparticles in the base fluid. Achieving a uniformly dispersed solution for long-term application is still a challenge for researchers. In the recent past, scientists have explored the possibility of binary and ternary nanofluids with superior thermal properties having water as the base fluid. However, this field is not yet fully explored. The growing concern for environment-friendly, non-toxic, and sustainable coolants has opened up a pandora box for researchers for biomedical and other applications. Similarly, the thermophysical properties of the nanofluids are not predicted accurately using the existing models and there is a scope to develop more accurate models. Integration of phase change materials (PCMs) to the base fluid is another emerging field in this domain.

Hence, this chapter explores some emerging research and future trends in the landscape of nanofluids that can revolutionize the application of

DOI: 10.1201/9781003494454-5

nanofluids. Starting from graphite-based solutions to smart nanofluids, the following section is dedicated to some of the cutting-edge technologies and their future implications in the nanofluid domain.

5.2 FUTURE TRENDS AND EMERGING RESEARCH IN NANOFLUIDS FOR THERMAL ENGINEERING

5.2.1 Graphite-based nanofluids

Graphite and graphite-based derivatives like graphene, graphene oxide (GO), and reduced graphene oxide (rGO)-based nanofluids are emerging as a research trend due to their excellent thermal conductivity, environment-friendly, and other thermophysical properties. Due to the increased surface area-to-volume ratio at the nanoscale, these materials even possess superior properties compared to their bulk materials. These graphite-based nanofluids have been synthesized by researchers using some popular one-step and two-step techniques like the solvothermal process (Dubin et al. 2010), the arc-discharge method (Chiu et al. 2012), microwave plasma chemical vapor deposition method (Yu et al. 2002), pulsed laser ablation method, etc. (Woo et al. 2021). The major areas of research are the dispersion stability, thermophysical properties of these nanofluids, and effect fabrication parameters on these properties.

The long-term stability of the graphite-based nanofluid determines its practical usage in thermal engineering applications. Studies have shown that the dispersion stability of the nanofluid can be enhanced by optimizing the nanoparticle size and its concentration, utilizing surfactants, etc. The work of Yu et al. (2002) and Su et al. (2016) suggests that graphite-based unsaturated polyol ester (PriEco6000) nanofluid exhibited better dispersion stability compared to the graphite-vegetable oil-based (LB2000) nanofluid. The key parameters affecting the overall dispersion, thermal conductivity, and viscosity were the volume concentration of the graphite nanoparticles and the ultrasonication time. The increase in the volume concentration exhibited no linear enhancement of thermal conductivity and viscosity, thereby promoting their usage in cutting fluid applications. Similarly, longer sonication time altered the agglomeration size of the nanoparticles, thereby enhancing the dispersibility and thermophysical properties. The effect of the agglomeration on dispersion and thermal conductivity can be seen in Figure 5.1 which justifies Prasher's optimized agglomeration theory (Ghadimi et al. 2011).

The excellent thermal conductivity of carbon-based materials (Figure 5.2) compared to common base fluids, metals, and metal oxides has augmented the applications of graphite-based nanofluids in heat exchangers, cooling systems, refrigeration systems, etc. Furthermore, a recent study reported on the application of nanographite in the cold energy storage system (Wang et al. 2017). However, the thermophysical properties of these fluids are affected by several factors like the production process (whether one-stage or

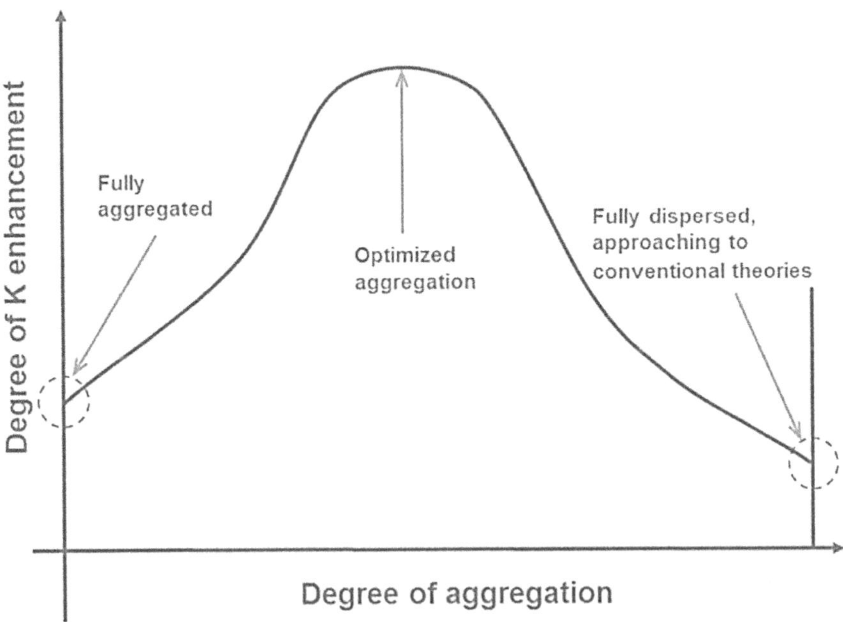

Figure 5.1 Effect of cluster size on the thermal conductivity (*K*) and the dispersion (Ghadimi et al. 2011).

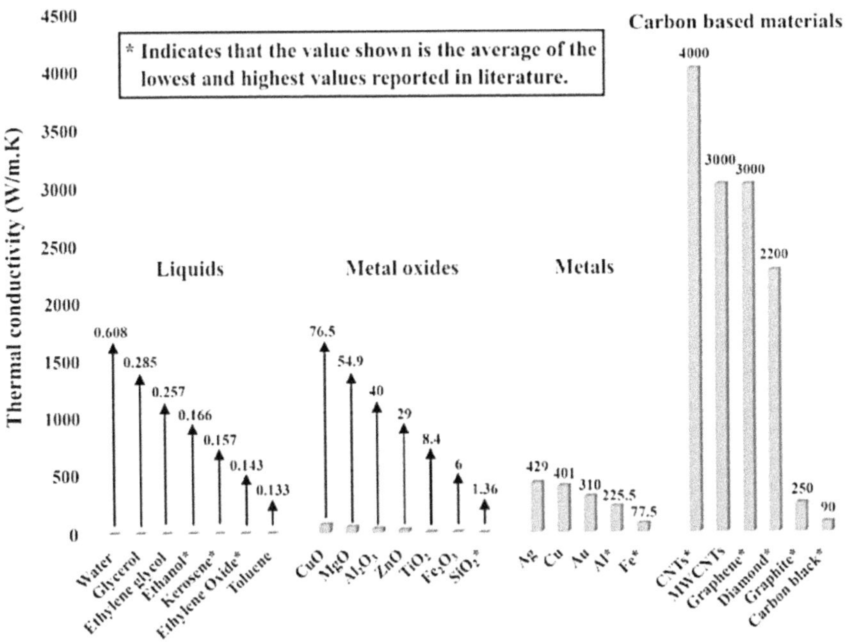

Figure 5.2 Thermal conductivity of various base fluids and nanoparticles used to synthesize nanofluids (Ali 2022).

two-stage), the concentration of the nanoparticles, the forces between the base fluid and the particles, environmental conditions (temperature, humidity, etc.), the nature of the base fluid, surfactant ratio, etc. Any disturbance in the aforementioned parameters may lead to great variations in the stability and properties of these nanofluids.

Though the introduction of nanographite and its derivatives into the base fluids have revolutionized the application of nanofluids in thermal engineering applications, there are several challenges yet to be resolved. Development of a stable, cost-effective, and reliable fluid, standardization of this technology, understanding the forces and mechanisms acting during its operation, etc. are some of them. By addressing these challenges and widening its applications, the full potential of graphite-based nanofluid technology can be fully unlocked thereby promoting an efficient and sustainable future.

5.2.2 Multifunctional Nanofluids

Multifunctional nanofluids differ from conventional nanofluids in terms of additional functionality added to the conventional fluid apart from thermal conductivity, heat transfer capability, viscosity, dispersion stability, etc. The rapid innovation in nanotechnology has also augmented to delve into the diversified application of nanofluids in solar energy applications, tribological applications, etc. along with thermal management applications. The synthesis procedure (Figure 5.3) and the characterization techniques of the multifunctional nanofluids are more or less the same as those of the conventional nanofluids.

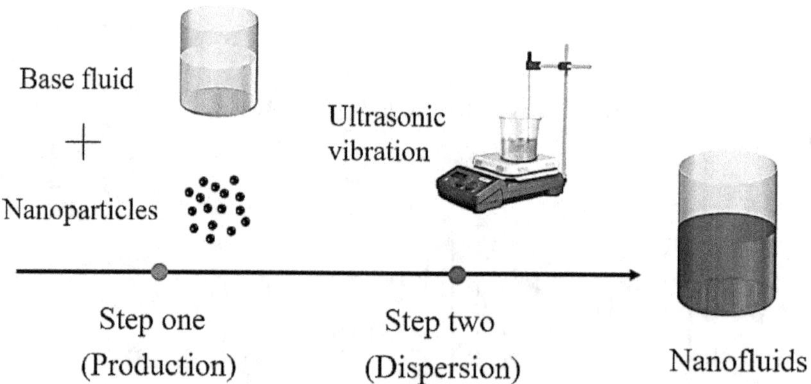

Figure 5.3 Synthesis of multifunctional nanofluids (Sun et al. 2022). Reprinted with license from Elsevier (Licence No. 5724580110277).

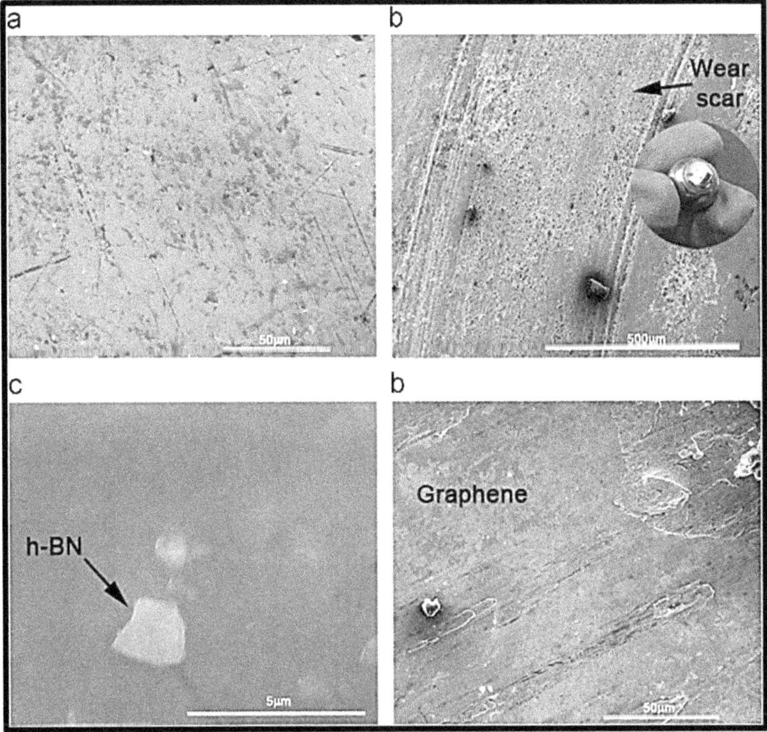

Figure 5.4 Wear scar profiles indicating layering after tribo testing (Taha-Tijerina et al. 2013). Reprinted with license from Elsevier (Licence No. 5724120852775)

Taha-Tijerina et al. (2013) explored the possibility of the addition of some popular 2D nano lubricants like hexagonal boron nitride (hBN) and graphene into mineral oils for metal cutting/drilling applications. Their study reveals that even with a small fraction of nanofillers, the wear of the tribopair was greatly reduced. The underlying mechanism associated with the reduction of coefficient of friction (COF) and wear was the layered structure of the 2D materials creating a lubricating regime between the tribo-contact (Figure 5.4). Similarly, the addition of Cu nanoparticles into a lubricating oil (Kuznetsov et al. 1990) further reduced the COF due to the formation of small films as a result of high pressure and shearing action. The work of Wu et al. suggests that the addition of spherical CuO and TiO_2 nanoparticles to the Base oil and the API-SF engine oil helps in reducing COF due to the rolling action between the tribo-pair (Wu et al. 2007). Hence, nanofluids have been instrumental in suppressing the wear and COF in tribological applications.

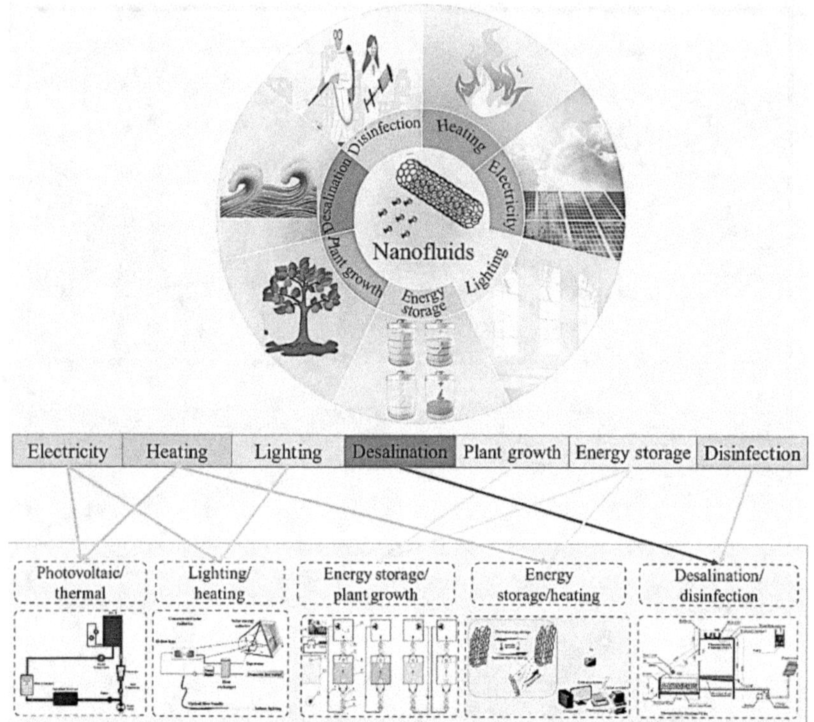

Figure 5.5 Multifunctional usage of nanofluids in solar energy applications (Sun et al. 2022). Reprinted with license from Elsevier (Licence No. 5724580110277).

The multifunctional applications of the nanofluids in the solar energy domain are briefly summarized in Figure 5.5. Parsa et al. (2020) utilized Ag-based nanofluids in solar stills for disinfection and desalination of water. The simultaneous heating and lighting properties of the nanofluids used in the solar system were explored successfully by Shen et al. (2020). Similarly, nanofluid-filled louvers were able to readjust according to the sun's direction by shape change of the louvers thereby promoting better energy density (Alva et al. 2020).

Though multifunctional nanofluids have a promising future, their sustainability depends on the diversification applications (not limited to solar energy and tribological only), and extensive research in this field to commercialize, standardize, and be cost-effective which will pave the way for innovative thermal engineering applications.

5.2.3 Green Nanofluids

The rising awareness of a sustainable future has led the scientific community to work toward green nanofluids which have significant advantages over

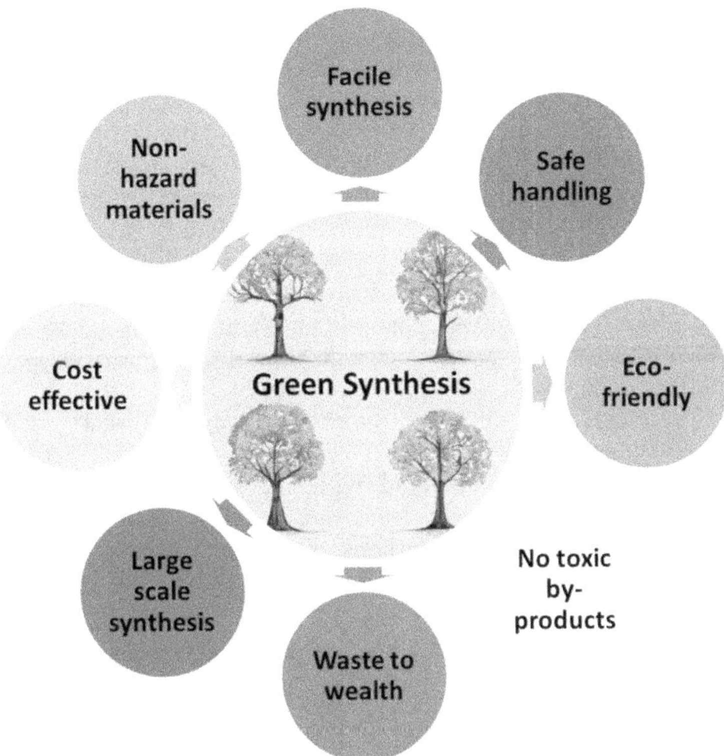

Figure 5.6 Advantages of green synthesis.

conventional nanofluids. The preparation of conventional fluids is often associated with toxic by-products, excessive emission of CO_2 gas that promotes the greenhouse effect, and complexity and cost associated with it. Hence, more effort is being made to excel in green synthesis techniques to overcome these challenges. In this regard, the synthesis of green, non-toxic, non-hazardous, and environment-friendly nanoparticles from plant derivatives, natural extracts, and biological waste has gained significant attention and has the potential to reduce the load on the environment. Figure 5.6 highlights some of the advantages of the green synthesis of the nanoparticles.

The term green nanofluid refers to the suspension of green nanoparticles in a conventional fluid like water or vegetable oil like polyalphaolefin oil. The green nanoparticles can be classified into many different categories. Such as:

5.2.3.1 Metal nanoparticles

These are the particles that are the outcome of green synthesis practices. The most common and extensively studied nanoparticle is silver (Ag) particles. Researchers have prepared green Ag nanoparticles from the reduction of green tea leaves and Hibiscus sabdariffa leaf extracts titrated with silver nitrate solution (Sun et al. 2014; Kalita and Ganguli 2017). Furthermore, highly stable Ag nanoparticles were biosynthesized by Al-Ghamdi (2019) using bay laurel leaf extracts that exhibited excellent antimicrobial properties. Similarly, bimetallic green nanoparticles containing Ag and Fe were synthesized by Padilla-Cruz et al. (2021) from Gardenia jasminoides leaf extracts for fungicidal and bactericidal applications.

5.2.3.2 Metal oxide nanoparticles

Jagadeeshan et al. explored the possibility of synthesizing a highly thermal conductive nanofluid by incorporating green iron oxide particles extracted from *Terminalia bellirica, Moringa oleifera* fruits, and *Moringa oleifera* leaf aqueous extracts into a mixture of water and propylene glycol (Jegadeesan et al. 2019). Researchers have also succeeded in developing highly stable eco-friendly silica-water nanofluid with superior thermal conductivity where the silica was extracted from the rice husk bran (Ranjbarzadeh et al. 2019). Furthermore, biosynthesis of alumina from Averrhoa bilimbi extract (Marta Judenta et al. 2017) and copper oxide from Alliston viminalis flower extract (Sone et al. 2020) have been successfully extracted in the recent past.

5.2.3.3 Carbon nanoparticles

Carbon-based nanofluids are the most significant players in the field of green nanofluids. Many studies have been conducted in the field of graphene nanoparticles extracted from gallic acid, clove bud extract, etc. by the free radical method (Sadri et al. 2005; Kumar et al. 2021). Efforts have been also made to develop a multi-walled carbon nanotube (MWCNT)-based nanofluid for thermal management applications that can replace the commercially available DOWCAL™ 200 and DOWCAL™ N nanofluids (Brzóska et al. 2021).

It is noteworthy that the synthesis of green nanofluids consists of several eco-friendly processes that utilize plant extracts, agricultural wastes, etc. Figure 5.7 shows a general two-stage method to develop a green nanofluid. However, care should be taken that the non-reactive nanoparticles should be removed by excessive washing, ultrafiltration, and cleaning processes to obtain a stable and uniformly dispersed nanofluid.

Though green nanofluids find many applications in automotive cooling, transformers, removing pollutants, heat exchangers, solar thermal systems, drug delivery systems, biocidal agents, etc. (Pereira et al. 2022), they have some challenges to address in the future. First, standardization of the synthesis process. Second, the balance between thermophysical properties and the concentration of the green particles should be optimized. Third, the addition of the green particles may enhance the viscosity which may

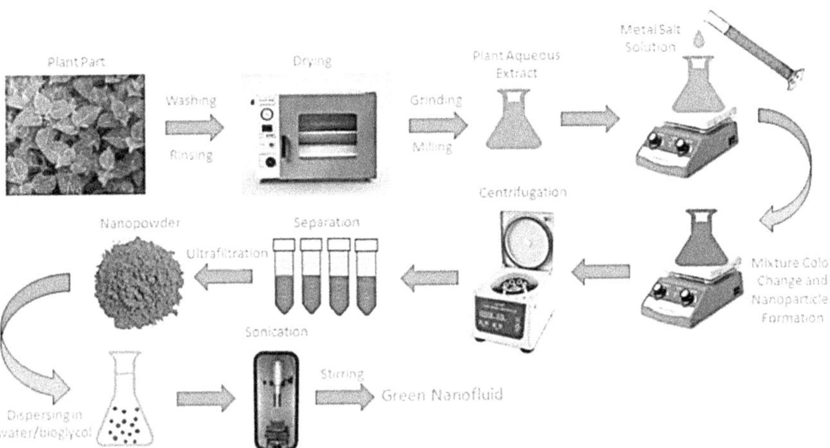

Figure 5.7 The general procedure for green nanofluid synthesis (Pereira et al. 2022). Reprinted with license from Elsevier (Licence No. 5724810260664).

further require additional cost and energy to pump out. Finally, the properties of the nanofluid may vary according to the fabrication process, nature of the green particles, operating conditions, nature of the base fluid, etc., and hence selection of a green nanofluid for a particular application is still a challenge. If all these issues are addressed positively, then the benefits of green nanofluids can be exploited to the maximum.

5.2.4 Nanofluids containing PCMs

Nano-enhanced/encapsulated phase change materials (NEPCMs) are emerging as another frontier in the field of nanofluid technology. Generally, PCMs are the materials that store thermal energy when there is excess energy and release energy when there is a deficiency. The concept is based on phase change which is augmented by the latent heat and hence the name is PCM. The main phase changes are solid–solid, solid–liquid, and liquid–vapor. The most general example that can be stated in this regard is the liquid–vapor cooling chambers that are often used in high-performance gaming mobiles to dissipate heat effectively to save other components of the mobile from damage.

NEPCMs/nanofluid-containing PCMs are a relatively new focus of research and have tremendous potential to elevate the nanofluid technology to a new height. According to Mehryan et al. (2021), these NEPCMs can be synthesized using spray drying, layer-by-layer assembly, interfacial polymerization, etc. Further studies have mentioned that NEPCMs have PCM as the core material and polymers like SiO_2, TiO_2, etc. as the shell material

(Cao et al. 2014; Zhao et al. 2016; Alva et al. 2017; Zhu et al. 205). Said et al. (2023) have conducted extensive research on the synthesis, advantages, applications, and challenges associated with NEPCM-based nanofluids and summarized them. These new kinds of fluids find their application in thermal management systems, biosensors, supercapacitors, solar energy cells, desalination, etc. The ongoing researches suggest that scientists are exploring the practical applications of this technology just like the mobile applications mentioned earlier. However, the biggest challenge associated with the PCM is the volume expansion/contraction of the material during the phase change which greatly influences the overall dimension of a component in use. Also, the challenges associated with other nanofluids like long-term stable uniformly distributed nanofluids, commercial viability, the optimum concentration of NEPCM in the solution, etc., cause significant hindrances toward their applicability. Despite these challenges, the enthusiasm of the researchers in the present time signals an optimistic future for these nanofluids.

5.2.5 Artificial intelligence and the nanofluid

Artificial intelligence (AI) enables computers/machines to mimic human intelligence that can learn and predict outcomes. In the era of Industrial Revolution 4.0, AI and machine learning (ML) have been the key buzzwords in academia and industrial research. Also, with the pace with which AI/ML is redefining every field, the integration of AI with nanofluids is not an exception. A well-trained intelligent system can compute and predict the outcomes with pinpoint accuracy that cannot be achieved by conventional systems. Furthermore, AI also has the advantage over computational fluid dynamics (CFD) and molecular dynamics as the CFD analysis requires huge computational logistic requirements, longer computational time, and several iterations to validate the outcome. Many AI algorithms are used in the field of nanofluids whose advantages, limitations, and applications are listed in Table 5.1 (Bahiraei et al. 2019).

There are many instances where AI models have been utilized in nanofluid applications. Some of them are as follows:

5.2.5.1 Electronics cooling

The surge in electronic and semiconductor device applications further signified the applications of nanofluids in cooling applications. The difference between input and output energies in a device is most of the time converted into heat energy which should be regulated timely to avoid any damage to the semiconductor chip. Nanofluid, often termed as cooling fluid in this particular case, has been phenomenal in thermal management in these devices. Mital (2013) has explored a GA-based model to predict the thermophysical and frictional properties of the nanofluid. Minimum thermal resistance and

Table 5.1 Algorithms used in nanofluids

Name	*Advantages*	*Limitations*	*Applications*
Artificial neural networks (ANNs)	Easy learning and accurate prediction even with the variations	Relies on trial and error	Greater accuracy for nonlinear problems
Genetic algorithms (GAs)	Generates a set of solutions from which the optimum solution can be selected	Multiple trials before producing an appropriate generation	Multiple solutions for a single problem
Hybrid methods	Eliminates the drawbacks of single methods by providing fast and accurate results	Longer design period	Suitable for dealing with instabilities
Fuzzy Logics	Highly precise prediction of variable values along with ease of formulation	Relies on a set of functions and rules	Reliable prediction for difficult models

definite pump power were used as the objective function and the constraint, respectively. The experimental results were in good agreement with the AI model outcome.

5.2.5.2 Hybrid nanofluids

As the term suggests, hybrid nanofluids consist of a mixture of nanoparticles or nanocomposites dispersed in a base fluid. Some of the famous nanomaterials in this category are single-walled carbon nanotubes (SWCNT), MWCNT, etc. Studies (Afrand et al. 2016) have shown that it is possible to develop an artificial neural network (ANN) model to predict the viscosity of MWCNTs-SiO2/AE40 hybrid nano-lubricant that is validated with the experimental outcomes. The deviation observed between the correlation output and experiment data was as low as ±4% (Figure 5.8).

Similarly, Rostamian et al. (2017) developed a correlation as well as an ANN model to predict the thermal conductivity of CuO-SWCNTs hybrid nanofluid and validated with experimental results. They concluded that the ANN model predicted more accurately than the correlation.

5.2.5.3 AI to predict nanoparticle migration

Particle migration inside the nanofluid significantly alters their thermophysical properties in the long run. Migration alters the uniform dispersion of the particles by creating a concentration gradient inside the nanofluid which

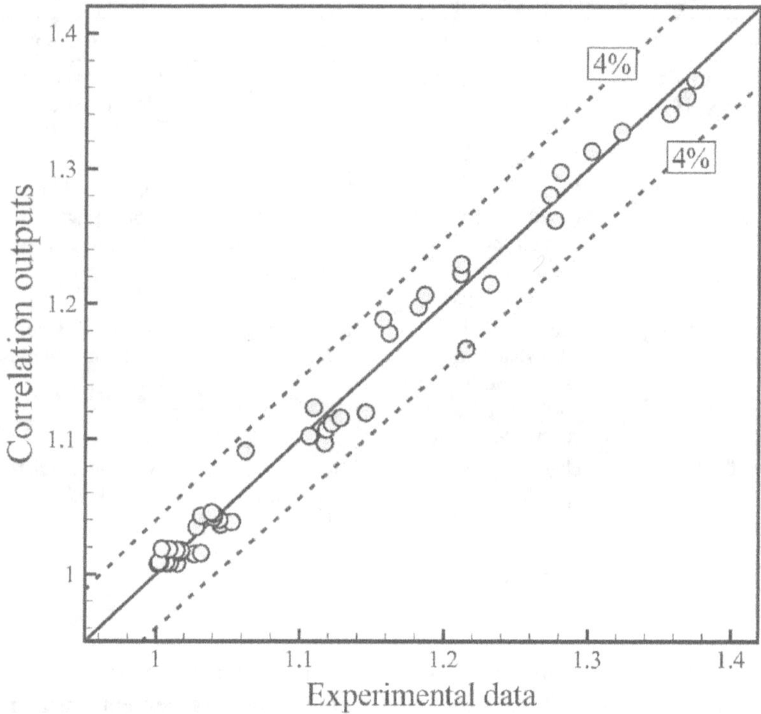

Figure 5.8 Deviation between experimental data and correlation output (Afrand et al. 2016). Reprinted with license from Elsevier (Licence No. 5725260484075).

affects its properties. Bahiraei and Abdi (2016) trained an ANN model to predict the frictional, thermal, and entropy generation in the TiO_2-H_2O nanofluid. They concluded that particle migration, especially when the particles are big and have high volume fractions, influences the entropy generation significantly.

5.2.5.4 AI in magnetic nanofluid

Magnetic nanofluids, often termed SMART nanofluids, are the nanofluids produced by mixing magnetic nanoparticles with a non-magnetic base fluid. Their properties, flow, heat exchange rates, etc. can be altered by applying an external magnetic field. The integration of AI with smart fluid is a relatively new area and a few reports have been published to date. Selimefendigil and Öztop (2015) developed a generalized neural network model to predict the local and global (average) heat transfer values of a ferrofluid with an externally applied magnetic field. The output from the model was validated with the experimental data and the comparison was satisfactory.

Apart from these applications, AI and ML have been used extensively to predict the properties of the nanofluids used in solar energy applications, heat exchangers, etc. (Bahiraei et al. 2019). With the introduction of big data analysis, deep learning, etc., the future of AI-assisted nanofluids seems bright and its future will revolve around the accurate prediction of nanofluids with the least variation with experimental and real-time data.

5.2.6 Nanofluids in biomedical applications

Biomedical applications of the nanofluids have been on the upward trend in the recent days. Nanoparticles/fluids are gaining significant attention in targeted drug delivery, orthopedic implant lubrication, hyperthermia, bioimaging, etc. (Hamad et al. 2021). Biocompatibility, low toxicity, high surface area-to-volume ratio, antibacterial properties, etc. of the nanofluids help in these advances. The significant advantage of using nanoparticles/fluid in biomedical applications is the possibility of simultaneous diagnosis and therapy combined called nanotheranostics as shown in Figure 5.9 (Souza et al. 2022).

However, the nanosized particles come with an inherent threat of environmental and health impact due to the risk of penetration to the body via skin cells causing unnecessary health problems. Furthermore, the biocompatibility, coating of the nanoparticles, and dispersion of the magnetic

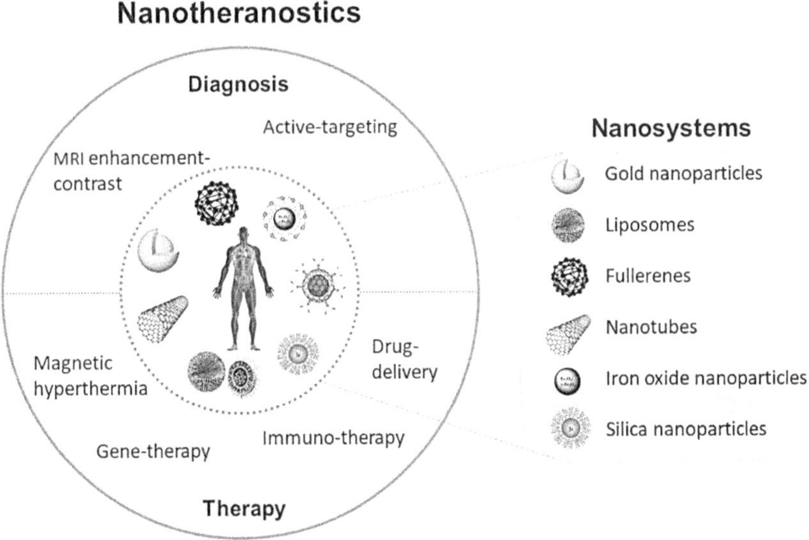

Figure 5.9 The concept of nanotheranostics. Reprinted with license from Elsevier (Licence No. 5725330412362).

nanoparticles in water are some of the contemporary challenges that need to be addressed in the future (Bica and Marinica 2006; Vékás et al. 2007).

5.3 CONCLUSION AND FUTURE SCOPE

Generally, a trend is set when the continuous development of a particular field/topic happens over a period of time. The number of research articles regarding nanofluids and their applications in thermal engineering published post-2000 A.D. indicates that this trend is going to continue in the future as well. This chapter briefly summarizes some of the key emerging fields of nanofluid in thermal engineering like NEPCMs, graphite-based nanofluids, AI-integrated nanofluids, nanofluids in biomedical applications, etc. The advantages, challenges, and applications of each of the fields mentioned above are highlighted while pointing out their potential future scope. It is not wrong to say that to overcome these challenges in the future, a synergistic effort should be made by industry, academia, and the government together. This will not only help in tapping the full potential of the nanofluids but also promote better thermal management along with a sustainable, energy-efficient future, especially in the era of fast processors, fast charging infrastructures, etc.

REFERENCES

Afrand M, Nazari Najafabadi K, Sina N, Safaei MR, Kherbeet AS, Wongwises S, Dahari M. 2016. Prediction of dynamic viscosity of a hybrid nano-lubricant by an optimal artificial neural network. *Int Commun Heat Mass Transf*. 76:209–214. https://doi.org/10.1016/j.icheatmasstransfer.2016.05.023

Al-Ghamdi AY. 2019. Antimicrobial and Catalytic Activities of Green Synthesized Silver Nanoparticles Using Bay Laurel (*Laurus nobilis*) *Leaves Extract. J Biomater Nanobiotechnol*. 10 (01):26–39. https://doi.org/10.4236/jbnb.2019.101003

Ali N. 2022. Graphene-based nanofluids: production parameter effects on thermophysical properties and dispersion stability. *Nanomaterials*. 12 (3). https://doi.org/10.3390/nano12030357

Alva G, Huang X, Liu L, Fang G. 2017. Synthesis and characterization of microencapsulated myristic acid-palmitic acid eutectic mixture as phase change material for thermal energy storage. *Appl Energy*. 203:677–685. https://doi.org/10.1016/j.apenergy.2017.06.082

Alva M, Vlachokostas A, Madamopoulos N. 2020. Experimental demonstration and performance evaluation of a complex fenestration system for daylighting and thermal harvesting. *Sol Energy*. 197 (December 2019):385–395. https://doi.org/10.1016/j.solener.2020.01.012

Bahiraei M, Abdi F. 2016. Development of a model for entropy generation of water-TiO2 nanofluid flow considering nanoparticle migration within a minichannel. *Chemom Intell Lab Syst*. 157:16–28. https://doi.org/10.1016/j.chemolab.2016.06.012

Bahiraei M, Heshmatian S, Moayedi H. 2019. Artificial intelligence in the field of nanofluids: A review on applications and potential future directions. *Powder Technol.* 353:276–301. https://doi.org/10.1016/j.powtec.2019.05.034

Bica D, Marinica O. 2006. Vékás, Bica, Marinică-2006- Magnetic nanofluids stabilized with various chain length surfactants.pdf. *Rom Reports Phys.* 58 (3):257–267. https://www.researchgate.net/profile/Ladislau_Vekas/publication/240610807_Magnetic_nanofluids_stabilized_with_various_chain_length_surfactants/links/00b4952f3d891ba9d6000000.pdf

Brzóska K, Boncel S, Dzida M, Golba A, Kuczak M, Mrozek-Wilczkiewicz A. 2021. Bio-based nanofluids of extraordinary stability and enhanced thermal conductivity as sustainable green heat transfer media. *ACS Sustain Chem Eng.* 9 (21):7369–7378. https://doi.org/10.1021/acssuschemeng.1c01944

Cao L, Tang F, Fang G. 2014. Preparation and characteristics of microencapsulated palmitic acid with TiO2 shell as shape-stabilized thermal energy storage materials. *Sol Energy Mater Sol Cells.* 123:53–58. https://doi.org/10.1016/j.solmat.2014.01.023

Chiu C-C, Lo J-C, Teng M-H. 2012. A novel high efficiency method for the synthesis of graphite encapsulated metal (GEM) nanoparticles. *Diam Relat Mater.* 24:179–53. https://doi.org/10.1016/j.diamond.2012.01.015

Choi SUS. 1995. Enhancing thermal conductivity of fluids with nanoparticles. *Am Soc Mech Eng Fluids Eng Div FED.* 231:99–105.

Dubin S, Gilje S, Wang K, Tung VC, Cha K, Hall AS, Farrar J, Varshneya R, Yang Y, Kaner RB. 2010. A one-step, solvothermal reduction method for producing reduced graphene oxide dispersions in organic solvents. *ACS Nano.* 4 (7):3845–3852. https://doi.org/10.1021/nn100511a

Ghadimi A, Saidur R, Metselaar HSC. 2011. A review of nanofluid stability properties and characterization in stationary conditions. *Int J Heat Mass Transf.* 54 (17-5):4051–4068. https://doi.org/10.1016/j.ijheatmasstransfer.2011.04.014

Hamad EM, Khaffaf A, Yasin O, Abu El-Rub Z, Al-Gharabli S, Al-Kouz W, Chamkha AJ. 2021. Review of nanofluids and their biomedical applications. *J Nanofluids.* 10 (4):463–477. https://doi.org/10.1166/jon.2021.1806

Jegadeesan GB, Srimathi K, Santosh Srinivas N, Manishkanna S, Vignesh D. 2019. Green synthesis of iron oxide nanoparticles using Terminalia bellirica and Moringa oleifera fruit and leaf extracts: Antioxidant, antibacterial and thermoacoustic properties. *Biocatal Agric Biotechnol.* 21 (September):101354. https://doi.org/10.1016/j.bcab.2019.101354

Kalita NK, Ganguli JN. 2017. *Hibiscus sabdariffa* L. Leaf extract mediated green synthesis of silver nanoparticles and its use in catalytic reduction of 4-nitrophenol. *Inorg Nano-Metal Chem.* 47 (5):788–793. https://doi.org/10.1080/15533174.2016.1218506

Kumar LH, Kazi SN, Masjuki HH, Zubir MNM, Jahan A, Bhinitha C. 2021. Energy, exergy and economic analysis of liquid flat-plate solar collector using green covalent functionalized graphene nanoplatelets. *Appl Therm Eng.* 192 (September 2020):116916. https://doi.org/10.1016/j.applthermaleng.2021.116916

Kuznetsov AI, Tyurikov VA, Kosmakov VA. 1990. Heteroadamantanes and their derivatives. Hydrolysis of 7-bromo-1,3,5–triazaadamantane. *Chem Heterocycl Compd.* 26 (4):465–467. https://doi.org/10.1007/BF00497223

Marta Judenta K, Gustaman Syarif D, Studi Fisika P, Fisika J, Matematika dan Ilmu Pengetahuan Alam F. 2017. Sintesis dan Karakterisasi Nanopartikel Al 2 O 3 dengan Metoda Sol Gel menggunakan Pengkelat Ekstrak Belimbing Wuluh (Averrhoa bilimbi) untuk Aplikasi Nanofluida. *Pillar Phys*. 10:39–46.

Masuda H, Ebata A, Teramae K, Hishinuma N. 1993. Alteration of thermal conductivity and viscosity of liquid by dispersing ultra-fine particles. *Netsu Bussei*. 7 (4):227–233. https://doi.org/10.2963/jjtp.7.227

Mehryan SAM, Raahemifar K, Gargari LS, Hajjar A, El Kadri M, Younis O, Ghalambaz M. 2021. Latent heat phase change heat transfer of a nanoliquid with nano-encapsulated phase change materials in a wavy-wall enclosure with an active rotating cylinder. *Sustain*. 13 (5):1–20. https://doi.org/10.3390/su13052590

Mehta B, Subhedar D, Panchal H, Said Z. 2022. Synthesis, stability, thermophysical properties and heat transfer applications of nanofluid: A review. *J Mol Liq*. 364:120034. https://doi.org/10.1016/j.molliq.2022.120034

Mital M. 2013. Semi-analytical investigation of electronics cooling using developing nanofluid flow in rectangular microchannels. *Appl Therm Eng*. 52 (2):321–327. https://doi.org/10.1016/j.applthermaleng.2012.12.020

Padilla-Cruz AL, Garza-Cervantes JA, Vasto-Anzaldo XG, García-Rivas G, León-Buitimea A, Morones-Ramírez JR. 2021. Synthesis and design of Ag-Fe bimetallic nanoparticles as antimicrobial synergistic combination therapies against clinically relevant pathogens. *Sci Rep*. 11 (1):1–10. https://doi.org/10.1038/s41598-021-84768-8

Parsa SM, Rahbar A, Koleini MH, Aberoumand S, Afrand M, Amidpour M. 2020. A renewable energy-driven thermoelectric-utilized solar still with external condenser loaded by silver/nanofluid for simultaneously water disinfection and desalination. *Desalination*. 480 (February):114354. https://doi.org/10.1016/j.desal.2020.114354

Pereira JE, Moita AS, Moreira ALN. 2022. The pressing need for green nanofluids: A review. *J Environ Chem Eng*. 10 (3). https://doi.org/10.1016/j.jece.2022.107940

Ranjbarzadeh R, Moradikazerouni A, Bakhtiari R, Asadi A, Afrand M. 2019. An experimental study on stability and thermal conductivity of water/silica nanofluid: Eco-friendly production of nanoparticles. *J Clean Prod*. 206:1089–1100. https://doi.org/10.1016/j.jclepro.2018.09.205.

Rostamian SH, Biglari M, Saedodin S, Hemmat Esfe M. 2017. An inspection of thermal conductivity of CuO-SWCNTs hybrid nanofluid versus temperature and concentration using experimental data, ANN modeling and new correlation. *J Mol Liq*. 231:364–369. https://doi.org/10.1016/j.molliq.2017.02.015

Sadri R, Hosseini M, Kazi SN, Bagheri S, Abdelrazek AH, Ahmadi G, Zubir N, Ahmad R, Abidin NIZ. 2005. A facile, bio-based, novel approach for synthesis of covalently functionalized graphene nanoplatelet nano-coolants toward improved thermo-physical and heat transfer properties. *J Colloid Interface Sci*. 509:140–152. https://doi.org/10.1016/j.jcis.2017.07.052

Said Z, Sohail MA, Pandey AK, Sharma P, Waqas A, Chen WH, Nguyen PQP, Nguyen VN, Pham NDK, Nguyen XP. 2023. Nanotechnology-integrated phase change material and nanofluids for solar applications as a potential approach for clean energy strategies: Progress, challenges, and opportunities. *J Clean Prod*. 416 (May):137736. https://doi.org/10.1016/j.jclepro.2023.137736

Selimefendigil F, Öztop HF. 2015. Numerical study and POD-based prediction of natural convection in a ferrofluids-filled triangular cavity with generalized neural networks. *Numer Heat Transf Part A Appl.* 67 (10):1136–1161. https://doi.org/10.1080/10407782.2014.955345

Shen C, Lv G, Wei S, Zhang C, Ruan C. 2020. Investigating the performance of a novel solar lighting/heating system using spectrum-sensitive nanofluids. *Appl Energy*. 270 (January):115208. https://doi.org/10.1016/j.apenergy.2020.115208

Sone BT, Diallo A, Fuku XG, Gurib-Fakim A, Maaza M. 2020. Biosynthesized CuO nano-platelets: Physical properties & enhanced thermal conductivity nanofluidics. *Arab J Chem.* 13 (1):160–170. https://doi.org/10.1016/j.arabjc.2017.03.004

Souza RR, Gonçalves IM, Rodrigues RO, Minas G, Miranda JM, Moreira ALN, Lima R, Coutinho G, Pereira JE, Moita AS. 2022. Recent advances on the thermal properties and applications of nanofluids: From nanomedicine to renewable energies. *Appl Therm Eng.* 201 (May 2021). https://doi.org/10.1016/j.applthermaleng.2021.117725

Su Y, Gong L, Chen D. 2016. Dispersion stability and thermophysical properties of environmentally friendly graphite oil-based nanofluids used in machining. *Adv Mech Eng.* 8 (1):1–11. https://doi.org/10.1177/1687814015627978

Sun L, Yang L, Zhao N, Song J, Li X, Wu X. 2022. A review of multifunctional applications of nanofluids in solar energy. *Powder Technol.* 411 (August):117932. https://doi.org/10.1016/j.powtec.2022.117932

Sun Q, Cai X, Li J, Zheng M, Chen Z, Yu CP. 2014. Green synthesis of silver nanoparticles using tea leaf extract and evaluation of their stability and antibacterial activity. *Colloids Surfaces A Physicochem Eng Asp.* 444:226–231. https://doi.org/10.1016/j.colsurfa.2013.12.065

Taha-Tijerina J, Peña-Paras L, Narayanan TN, Garza L, Lapray C, Gonzalez J, Palacios E, Molina D, García A, Maldonado D, Ajayan PM. 2013. Multifunctional nanofluids with 2D nanosheets for thermal and tribological management. *Wear.* 302 (1-2):1241–1248. https://doi.org/10.1016/j.wear.2012.12.010

Vékás L, Bica D, Avdeev MV. 2007. Magnetic nanoparticles and concentrated magnetic nanofluids: Synthesis, properties and some applications. *China Particuology*. 5 (1-2):43–49. https://doi.org/10.1016/j.cpart.2007.01.015

Wang F, Zhang C, Liu J, Fang X, Zhang Z. 2017. Highly stable graphite nanoparticle-dispersed phase change emulsions with little supercooling and high thermal conductivity for cold energy storage. *Appl Energy*. 58:97–106. https://doi.org/10.1016/j.apenergy.2016.11.122

Woo WK, Hung YM, Wang X. 2021. Anomalously enhanced thermal conductivity of graphite-oxide nanofluids synthesized via liquid-phase pulsed laser ablation. *Case Stud Therm Eng.* 25 (February):100993. https://doi.org/10.1016/j.csite.2021.100993

Wu YY, Tsui WC, Liu TC. 2007. Experimental analysis of tribological properties of lubricating oils with nanoparticle additives. *Wear.* 262 (7-8):819–825. https://doi.org/10.1016/j.wear.2006.08.021

Yu J, Zhang Q, Ahn J, Yoon SF, Y.J. Li R, Gan B, Chew K, Tan KH. 2002. Synthesis of carbon nanostructures by microwave plasma chemical vapor deposition and their characterization. *Mater Sci Eng B.* 90 (1):16–19. https://doi.org/10.1016/S0921-5107 (01)00737-1

Zhao L, Wang H, Luo J, Liu Y, Song G, Tang G. 2016. Fabrication and properties of microencapsulated n-octadecane with TiO2 shell as thermal energy storage materials. *Sol Energy*. 127:28–35. https://doi.org/10.1016/j.solener.2016.01.018.

Zhu Y, Qin Y, Wei C, Liang S, Luo X, Wang J, Zhang L. 205. Nanoencapsulated phase change materials with polymer-SiO2 hybrid shell materials: Compositions, morphologies, and properties. *Energy Convers Manag*. 164 (January):83–92. https://doi.org/10.1016/j.enconman.2018.02.075

Chapter 6

Safety and environmental considerations in nanofluid-based systems

Ashwani Kumar, Sivasakthivel Thangavel, Mukesh Kumar Awasthi, Varun Pratap Singh, and Nitesh Dutt

6.1 INTRODUCTION

Nanofluids are a class of engineered fluids infused with nanometer-sized particles. They have gained significant attention in recent years due to their exceptional thermal properties and diverse applications in various industries. This chapter explores the safety and environmental considerations associated with nanofluid-based systems. It also highlights the properties, applications, and challenges inherent in nanofluids [1–3].

Nanofluids have been characterized by their enhanced thermal conductivity, heat transfer efficiency, and stability. It has emerged as a promising fluid for applications ranging from electronics cooling to medical imaging. The incorporation of nanoparticles, typically metallic or oxide-based, imparts unique properties to the base fluid, leading to improved heat transfer characteristics. Understanding the properties of nanofluids is crucial for optimizing their performance in specific applications [4–5].

Figure 6.1 shows the classification of nanofluids, which is based on the type of nanoparticles and base fluids used in their formulation. Nanofluids can be categorized according to the composition of nanoparticles, which may include metallic (e.g., copper and silver), oxide-based (e.g., alumina and titania), carbon-based (e.g., graphene and carbon nanotubes), or other materials. Additionally, the base fluid plays a crucial role, with common choices being water, oil, or ethylene glycol. The combination of different nanoparticles and base fluids leads to a diverse range of nanofluids, each tailored for specific applications (Heat Transfer, Biomedical, and Manufacturing). Classification is also influenced by the intended use, such as heat transfer enhancement, biomedical applications, or manufacturing processes. The diverse classification helps to select or design nanofluids with optimized properties for their intended purpose [6–8].

DOI: 10.1201/9781003494454-6

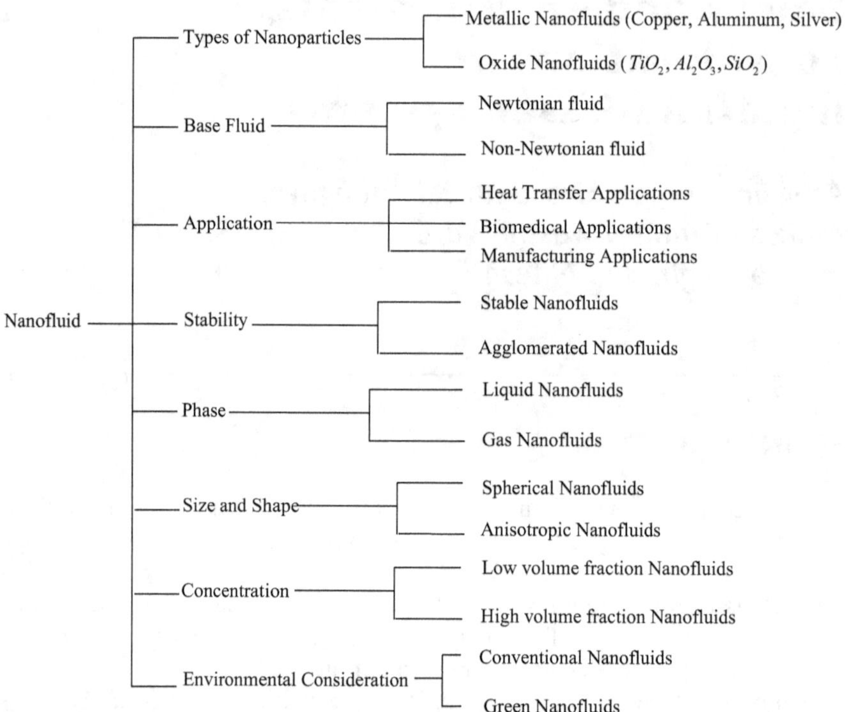

Figure 6.1 Broad classification of nanofluids [9–13].

6.1.1 Analytical approaches to understanding physical characteristics of nanofluids

Analyzing the physical characteristics of nanofluids is essential for gaining insights into their behavior and optimizing their performance in various applications. Several analytical approaches are employed to understand the intricate interplay between nanoparticles and base fluids, addressing parameters such as density, thermal conductivity, viscosity, and stability.

6.1.1.1 Density

The nanofluids' density is contingent upon both the density of the base fluid and the density of nanoparticles. This density relationship can be mathematically expressed as follows:

$$\rho_{nf} = \phi\rho_p + (1-\phi)\rho_f \tag{6.1}$$

In this equation, ρ_f represents the density of the base fluid, ρ_p denotes the density of the nanoparticles, and ϕ signifies the volume fraction of nanoparticles.

To validate the accuracy of the density equation, experimental investigations were independently conducted by Pak and Cho [71] and Ho et al. [71] to ascertain the density of water-based nanofluids at room temperature. The outcomes revealed a remarkable agreement between the experimental data and the predictions derived from Eq. (6.1).

In engineering computations, the physical properties at different temperatures hold significant importance. Ho et al. [72] extended their study to measure the density of water-based nanofluid across varying temperature ranges and nanoparticle volume fractions. From this investigation, a correlation for the density of the nanofluid was developed, incorporating temperature and nanoparticle volume fraction: $\rho_n = 1001.064 + 2738.6191\phi - 0.2095T$; $0.04 \geq \phi \geq 0$, $40^{\circ}\mathrm{C} \geq T \geq 5^{\circ}\mathrm{C}$. This newly formulated correlation, based on the experimental data from Ho et al. [30], exhibits high precision with a regression coefficient (R^2) of 99.97% and a maximum relative error of 0.22%.

6.1.1.2 Thermal conductivity measurements

Nanofluids, colloidal suspensions of nanoparticles in a base fluid, exhibit enhanced thermal conductivity compared to their base fluids. This unique property has garnered significant attention for applications in heat transfer, electronic cooling, and energy systems. Understanding and quantifying the thermal conductivity of nanofluids is crucial for optimizing their performance in various technological domains.

The thermal conductivity (k_{nf}) of nanofluids can be theoretically expressed using various models. One commonly employed model is the Maxwell-Eucken equation, which considers the effect of particle concentration on thermal conductivity: $k_{nf} = k_f\left(1 + 2.5\phi + 1.5\phi^2\right)$

Here, k_f represents the thermal conductivity of the base fluid, and ϕ denotes the volume fraction of nanoparticles. The Maxwell-Eucken equation provides insights into how the addition of nanoparticles influences the overall thermal conductivity of the nanofluid.

Experimental investigations play a pivotal role in validating theoretical models and providing accurate data for nanofluid thermal conductivity. Researchers employ techniques such as the transient hot-wire method, guarded hot-plate apparatus, and laser flash analysis to measure thermal conductivity accurately. The experimental data is then compared with theoretical predictions to assess the validity of the models under specific conditions.

The thermal conductivity of nanofluids is influenced by various nanoparticle properties, including size, shape, and material. Larger nanoparticles generally contribute more significantly to thermal conductivity enhancement, while specific shapes and materials may exhibit unique effects. The impact of these factors can be incorporated into models to tailor nanofluid formulations for specific applications [14–18].

Table 6.1 Summary of thermal conductivity data

Nanoparticle volume fraction (ϕ)	*Thermal conductivity enhancement*
0.1	5%
0.3	12%
0.5	20%
0.7	28%
0.9	35%

Note: Here is a summary table illustrating the thermal conductivity enhancement of nanofluids at different nanoparticle concentrations.

Table 6.1 demonstrates how increasing the volume fraction of nanoparticles leads to a proportional enhancement in thermal conductivity, showcasing the potential for tailored nanofluid formulations [19–21].

6.1.1.3 Viscosity

The viscosity of nanofluids, which are suspensions of nanoparticles in a base fluid, is a critical parameter influencing their flow behavior and performance in various applications such as heat transfer and lubrication systems. Understanding the viscosity of nanofluids is essential for optimizing their use in different technological domains.

The viscosity of nanofluids is influenced by various characteristics of nanoparticles, including size, shape, and surface chemistry. Smaller nanoparticles may exhibit less impact on viscosity, while specific shapes and surface modifications can result in unique effects. The interplay of these factors contributes to the overall flow behavior of nanofluids (Table 6.2).

Several theoretical models have been proposed to describe the viscosity of nanofluids. One is widely used, which considers the impact of nanoparticle concentration on the viscosity of the nanofluid:

The viscosity expression for nanofluids can be formulated as

$$\mu_{nf} = \mu_f \left(1 - \frac{\phi_{ag}}{\phi_m}\right)^{-[\delta]\phi_m} \tag{6.2}$$

where μ_f represents the density of the base fluid, ϕ_m indicates the maximum volume fraction achievable for spheroidal nanoparticles, and δ signifies the shape parameter of the nanoparticles. According to Agarwal et al. [70], the equivalent volume fractions of the nanoparticles, denoted as ϕ_{ag}, and the fractal index d are utilized to articulate the aggregate volume fraction of the nanoparticles, expressed as ϕ_{mod} in Eq. (6.3).

$$\phi_{ag} = \phi_{\text{mod}} \left(\frac{r_a}{r}\right)^{3-d} \tag{6.3}$$

Table 6.2 Summary of viscosity data

Nanoparticle volume fraction (ϕ)	*Viscosity increase*
0.1	5%
0.3	15%
0.5	25%
0.7	35%
0.9	45%

Note: Here is a summary table illustrating the viscosity behavior of nanofluids at different nanoparticle concentrations.

In this context, the equivalent volume fractions of the nanoparticles, represented by ϕ_{mod}, can be articulated concerning the volume fraction (ϕ), the thickness of the interfacial layer (γ), and the lengths of the semi-major and semi-minor axes (b and a), respectively, as indicated in Eq. (6.4).

$$\phi_{mod} = \phi\left(1+\frac{\gamma}{a}\right)\left(1+\frac{\gamma}{b}\right)^2 \tag{6.4}$$

Experimental determinations of nanofluid viscosity are crucial for validating theoretical models and providing accurate data for practical applications. Techniques such as rotational viscometers and rheometers are commonly employed to measure viscosity under varying conditions. Experimental results are then compared with theoretical predictions to assess the reliability of the models.

6.1.1.4 Stability

The stability of nanofluids, referring to their ability to maintain a homogeneous and well-dispersed state over time, is a critical aspect influencing their practical applications. Stability is crucial for ensuring consistent and reliable performance in various fields, including heat transfer systems, biomedical applications, and advanced manufacturing.

Nanofluid stability is often described using theoretical models that consider the forces acting on nanoparticles in the fluid. One commonly employed model is the DLVO theory, which accounts for the van der Waals attraction and repulsive forces between nanoparticles:

$$F_{DLVO} = F_{van\ der\ waals} + F_{electric\ double\ layer} \tag{6.5}$$

Here, $F_{van\ der\ waals}$ represents the attractive van der Waals forces, and $F_{electric\ double\ layer}$ denotes the repulsive forces due to the electric double layer surrounding nanoparticles. The balance between these forces influences the stability of nanofluids.

Table 6.3 Stability assessment

Nanoparticle concentration (ϕ)	*Stability rating*
0.1	Stable
0.3	Moderately stable
0.5	Unstable
0.7	Highly unstable
0.9	Agglomerated

Note: Here is a summary table illustrating the stability assessment of nanofluids under different conditions.

Experimental methods play a crucial role in assessing the stability of nanofluids. Techniques such as dynamic light scattering and zeta potential measurements provide insights into the size distribution and surface charge of nanoparticles. A stable nanofluid is characterized by minimal agglomeration or sedimentation over time (Table 6.3).

Surface modification of nanoparticles through the use of surfactants or stabilizing agents is a common approach to enhance stability. The addition of surface modifiers alters the inter-particle forces and reduces agglomeration tendencies, contributing to improved stability. The choice of stabilizing agents is often application specific, and their effectiveness is evaluated through experimental assessments [22–26].

6.1.1.5 Heat capacity of nanofluids

The heat capacity of nanofluids, a crucial thermophysical property, plays a significant role in applications related to heat transfer, energy storage, and thermal management. Understanding the heat capacity behavior of nanofluids is essential for optimizing their performance in diverse engineering and scientific domains.

The heat capacity (C_{nf}) of nanofluids can be theoretically expressed by considering the heat capacities of both the base fluid (C_f) and the nanoparticles (C_p) along with the volume fraction (ϕ) of the nanoparticles:

$$C_{nf} = (1 - \phi)C_f + \phi C_p \tag{6.6}$$

In this equation, $(1 - \phi)C_f$ represents the heat capacity of the base fluid, and ϕC_p represents the contribution from the nanoparticles. This model provides insights into how the addition of nanoparticles influences the overall heat capacity of the nanofluid.

Experimental measurements are critical for validating theoretical models and providing accurate heat capacity data for nanofluids. Techniques such as differential scanning calorimetry and specific heat calorimetry are commonly employed to measure heat capacities under varying conditions.

Table 6.4 Heat capacity trends

Nanoparticle concentration (ϕ)	*Heat capacity increase (%)*
0.1	3%
0.3	8%
0.5	15%
0.7	22%
0.9	30%

Note: Here is a summary table illustrating the heat capacity trends in nanofluids at different nanoparticle concentrations.

The experimental data allows for the verification and refinement of theoretical predictions [27–34] (Table 6.4).

The heat capacity of nanofluids is influenced by various nanoparticle properties, including size, shape, and material. Larger nanoparticles tend to contribute more significantly to heat capacity enhancement. The choice of nanoparticle material can also have a substantial impact, as different materials have distinct thermal properties.

6.1.1.6 Natural heat convection in nanofluids

Natural heat convection, a fundamental mode of heat transfer driven by buoyancy forces, is a critical phenomenon in various engineering applications. When nanofluids, which are colloidal suspensions of nanoparticles in a base fluid, are involved, their unique thermophysical properties can significantly influence the natural convection process. Understanding and optimizing natural heat convection in nanofluids have implications for applications in thermal management, renewable energy systems, and beyond.

The heat transfer rate due to natural convection in nanofluids can be theoretically expressed using the Nusselt number (Nu), Grashof number (Gr), and Prandtl number (Pr):

$$\mathrm{Nu} = 0.59(\mathrm{Gr}\,\mathrm{Pr})^{0.25} \tag{6.7}$$

Here, Nu represents the Nusselt number, which characterizes the enhancement of heat transfer due to natural convection, Gr is the Grashof number representing the ratio of buoyancy to viscous forces, and Pr is the Prandtl number representing the ratio of momentum diffusivity to thermal diffusivity. The Nusselt number correlates the convective heat transfer rate to the relevant fluid and geometric properties.

Experimental investigations are crucial for validating theoretical models and providing insights into the actual performance of nanofluids in natural

Table 6.5 Nusselt number enhancement

Nanoparticle concentration (ϕ)	*Nusselt number enhancement (%)*
0.1	10
0.3	20
0.5	30
0.7	40
0.9	50

Note: Here is a summary table illustrating the enhancement of the Nusselt number in nanofluids compared to the base fluid under natural convection.

heat convection. Researchers use experimental setups, such as vertical or inclined heated surfaces, to study the heat transfer enhancement in nanofluid systems. The experimental data aids in refining theoretical predictions and developing correlations for specific nanofluid compositions [35–39] (Table 6.5).

The impact of nanoparticle characteristics, such as size, shape, and material, on natural heat convection in nanofluids is a subject of ongoing research. Larger nanoparticles may alter the fluid dynamics, influencing the Nusselt number enhancement. The choice of nanoparticle material can also affect thermal conductivity, influencing the overall heat transfer rate (Table 6.6).

The integration of nanofluids into fluid-based systems extends beyond heat transfer applications, encompassing a diverse range of industries such as energy, healthcare, and manufacturing. Table 6.6 shows the thermo-mechanical properties and different industrial applications of the nanofluids. The careful control of particle size, shape, and surface chemistry is crucial for tailoring the thermal and rheological properties of nanofluids. The widespread adoption of nanofluid-based systems prompts an evaluation of their environmental impact. Environmental considerations are paramount in ensuring the responsible development and deployment of these advanced fluids. It is important to understand the potential toxicity of nanoparticles, their impact on ecosystems, and the end-of-life disposal of nanofluid-based products is crucial for mitigating any adverse effects on the environment [11]. The emergence of "green" nanofluids, characterized by environmentally friendly nanoparticles and base fluids, offers a sustainable solution to the environmental concerns associated with conventional nanofluids. By prioritizing environmental sustainability, researchers and industries can contribute to the responsible advancement of nanofluid-based systems [40–45].

Table 6.6 Nanofluid properties and industrial applications [1–8]

Nanofluid type	*Nanoparticles*	*Base fluid*	*Thermal conductivity (W/m·K)*	*Viscosity (cP)*	*Specific heat (J/g·K)*	*Mechanical properties*	*Industrial applications*
Metallic nanofluids	Copper, aluminum	Water, oil	10–200	Low to moderate	Moderate to high	Enhanced thermal conductivity, good stability	Electronics cooling, automotive applications, heat exchangers
Oxide nanofluids	Titanium dioxide, Al_2O_3	Water, ethylene glycol	5–80	Low to moderate	Moderate to high	Stability, resistance to corrosion	Solar thermal systems, refrigeration, lubrication
Carbon-based nanofluids	Carbon nanotubes, graphene	Water, oil	100–1,000	Low to moderate	Moderate to high	Excellent thermal conductivity, lightweight	Aerospace applications, nanocomposites, energy storage systems
Polymer nanofluids	Polymeric nanoparticles	Various polymers	0.1–2	Low to moderate	Low to moderate	Improved stability, viscosity control	Biomedical applications, drug delivery, rheology modifiers
Phase change nanofluids	PCM nanoparticles	Various PCMs	0.5–30	Low to moderate	High	Latent heat storage, temperature regulation	Thermal energy storage, building materials, HVAC systems
Hybrid nanofluids	Combination of above	Various combinations	Varies	Varies	Varies	Tailored properties for specific applications	Versatile applications, customized solutions for unique requirements

6.2 NANOFLUID CHARACTERIZATION TECHNIQUES

Nanofluid characterization techniques play an important role in understanding the structural and chemical attributes of nanoparticles. Few important techniques are shown in Figure 6.2 and discussed as follows:

- **Transmission Electron Microscopy:** This technique involves transmitting electrons through a thin specimen to form an image. It provides high-resolution details of the internal structure, size, and morphology of nanoparticles.
- **Scanning Electron Microscopy (SEM):** SEM scans the surface of a specimen with a focused beam of electrons, creating a detailed 3D image. It is valuable for studying the surface morphology and size distribution of nanoparticles.
- **X-ray Diffraction (XRD):** XRD is employed to analyze the crystal structure of nanoparticles. When X-rays interact with the crystal lattice of a material, they diffract, producing a diffraction pattern. This pattern is used to determine the crystallographic structure, phase composition, and crystallite size of nanoparticles.
- **UV-Visible Spectroscopy:** This technique measures the absorption of ultraviolet and visible light by nanoparticles. The absorption spectra provide information about the electronic transitions, concentration, and size of nanoparticles.
- **Infrared Spectroscopy (IR):** IR spectroscopy examines the vibrations of chemical bonds in nanoparticles. It is particularly useful for identifying functional groups on the nanoparticle's surface, providing information on surface chemistry.
- **Raman Spectroscopy:** Raman spectroscopy involves the scattering of monochromatic light by nanoparticles. It provides information on molecular vibrations and crystal structures, aiding in the identification of chemical compounds and phases.
- **Nuclear Magnetic Resonance (NMR) Spectroscopy:** NMR can be used to study the interaction of nanoparticles with surrounding molecules. It provides insights into the surface functionalization and chemical composition of nanoparticles.

These characterization techniques are often used in combination to obtain accurate information about nanoparticles. For example, electron microscopy provides detailed imaging, while XRD offers structural information and various spectroscopic techniques give insights into chemical composition and surface characteristics. The methods allow developing a clear understanding of the properties of nanoparticles, aiding in their successful application in nanofluid development and other fields [46–51].

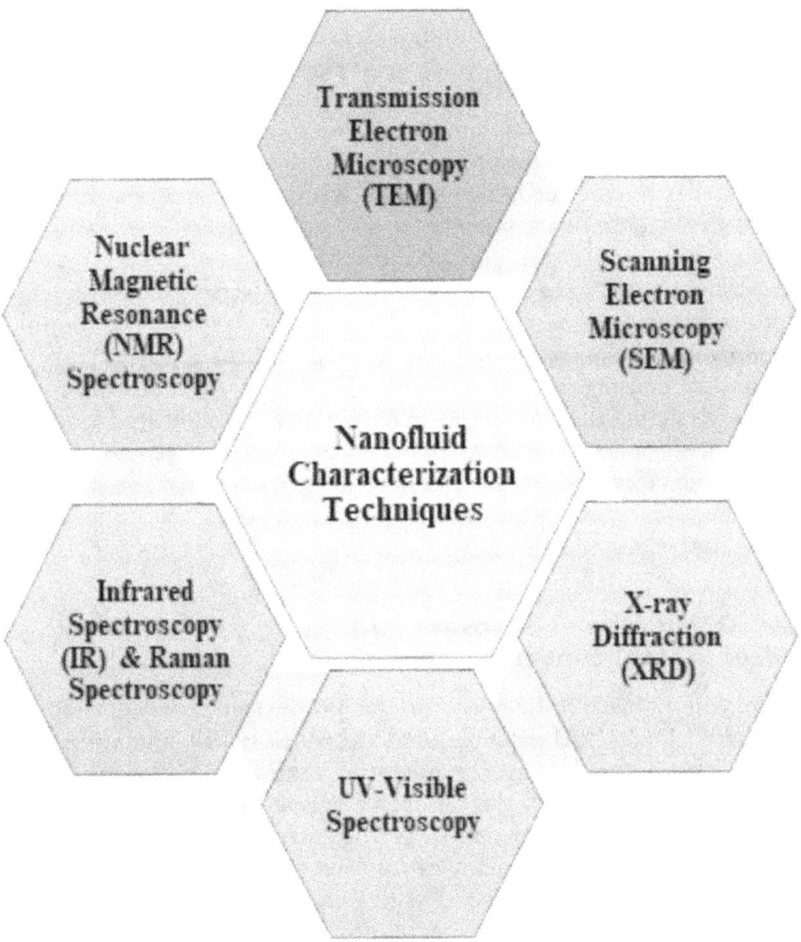

Figure 6.2 Nanofluid characterization techniques.

6.3 TOXICITY OF NANOPARTICLES AND HUMAN HEALTH IMPACT

Nanoparticles, due to their unique size and surface properties, can exhibit distinct biological interactions that raise concerns about their potential toxicity and impact on human health. The small size of nanoparticles allows them to penetrate biological barriers, including cell membranes, leading to interactions with cellular components. Various factors contribute to the toxicity of nanoparticles, such as their composition, shape, surface charge, and chemical reactivity.

The potential toxicity of nanoparticles arises from several mechanisms, including oxidative stress, inflammation, genotoxicity, and cellular uptake. Nanoparticles can generate reactive oxygen species, causing oxidative stress and damage to cellular structures. Additionally, certain nanoparticles may induce an inflammatory response in tissues, impacting immune function and exacerbating pre-existing health conditions. The ability of nanoparticles to enter cells and interact with genetic material raises concerns about genotoxicity, potentially leading to adverse effects on DNA integrity.

Exposure to toxic nanoparticles can have diverse effects on human health, ranging from respiratory and cardiovascular issues to neurological impacts. Inhaled nanoparticles, for example, can reach the respiratory system, causing inflammation and potentially leading to respiratory diseases. Systemic distribution of nanoparticles through the bloodstream raises concerns about cardiovascular effects. Moreover, the ability of certain nanoparticles to cross the blood-brain barrier raises questions about their impact on neurological function, with potential implications for neurodegenerative diseases [11–12].

6.3.1 Occupational exposure and environmental concerns

Workers in industries involved in the production and application of nanoparticles may face occupational exposure risks. Inhalation, ingestion, and dermal exposure are potential routes for nanoparticle entry into the body. Moreover, the release of nanoparticles into the environment raises ecological concerns, as they can accumulate in soil and water, affecting various organisms and ecosystems. Understanding the potential health risks associated with nanoparticle exposure is crucial for establishing guidelines and regulations to protect both workers and the environment.

Recognizing the potential health hazards, regulatory bodies worldwide are actively working on guidelines for the safe handling and use of nanoparticles. Establishing occupational exposure limits, ensuring proper labeling, and implementing risk assessment frameworks are essential components of regulatory measures. Ongoing research aims to further elucidate the mechanisms of nanoparticle toxicity, identify safe design principles, and develop effective protective measures to mitigate potential health risks associated with the expanding use of nanotechnology in various industries. As the field evolves, interdisciplinary collaboration between scientists, regulators, and industry stakeholders is imperative to ensure the responsible development and application of nanomaterials [52–58].

6.4 SAFETY NORMS FOR NANOFLUID

Safety and environmental regulations related to nanofluid-based systems are evolving, and various countries, including India, have been working on guidelines and standards related to nanofluid testing, use, and disposal as shown in Figure 6.3.

- **OECD Guidelines for the Testing of Chemicals:** The Organisation for Economic Co-operation and Development (OECD) provides guidelines for the testing of chemicals, including nanomaterials. These guidelines cover aspects such as toxicity testing and environmental fate assessment.
- **ISO Standards:** The International Organization for Standardization (ISO) has been working on standards related to nanotechnologies, including those relevant to safety and environmental aspects. Examples include ISO 13014 on characterization of nanoparticles and ISO/TR 12885 on nanotechnologies – occupational risk management.
- **EU Regulations:** The European Union has specific regulations concerning nanomaterials. REACH (Registration, Evaluation, Authorization, and Restriction of Chemicals) regulations, for instance, may cover the registration and use of nanomaterials.

In India, the safety and environmental rules and regulations related to nanofluid-based systems are governed by several laws and agencies. Some of the key regulations include:

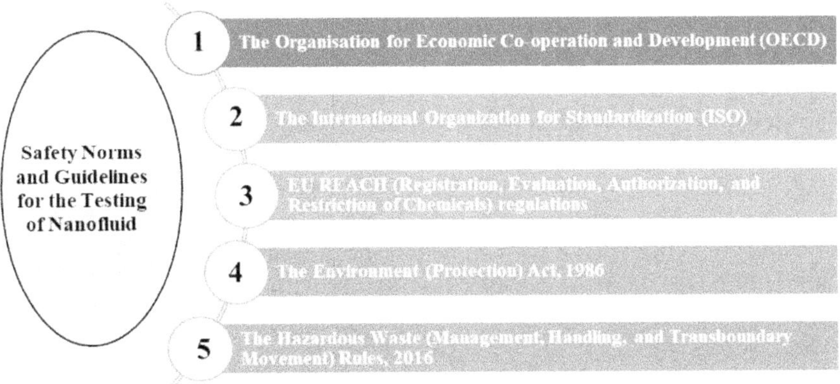

Figure 6.3 Safety norms and guidelines for the testing of nanofluid.

- **The Environment (Protection) Act, 1986:** This law empowers the Ministry of Environment, Forest and Climate Change (MoEFCC) to regulate the production, import, export, and use of hazardous substances, including nanomaterials.
- **The Hazardous Waste (Management, Handling, and Transboundary Movement) Rules, 2016:** These rules require the registration and authorization of facilities that generate, store, and dispose of hazardous wastes, including nanomaterials.
- **The Food Safety and Standards Act:** This regulation sets out the procedures and requirements for the environmental safety assessment of products, including those that use nanotechnology.

In addition to these, the regulation of nanotechnology-based products in India falls under the purview of various agencies including the Department of Biotechnology (DBT), the Ministry of Environment, Forest and Climate Change (MoEFCC), the Food Safety and Standards Authority of India (FSSAI), and the Indian Council of Agricultural Research (ICAR) [8].

Internationally, various countries have also established regulations and standards for the safe use of nanomaterials. For example, Japan has used the Ministry of Economy, Trade and Industry (METI) to gather information about the nano industry and assess the negative effects of nanomaterials by the environment ministry (ME). Additionally, the International Organization for Standardization (ISO) has developed standards for the safe use of nanomaterials [9]. These regulations and standards are aimed at ensuring the safe and responsible use of nanotechnology and nanomaterials, as well as protecting the environment and human health [59–64].

6.5 SAFETY AND ENVIRONMENTAL IMPACT OF NANOFLUID

The safety and environmental impact of nanofluids have been the subject of analysis, and it has been concluded that current regulatory regimes are generally adequate for ensuring their safe use. However, it is essential to adhere to safety, health, and environmental regulations when using nanofluids. This includes conducting environmental, health, and safety assessments of nanofluid-based systems to ensure compliance with stringent standards and to identify and mitigate any potential risks [65–70].

Some proactive safety measures that can be taken when using nanofluids are shown in Figure 6.4 and include the following:

- **Engineering Controls:** Implementing engineering measures such as enclosed systems and ventilation to minimize exposure during the production and use of nanofluids [24].

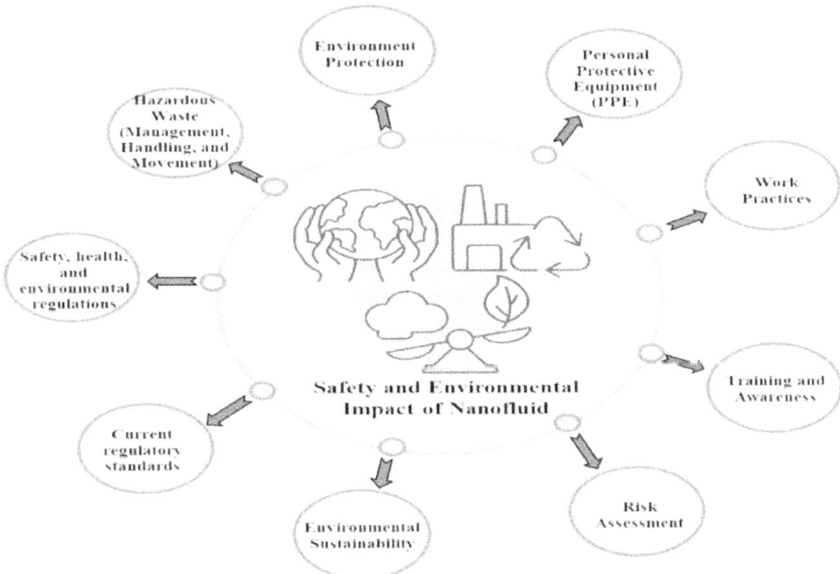

Figure 6.4 Safety and environmental impact of nanofluid.

- **Personal Protective Equipment (PPE):** Providing and mandating the use of appropriate PPE, such as gloves, lab coats, and respiratory protection, to prevent direct contact with nanofluids [24].
- **Work Practices:** Developing and enforcing safe work practices, including proper handling, storage, and disposal procedures for nanofluids to minimize potential exposure and environmental impact.
- **Training and Awareness:** Providing comprehensive training and raising awareness among workers about the potential risks and safe handling practices associated with nanofluids.
- **Risk Assessment:** Conducting thorough risk assessments to identify and mitigate potential hazards associated with the use of nanofluids, and implementing control measures based on the assessment findings.

By adhering to these safety measures, the potential risks associated with the use of nanofluids can be effectively managed, ensuring the protection of human health and the environment.

6.6 GREEN NANOFLUIDS FOR SUSTAINABLE DEVELOPMENT

Green nanofluids represent a sustainable approach to harnessing the unique properties of nanoparticles while minimizing environmental impact. Green nanofluids share similar thermal and rheological properties with conventional nanofluids but stand out due to the use of environmentally benign nanoparticles and base fluids. The choice of biocompatible and biodegradable materials ensures that these fluids meet stringent environmental standards. Additionally, their thermal conductivity, heat transfer efficiency, and stability are tailored to specific applications while prioritizing sustainability.

Characterizing green nanofluids involves employing techniques such as electron microscopy, XRD, and spectroscopy. These methods offer insights into the structural and chemical attributes of nanoparticles, ensuring that the chosen materials align with green principles. Analyzing the environmental impact, toxicity, and biodegradability of both nanoparticles and base fluids is also crucial in the characterization process [71–74].

6.6.1 Advantage

The main advantages of Green Nanofluids are as follows:

- **Environmental Sustainability:** The use of eco-friendly nanoparticles and base fluids reduces the environmental footprint associated with conventional nanofluids.
- **Biocompatibility:** Green nanofluids find applications in fields such as biomedicine and food processing due to the biocompatibility of their components.
- **Renewable Resources:** Utilizing nanoparticles derived from renewable resources contributes to the overall sustainability of the nanofluid.

6.6.2 Limitations

The limitation of Green Nanofluids is:

- **Limited Material Options:** The availability of biocompatible and biodegradable nanoparticles may be limited compared to conventional materials, restricting the range of applications.
- **Potential Performance Trade-offs:** Green nanofluids may exhibit slightly lower thermal conductivity or stability compared to their conventional counterparts, requiring careful optimization for specific applications.
- **Cost Considerations:** The production of green nanofluids, especially those using specialized biodegradable materials, may be costlier than conventional counterparts.
- **Challenges in Synthesis:** The synthesis of nanoparticles from renewable resources might pose challenges in achieving uniform size, shape, and stability.

Conventional nanofluids are typically prepared using various base fluids and metallic/non-metallic nanoparticles. They enhance the thermal conductivity of the base fluid but may involve toxic nanoparticles and non-biodegradable materials. Industrial examples of conventional nanofluids include their use in engine cooling systems and heat transfer fluids. On the other hand, green nanofluids are prepared using environmentally friendly base fluids and biodegradable nanoparticles. They also enhance the thermal conductivity of the base fluid but involve biodegradable and non-toxic nanoparticles. Industrial examples of green nanofluids include their use in solar energy systems and HVAC systems. Green nanofluids represent a conscientious approach to nanofluid development, aligning with the growing emphasis on sustainability. While their advantages include environmental friendliness and biocompatibility, considerations such as limited material options, potential trade-offs in performance, and cost implications need to be carefully weighed [75–78].

6.7 DISCUSSION

In conclusion, the role of safety and environmental issues in nanofluid-based systems is a consideration. Conventional nanofluids, although effective in enhancing thermal conductivity, raise concerns due to the potential use of toxic nanoparticles and non-biodegradable materials. The shift toward green nanofluids signifies a commendable effort to fulfill these concerns by incorporating environmentally friendly base fluids and biodegradable nanoparticles. This shift aligns with the global emphasis on sustainability, aiming to reduce the environmental footprint associated with nanofluid technologies. While green nanofluids have added advantages such as biocompatibility and reduced toxicity, careful evaluation of material options, potential trade-offs in performance, and cost implications remains crucial. Striking a balance between performance and sustainability is paramount in the pursuit of safer and environmentally responsible nanofluid-based systems, ensuring that technological advancements align with the broader goals of ecological consciousness and human well-being.

REFERENCES

1. M. Karkare (2013). *Nanotechnology: Fundamentals and Applications.* IK International Pvt Ltd, Daryaganj, New Delhi.
2. K. V. Wong and O. De Leon (2010). Applications of nanofluids: Current and future. *Advances in Mechanical Engineering*, 519659. doi:10.1155/2010/519659.

3. V. Verma, C. S. Meena, S. Thangavel, A. Kumar, T. Choudhary, G. Dwivedi (2023). Ground and solar assisted heat pump systems for space heating and cooling applications in the northern region of India: A study on energy and CO2 saving potential. *Sustainable Energy Technologies and Assessments*, 59, 103405. doi:10.1016/j.seta.2023.103405.
4. M. K. Awasthi, N. Dutt, A. Kumar, S. Kumar (2023). Electrohydrodynamic capillary instability of Rivlin-Ericksen viscoelastic fluid film with mass and heat transfer. *Heat Transfer*, 1–19. doi:10.1002/htj.22944
5. N. Dutt, A. J. Hedau, A. Kumar, M. K. Awasthi, V. P. Singh, G. Dwivedi (2023). Thermo-hydraulic performance of solar air heater having discrete D-shaped ribs as artificial roughness. *Environmental Science and Pollution Research*. doi:10.1007/s11356-023-28247-9.
6. N. Nabhani, A. Tofighi (2010, April). The assessment of health, safety and environmental risks of nano-particles and how to control their impacts. In: *SPE International Conference and Exhibition on Health, Safety, Environment, and Sustainability?* (SPE-127261). SPE.
7. P. J. Borm, D. Robbins, S. Haubold, T. Kuhlbusch, H. Fissan, K. Donaldson, R. Schins, V. Stone, W. Kreyling, J. Lademann, J. Krutmann (2006). The potential risks of nanomaterials: A review carried out for ECETOC. *Particle and Fibre Toxicology*, 3, 1–35.
8. P. K. Kushwaha, N. K. Sharma, A. Kumar, C. S. Meena (2023). Recent advancements in augmentation of solar water heaters using nanocomposites with PCM: Past, present, and future. *Buildings*, 13, 79. doi:10.3390/buildings13010079.
9. V. P. Singh, S. Jain, A. Karn, A. Kumar (2022). Experimental assessment of variation in open area ratio on thermohydraulic performance of parallel flow solar air heater. *Arabian Journal for Science and Engineering*. doi:10.1007/s13369-022-07525-7.
10. C. S. Meena, A. N. Prajapati, A. Kumar, M. Kumar (2022). Utilization of solar energy for water heating application to improve building energy efficiency: An experimental study. *Buildings*, 12, 2166. doi:10.3390/buildings12122166.
11. P. C. Ray, H. Yu, P. P. Fu (2009). Toxicity and environmental risks of nanomaterials: Challenges and future needs. *Journal of Environmental Science and Health Part C*, 27(1), 1–35.
12. A. Boxall, K. Tiede, Q. Chaudhry, R. Aitken, A. Jones, B. Jefferson, J. Lewis (2007). Current and future predicted exposure to engineered nanoparticles. In: *Safety of Nanomaterials Interdisciplinary Research Centre Report*, pp. 1–13.
13. R. D. Handy, R. Owen, E. Valsami-Jones (2008). The ecotoxicology of nanoparticles and nanomaterials: Current status, knowledge gaps, challenges, and future needs. *Ecotoxicology*, 17, 315–325.
14. V. P. Singh, S. Jain, A. Kumar, S. Mishra, N. K. Sharma (2022). Heat transfer and friction factor correlations development for double pass solar air heater artificially roughened with perforated multi-V ribs. *Case Studies in Thermal Engineering*, 39, 102461. doi:10.1016/j.csite.2022.102461.
15. V. P. Singh, S. Jain, A. Karn, A. Kumar, G. Dwivedi, C. S. Meena (2022). Recent developments and advancements in solar air heaters: A detailed review. *Sustainability*, 14, 12149. doi:10.3390/su141912149.

16. C. S. Meena, A. Kumar, S. Jain, A. U. Rehman, S. Mishra (2022). Innovation in green building sector for sustainable future. *Energies*, 15, 6631. doi:10.3390/en15186631.
17. V. Vishwakarma, S. S. Samal, N. Manoharan (2010), Safety and risk associated with nanoparticles: A review. *Journal of Minerals & Materials Characterization & Engineering*, 9, 455–459.
18. R. Kumari, K. Suman, S. Karmakar, V. Mishra, S. Gunjan Lakra, G. Kumar Saurav, B. K. Mahto (2023). Regulation and safety measures for nanotechnology-based agri-products, *Front Genome Edition*, 5, 1200987. doi:10.3389/fgeed.2023.1200987.
19. V. P. Singh, S. Jain, A. Karn, A. Kumar, G. Dwivedi, C. S. Meena, R. Cozzolino (2022). Mathematical modeling of efficiency evaluation of double-pass parallel flow solar air heater. *Sustainability*, 14, 10535. doi:10.3390/su141710535.
20. C. S. Meena, A. Kumar, S. Roy, A. Cannavale, A. Ghosh (2022). Review on boiling heat transfer enhancement techniques. *Energies*, 15, 5759. doi:10.3390/en15155759.
21. V. P. Singh, S. Jain, A. Kumar (2022). Establishment of correlations for the thermo-hydraulic parameters due to perforation in a multi-V rib roughened single pass solar air heater. *Experimental Heat Transfer*. doi:10.1080/08916152.2022.2064940.
22. D. Ramos, L. Almeida (2022). Overview of standards related to the occupational risk and safety of nanotechnologies. *Standards*, 2(1), 83–89. doi:10.3390/standards2010007.
23. W. Trybula, D. Newberry (2013). Nanotechnology risk assessment. In: Asmatulu, R. (ed) *Nanotechnology Safety*, Elsevier, Amsterdam, pp. 195–206. ISBN 9780444594389. doi: 10.1016/B978-0-444-59438-9.00014-X.
24. Y. K. Prajapati, P. K. Gupta, A. Kumar (2011). Free surface undulation and air entrainment in a rectangular tank. *International Journal of Applied Engineering Research*, 6(4), 409–419.
25. G. Pant, C. S. Meena, A. Saxena, A. Kumar, V. P. Singh, N. Dutt (2023). Study the temperature variation in alternate coils of insulated condenser cum storage tank: Experimental study. In: Sikarwar, B. S., Sharma, S. K., Jain, A., Singh, K. M. (eds) *Advances in Fluid and Thermal Engineering. FLAME 2022. Lecture Notes in Mechanical Engineering*. Springer, Singapore. doi: 10.1007/978-981-99-2382-3_52.
26. A. Saxena, A. N. Prajapati, G. Pant, C. S. Meena, A. Kumar, V. P. Singh (2023). Water consumption optimization of hybrid heat pump water heating system. In: Shukla, A. K., Sharma, B. P., Arabkoohsar, A., Kumar, P. (eds) *Recent Advances in Mechanical Engineering. FLAME 2022. Lecture Notes in Mechanical Engineering*. Springer, Singapore. doi:10.1007/978-981-99-1894-2_61.
27. M. A. Subhan, T. Subhan (2022). Safety and global regulations for application of nanomaterials. In: Rai, M., Nguyen, T. A. (eds) *Micro and Nano Technologies, Nanomaterials Recycling*. Elsevier, Amsterdam, pp. 83–107.
28. NNI. National Nanotechnology Initiative.Standards for Nanotechnology (2021). https://www.nano.gov/you/standards (accessed on 1 February 2024).

29. L. Almeida, D. Ramos (2022). Chapter: VII-governance/legislation/EU legal framework-REACH; occupational health aspects (EU); De Gruyter STEM. In: Cornier, J., Pursche, F., (eds) *Particle Technology and Textiles*. De Gruyter, Berlin. https://www.degruyter.com/document/isbn/9783110670776/html (accessed on 2 February 2024).
30. A. Kundu, A. Kumar, N. Dutt, C. S. Meena, V. P. Singh (2023). Introduction to thermal energy resources and their smart applications. In: Kumar, A.; Singh, V. P.; Meena, C. S.; Dutt, N. (eds) *Thermal Energy Systems: Design, Computational Techniques and Applications*. CRC Press, Boca Raton, pp. 1–15. doi:10.1201/9781003395768-1.
31. A. K. Dewangan, S. Q. Moinuddin, M. Cheepu, S. K. Sajjan, A. Kumar (2023). Thermal energy storage: Opportunities, challenges and future scope. In: Kumar, A.; Singh, V. P.; Meena, C. S.; Dutt, N. (eds) *Thermal Energy Systems: Design, Computational Techniques and Applications*. CRC Press, Boca Raton, FL, pp. 17–28. doi:10.1201/9781003395768-2.
32. A. Kundu, A. Kumar, N. Dutt, V. P. Singh, C. S. Meena (2023). Modelling and simulation of thermal energy system for design optimization. In: Kumar, A., Singh, V. P., Meena, C. S., Dutt, N. (eds) *Thermal Energy Systems: Design, Computational Techniques and Applications*. CRC Press, Boca Raton, FL, pp. 103–140. doi:10.1201/9781003395768-7.
33. A. Kumar, V. P. Singh, C. S. Meena, N. Dutt (2023). *Thermal Energy Systems: Design, Computational Techniques and Applications*. Taylor & Francis. doi:10.1201/9781003395768.
34. L. Almeida, D. Ramos (2017). Health and safety concerns of textiles with nanomaterials. *IOP Conference Series: Materials Science and Engineering*, 254, 102002.
35. D. Ramos, L. Almeida, M. Gomes (2019). Application of control banding to workplace exposure to nanomaterials in the textile industry. In: Arezes, P., Baptista, J. S., Barroso, M. P., Carneiro, P., Cordeiro, P., Costa, N., Melo, R. B., Miguel, A. S., Perestrelo, G. (eds) *Occupational and Environmental Safety and Health Studies in Systems, Decision and Control*. Springer, Cham, Vol. 202, pp. 105–113.
36. ANSI. American National Standards Institute. Risk and Safety USA. (2021). https://webstore.ansi.org/industry/nanotech/Risk-and-Safety (accessed on 5 February 2024).
37. N. Dutt, A. Hedau, M. K. Awasthi, A. Kumar, C. S. Meena (2024). Thermo-hydraulic performance investigation of solar air heater duct having staggered D-shaped ribs: Numerical approach, *Heat Transfer*. doi:10.1002/htj.22998.
38. A. Bansal, H. K. Bhardwaj, V. K. Sharma, A. Kumar (2022). Static and dynamic behavior analysis of Al-6063 alloy using modified Hopkinson bar. In: Sharma, V.K, Kumar, A., Gupta, M., Kumar, V., Sharma, D. K., Sharma, S. K. (eds) *Additive Manufacturing in Industry 4.0: Methods, Techniques, Modeling and Nano* Aspects. CRC Press, Boca Raton, pp. 107–124. doi:10.1201/9781003360001-6.
39. V. K. Sharma, V. Kumar, R. S. Joshi, A. Kumar (2022). Effect of REOs on tribological behavior of aluminum hybrid composites using ANN. In: Sharma, V.K; Kumar, A.; Gupta, M.; Kumar, V.; Sharma, D. K.; Sharma, S. K. (eds)

Additive Manufacturing in Industry 4.0: Methods, Techniques, Modeling and Nano Aspects. CRC Press, Boca Raton, pp. 153–168. doi:10.1201/9781003360001-9.

40. *ISO/TR* 13121:2011 (2011). *Nanotechnologies-Nanomaterial Risk Evaluation.* International Organization for Standardization, Geneva, Switzerland. https://www.iso.org/obp/ui/#iso:std:iso:tr:13121:ed-1:v1:en (accessed on 5 February 2024).
41. ISO/TS 12901-1:2012 (2012). *Nanotechnologies-Occupational Risk Management Applied to Engineered Nanomaterials-Part 1: Principles and Approaches.* International Organization for Standardization, Geneva, Switzerland. https://www.iso.org/standard/52125.html (accessed on 10 February 2024).
42. V. Kumar Sharma, A. Kumar, M. Gupta, V. Kumar, D. K. Sharma, S. Sharma (2022). *Additive Manufacturing in Industry 4.0: Methods, Techniques, Modeling, and Nano Aspects.* Taylor & Francis. doi:10.1201/9781003360001.
43. A. Kumar, A. Kumar, A. Kumar (2023). *Laser Based Technologies for Sustainable Manufacturing.* Taylor & Francis. doi:10.1201/9781003402398.
44. R. Kumar, A. Prasad, A. Kumar (2023). *Sustainable Smart Manufacturing Processes in Industry 4.0.* Taylor & Francis. doi:10.1201/9781003436072.
45. G. Chakraborty, V. Pandey, A. Prasad, A. Kumar. Introduction to sustainable manufacturing for industries 4.0. In: Kumar, R., Prasad, A., Kumar, A. (eds) *Sustainable Smart Manufacturing Processes in Industry 4.0.* CRC Press, Boca Raton, FL, pp. 1–17. doi:10.1201/9781003436072-1.
46. ISO/TR 13329:2012 (2012). *Nanomaterials-Preparation of Material Safety Data Sheet (MSDS). ISO. International Organization for Standardization.* Geneva, Switzerland. https://www.iso.org/standard/53705.html (accessed on 10 February 2024).
47. *ISO/TS* 12901-2:2014 (2014). *Nanotechnologies-Occupational Risk Management Applied to Engineered Nanomaterials-Part 2: Use of the Control Banding Approach.* International Organization for Standardization, Geneva, Switzerland. https://www.iso.org/standard/53375.html (accessed on 10 February 2024).
48. ISO/TR 12885:2018 (2018). *Nanotechnologies-Health and Safety Practices in Occupational Settings.* International Organization for Standardization, Geneva, Switzerland. https://www.iso.org/standard/67446.html (accessed on 10 February 2024).
49. M. Francis Luther King, A. Kumar, A. Kumar, A. Kumar (2023). Introduction to optics and laser-based manufacturing technologies. In: Kumar, A., Kumar, A., Kumar, A. (eds) *Laser-based Technologies for Sustainable Manufacturing.* CRC Press, Boca Raton, pp. 1–43. doi:10.1201/9781003402398-1.
50. P. P Patil, A. Kumar (2017). Dynamic structural and thermal characteristics analysis of oil lubricated multi speed transmission gearbox: Variation of load, rotational speed and convection heat transfer. *Iranian Journal of Science and Technology Transactions of Mechanical Engineering*, 41, 281–291, doi:10.1007/s40997-016-0063-z.
51. P. Patil, Y. Gori, A. Kumar, M. R. Tyagi (2021). Experimental analysis of tribological properties of polyisobutylene thickened oil in lubricated contacts. *Tribology International*, 159, 106983. doi:10.1016/j.triboint.2021.106983.

52. A. Kumar, A. Pathak, A. Kumar, A. Kumar (2023). Physics of laser-matter interaction in laser-based manufacturing. In: Kumar, A., Kumar, A., Kumar, A. (eds) *Laser-based Technologies for Sustainable Manufacturing*. CRC Press, Boca Raton, pp. 45–54. doi:10.1201/9781003402398-2.
53. R. Kumar, A. Kumar, L. Kant, A. Prasad, S. Bhoi, C. S. Meena, V. P. Singh, A. Ghosh (2023). Experimental and RSM-based process-parameters optimisation for turning operation of EN36B steel. *Materials*, 16, 339. doi:10.3390/ma16010339.
54. R. Purohit, A. Mittal, S. Dalela, V. Warudkar, K. Purohit, S. Purohit (2017). Social, Environmental and ethical impacts of nanotechnology. *Materials Today Proceedings*, 4, 5461–5467.
55. C. Martin, A. Nourian, M. Babaie, G. G. Nasr (2023). Environmental, health and safety assessment of nanoparticle application in drilling mud: Review. *Geoenergy Science and Engineering*, 226, 211767. doi:10.1016/j.geoen.2023.211767.
56. S. Bhoi, A. Kumar, A. Prasad, C. S. Meena, R. B. Sarkar, B. Mahto, A. Ghosh (2022). Performance evaluation of different coating materials in delamination for micro- milling applications on high-speed steel substrate. *Micromachines*, 13, 1277. doi:10.3390/mi13081277.
57. A. Pathak, A. Kumar, A. Kumar, A. Kumar, M. Francis Luther King (2023). Application of laser technology in the mechanical and machine manufacturing industry. In: Kumar, A., Kumar, A., Kumar, A. (eds) *Laser-based Technologies for Sustainable Manufacturin*. CRC Press, Boca Raton, FL, pp. 107–155. doi:10.1201/9781003402398-6.
58. M. K. Awasthi, N. Dutt, A. Kumar, A. Hedau (2024). Introduction to mathematical and computational methods. In: Awasthi, M. K., Kumar, A., Dutt, N., Singh, S. (eds) *Computational Fluid Flow and Heat Transfer: Advances, Design, Control and Applications*. CRC Press, Boca Raton, FL, pp. 1–20.
59. M. K. Awasthi, A. Kumar, N. Dutt (2024). Modeling Rayleigh-Taylor Instability in Nanofluid Layers. In: Awasthi, M. K., Kumar, A., Dutt, N., Singh, S. (eds) *Computational Fluid Flow and Heat Transfer: Advances, Design, Control and Applications*. CRC Press, Boca Raton, FL, pp. 249–262.
60. M. K. Awasthi, A. Kumar, N. Dutt, and S. Singh (eds) (2024). *Computational Fluid Flow and Heat Transfer: Advances, Design, Control and Applications*. CRC Press, Boca Raton, FL.
61. E. B. Elcioglu, A. Turgut, S. M. S. Murshed (eds) (2022). *Fundamentals and Transport Properties of Nanofluids*. The Royal Society of Chemistry, pp. 437–451.
62. M. K. Awasthi, Z. Uddin, R. Asthana (2021). Temporal instability of a Power-law viscoelastic nanofluid layer. *The European Physical Journal Special Topics*, 230, 1427–1434.
63. V. P. Singh, A. Kumar, C. S. Meena, G. Dwivedi (2024). Introduction to energy efficient vehicles for sustainable transportation: Transitions and challenges. In: Singh, V. P., Kumar, A., Meena, C. S., Dwivedi, G. (eds) *Energy Efficient Vehicles: Technologies and Challenges*. CRC Press, Boca Raton, FL, pp. 1–18. doi: 10.1201/9781003464556-1.
64. R. K. Upadhyay, V. P. Singh, A. Kumar (2024). Sustainable transportation: Policy, planning and implementation. In: Singh, V. P., Kumar, A., Meena, C. S., Dwivedi, G. (eds) *Energy Efficient Vehicles: Technologies and Challenges*. CRC Press, Boca Raton, FL, pp. 74–95. doi: 10.1201/9781003464556-5.

65. V. P. Singh, A. Kumar (2024). Techno-economic and future aspects of the HEV-EV-FCV: Decarbonisation, digitalization and sustainability. In: Singh, V. P., Kumar, A., Meena, C. S., Dwivedi, G. (eds) *Energy Efficient Vehicles: Technologies and Challenges*. CRC Press, Boca Raton, FL, pp. 212–231. doi: 10.1201/9781003464556-11.
66. V. P. Singh, A. Kumar, C. S. Meena, G. Dwivedi (eds) (2024). *Energy Efficient Vehicles: Technologies and Challenges*. CRC Press, Boca Raton, FL, pp. 1–314. doi: 10.1201/9781003464556.
67. A. Kumar, Y. Singla and T. Namboodri (2024). Globalization and international issues in sustainable manufacturing. In: Shah, S., Nautiyal, H., Gugliani, G., Kumar, A., Namboodri, T., Singla, Y. K. (eds) *Sustainability in Smart Manufacturing: Trends, Scope, and Challenges*. CRC Press, Boca Raton, FL, pp. 1–18. doi: 10.1201/9781003467496-1.
68. S. Shah, H. Nautiyal, G. Gugliani, A. Kumar, T. Namboodri, Y. K. Singla (eds) (2024). *Sustainability in Smart Manufacturing: Trends, Scope, and Challenges*. CRC Press, Boca Raton, FL.
69. M. K. Awasthi, Dharamendra, D. Yadav (2022). Temporal instability of nanofluid layer in a circular cylindrical cavity. *The European Physical Journal Special Topics*, 231, 2773–2779.
70. S. Agarwal, M. K. Awasthi, A. K. Shukla (2023). Stability analysis of water-alumina nanofluid film at the spherical interface. *Proceedings of the Institution of Mechanical Engineers, Part E: Journal of Process Mechanical Engineering*. doi:10.1177/09544089221150733
71. B. C. Pak, Y. I. Cho (1999). Hydrodynamic and heat transfer study of dispersed fluids with submicron metallic oxide particles. *Expperimental Heat Transfer*, 11, 151–170.
72. C. J. Ho, W. K. Liu, Y. S. Chang, C. C. Lin (2010). Natural convection heat transfer of alumina-water nanofluid in vertical square enclosures: An experimental study. *International Journal of Thermal Sciences*, 49, 1345–1353.
73. V. Verma, S. Thangavel, N. Dutt, A. Kumar, R. Weerasinghe (2024). Recent development of Introduction to thermal energy storage: Solar, geothermal and hydrogen energy. In: Verma, V., Thangavel, S., Dutt, N., Kumar, A., Weerasinghe, R. (eds) *Highly Efficient Thermal Renewable Energy Systems: Design, Design, Optimization and Applications*. CRC Press, Boca Raton, FL, pp. 1–22. doi: 10.1201/9781003472629-1.
74. M. Ahmadizadeh, M. Heidari, S. Thangavel, E. A. Naamani, M. Khashehchi, V. Verma, A. Kumar. Technological advancements in sustainable and renewable solar energy systems. In: Verma, V., Thangavel, S., Dutt, N., Kumar, A., Weerasinghe, R. (eds) *Highly Efficient Thermal Renewable Energy Systems: Design, Design, Optimization and Applications*. CRC Press, Boca Raton, FL, pp. 23–39. doi: 1201/9781003472629-2.
75. M. Chitt, S. Thangavel, V. Verma, A. Kumar. Green hydrogen productions: Methods, designs and smart applications. In: Verma, V., Thangavel, S., Dutt, N., Kumar, A., Weerasinghe, R. (eds) *Highly Efficient Thermal Renewable Energy Systems: Design, Design, Optimization and Applications*. CRC Press, Boca Raton, FL, pp. 261–276. doi: 10.1201/9781003472629-16.
76. M. Khashehchi, S. Thangavel, P. Rahmanivahid, M. Heidari, T. Moazzeni, V. Verma, A. Kumar (2024). Solar desalination techniques: Challenges and opportunities. In: Verma, V., Thangavel, S., Dutt, N., Kumar, A., Weerasinghe, R. (eds) *Highly Efficient Thermal Renewable Energy Systems: Design, Design,*

Optimization and Applications. CRC Press, Boca Raton, FL, pp. 305–329. doi: 10.1201/9781003472629-19.

77. V. Verma, S. Thangavel, N. Dutt, A. Kumar, R. (2024). *Weerasinghe Editors; Highly Efficient Thermal Renewable Energy Systems: Design, Design, Optimization and Applications*. CRC Press, Boca Raton, FL, pp. 1–341. doi: 10.1201/9781003472629.
78. P. N. Reddy, V. Verma, A. Kumar, M. Awasthi 2023. CFD simulation and thermal performance optimization of flow in a channel with multiple baffles. *Journal of Heat and Mass Transfer Research*, 10(2), 257–268. doi:10.22075/JHMTR.2023.31108.1458.

Chapter 7

Heat exchanger design and performance improvement with nanofluids

Shailandra Kumar Prasad and Ashwini Kumar

7.1 INTRODUCTION OF HEAT EXCHANGE DESIGN

The introduction of nanofluids has revolutionized the technology of heat exchange. These advancements have paved the way for smaller, more efficient, and cost-effective heat-exchanging processes. These advancements have a scope that can be included in various applications. The traditional methods of heat exchange were limited to the direct or indirect mode of exchange. However, these systems could only heat and cool through divided media. The traditional mode of heat exchange designs required and emphasized heat exchange through plate devices or more compact heat exchangers. However, as the advancing times continued to bring more innovative prospects, it introduced the technology of micro or nano heat exchangers.

Our study uniquely explores the integration of nanofluids into double-tube heat exchangers, offering a fresh perspective on heat exchanger research. Unlike previous studies, we comprehensively analyse flow characteristics, emphasizing the intricate interplay between fluid properties and nanoparticles. The optimization of nanoparticle volume percentage addresses a crucial trade-off, advancing beyond existing research. We synthesize experimental and computational approaches, bridging theory and practice. Additionally, our collaborative framework involving researchers, engineers, and industry experts enhances the study's applicability. In summary, our work significantly contributes to heat exchanger design, introducing novel insights and methodologies for improved performance and practical applications (Table 7.1).

The need for a smaller and more portable heat exchanger is the pivotal reason for scientific and technology developers to develop more compact models of heat exchange (Kaur and Singh 2021). Nanofluids are a combination of base fluid and a low concentration of nano-sized particles. They are made of metals or metal oxides which are used in different fields of human activity. This includes engineering devices in power and chemical engineering, medicine, electronics, and other various sectors (Chai and Tassou 2020). They have covered a great innovative scope in the field of medical science as well. Besides, as they are dependent on their chemical structure

DOI: 10.1201/9781003494454-7

Table 7.1 Heat exchanger types and applications

Heat exchanger type	*Description*	*Applications*
Shell and tube	Traditional design with a bundle of tubes inside a shell.	HVAC, chemical processing, power plants
Plate heat exchanger	Consists of multiple thin plates for enhanced heat transfer.	Food industry, refrigeration, HVAC
Finned tube heat exchanger	Tubes with extended surfaces (fins) for increased heat transfer area.	Air conditioning, automotive radiators
Microchannel heat exchanger	Compact design with small channels for improved heat transfer.	Electronics cooling, aerospace applications

Source: Self-developed.

and synthesis method, nanofluids are capable of improving thermophysical properties. The introduction of nano-heating technology has proved influential in improving the refrigerating effect.

The advanced industrial processes these days require more cost-effective and goal-oriented devices. Devices that can speed up their work and productivity and provide effective solutions to manage organizational investments (Sait et al. 2023). The introduction of nanofluids in the process of heat exchange has several benefits. It has optimized the utilization of energy resources, has reduced operational costs, and has also reduced the adverse environmental impacts. It is crucial to note that improved efficiency means a scope for higher productivity. Thus, the nanofluids have provided the industries a scope to evolve with supreme energy conservation techniques. They have made it a priority. The imperative for advanced heat exchanger technologies has grown, thus emphasizing the pivotal role that nanofluids play in fostering sustainable and economical industrial practice.

The objective of this chapter is to determine the impact of nanofluids on heat transfer characteristics. This chapter explores the design and its operational principles for incorporating these nanofluids in the heat-exchanging devices. This chapter seeks to focus on the impact of nanofluids on thermal conductivity. It will further delve into exploring its relation with the Nusselt number and establish its impact on the heat exchanger performance.

7.2 EVOLVEMENT IN HEAT-EXCHANGING TECHNOLOGIES

Traditional heat exchanger technologies include devices such as shell-and-tube and plate heat exchangers. They have been the most optimal source of heat suppliers to industrial thermal management. In the shell-and-tube devices, fluids were designed to circulate through a series of tubes within a shell (Sekulic and Shah 2023). It allowed them the scope for heat transfer between them internally. The plate heat exchangers brought a more compact scope. It employed stacked metal plates with fluid passages which were designed for efficient heat management.

As the traditional devices were very much dependent on their range of distribution, they were limited to factors like fouling and uneven fluid distribution. Again, they used to struggle to achieve optimal Nusselt numbers. The Nusselt Numbers is the ratio of convective to conductive heat transfer (Kilkovsky 2020). Along with it, problems arose on the thermal conductivity. The traditional heat exchangers failed to cope up with the industrial demands for higher productivity. Thus, the nanofluids were a sign of the emerging technology which was developed to enhance the heat exchanger performance. It fulfils this need by addressing the key parameters of heat exchange. The nanofluids technology has made the heat exchange process more efficient and sustainable.

The methodologies employed to quantify thermophysical enhancements involved rigorous experimentation and analysis. Heat transfer coefficient improvements were measured through controlled experiments in plate and double-tube heat exchangers, employing sensors and data acquisition systems to capture temperature differentials. Thermal conductivity enhancements were determined using established techniques such as transient hot-wire methods or comparative analysis of heat transfer rates in nanofluid and base fluid systems. These methods ensured reliable and precise measurements, contributing to the credibility of the observed enhancements in thermophysical qualities and providing valuable insights into the effectiveness of nanofluids in improving heat exchanger performance.

7.3 NANOFLUIDS: CHARACTERISTICS, PROPERTIES AND SIGNIFICANCE

A nanofluid is defined as a fluid element that contains the nanoparticles, that is, nanometre-sized particles. Nanotechnology has greatly improved the science of heat transfer. They allow heat transfer by improving the properties of the fluids which enable an easy flow of heat transfer (Smaisim et al. 2022). A high heat transfer device can be designed through the creation of innovative fluids. The composition of it involves dispersing the nanoparticles into metallic and non-metallic dimensions. These dimensions range from 1 to 100 nm. It is developed into a base fluid which includes water, oil, ethylene glycol, and other required components (Awais et al. 2021a). It helps in providing the heat exchange mediums with a versatile platform for various applications in the industrial, medical, and other sectors.

The study investigates the use of various nanofluids, each tailored to specific applications. Notable nanofluids include those containing carbon

Table 7.2 Properties of nanofluids

Nanoparticle type	*Base fluid*	*Concentration (vol. %)*	*Thermal conductivity enhancement*	*Viscosity increase*	*Stability*
Carbon nanotubes	Water	0.1–2.0	20–150%	Moderate	Good
Copper oxide	Ethylene glycol	1–5	10–70%	Low	Excellent
Aluminium oxide	Engine oil	0.5–4	5%–50%	Low	Fair

Source: Self-developed.

nanotubes dispersed in water with concentrations ranging from 0.1% to 2.0%, showcasing a thermal conductivity enhancement of 20%–150% and moderate viscosity. Copper oxide nanofluids in ethylene glycol, with concentrations between 1% and 5%, exhibit a thermal conductivity enhancement of 10–70% and low viscosity, demonstrating excellent stability. Additionally, aluminium oxide nanofluids in engine oil, featuring concentrations from 0.5% to 4%, showcase a thermal conductivity enhancement of 5%–50%, low viscosity, and fair stability. These nanofluids are strategically chosen based on their specific applications, with compositions and concentrations fine-tuned to optimize thermal conductivity and stability for enhanced heat transfer efficiency in plate and double-tube heat exchangers (Table 7.2).

The composition of the nanoparticles depends on the industry they are used for. The addition of nanoparticles increases the effectivity of thermal conductivity in the nanofluid particles. Thus, it helps in facilitating a more efficient heat transfer. The increase in thermal conductivity arises from the enhanced surface area and is thus intensified by the inter-particle interactions within the nanofluid elements (Said et al. 2022a). Nanofluids have a tendency to exhibit a modified attribute of heat transfer. These characteristics are aligned to interact between the nanoparticles and the original fluid matrix. This alteration provides and seeks to improve the process of heat transfer. The available coefficients measure the rate of heat transfer. Thus, it allows the nanofluids to achieve an enhancement in the overall heat exchanger performance (Said et al. 2022b). These processes promote a more effective heat exchange during the requirement of it.

It is important to acknowledge that the heat exchange performed by the use of nano-fluids differs in their need for temperature. Nanofluids are prone to show a high rate in their thermal conductivity, they are very popular in industrial; sectors which deal in high and volatile supply of electric use and heat procedures (Sofiah et al. 2021). Therefore, the integration of nanofluids into heat exchanger systems has improved the scope of usage of heat devices. They are capable of offering several advantages that can contribute to improving the overall performance of the heat exchange devices.

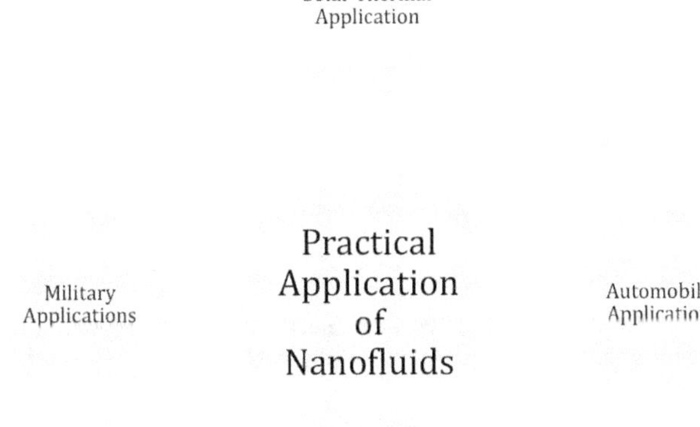

Figure 7.1 Practical applications of nanofluids (*Source*: self-developed).

The atomic design of the nanofluids makes it a sustainable source of heat exchange medium (Figure 7.1).

The properties of nanofluids have made them a popular source for designing heat exchange devices. Be it an industry of automobiles, or medical, military applications, electronic cooling devices or solar thermal plants, nanofluids have shown resilience in withstanding the test of every sector. The growing use of nanofluids is hinged on its various core properties. Its atomic integration provides it with various advantages. Incorporating the ore of the fluid matrix, the nanoparticles are able to improve the performance of heat exchangers (Younes et al. 2022). They provide scope for more efficient heat transfer. This results in an increase in the rate of heat transfer and improves temperature control. By leveraging the unique thermophysical properties of the nanofluids, heat exchangers seek to achieve higher efficiency levels (Sivaraj and Banerjee 2021). The increase in heat transfer rates enables for better utilization of the thermal energy. It thus contributes to energy savings and reduces the operational cost in industrial processes (Figure 7.2).

The Nusselt Number is a parameter that estimates the properties and influences of heat transfer devices. In the nanofluids, the improved Nusselt number shows how the improved heat transfer efficiency has led to

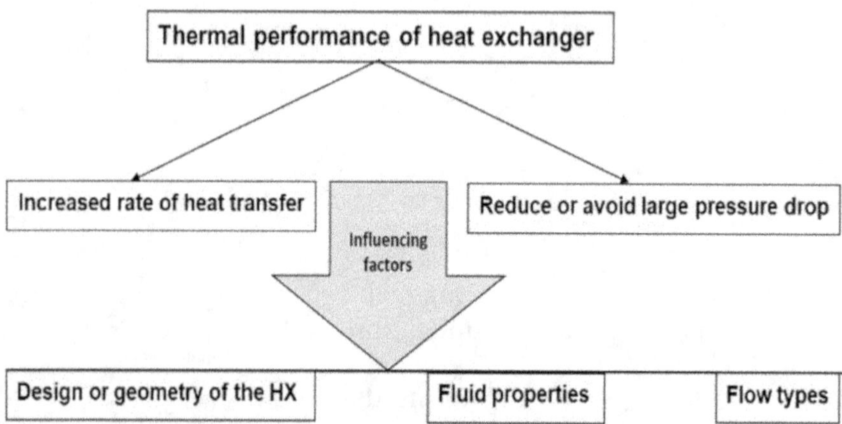

Figure 7.2 Application of nanofluids in improving performance (*Source*: self-developed).

improving the heat exchanger performance (Urmi et al. 2021). The superior capabilities of heat transfer mediums allow for a more compact and lightweight design. The nanofluids which are more compact and lightweight come out as excellent heat exchangers because they consume less electric value. It is very advantageous in applications where the space is compact. In the performance of nanofluids, weight considerations act as a critical factor. In applications of automotive or cooling systems or aerospace applications, it improves effectiveness and acts comparatively eco-friendly.

In addition to these, nanofluids also have the potential to enhance temperature. They allow for a uniform heat exchange. It also addresses the challenges which are associated with hotspots or uneven heat temperature distribution. This property is very valuable in applications where precise temperature control is essential. The nanofluids are crucial for maintaining the equipment's integrity and process efficiency.

7.4 FACTORS THAT IMPROVE HEAT TRANSFER

The use of copper and aluminium oxide boosts the overall thermal conductivity of the nanofluid. The primary mechanism behind the improved heat transfer in nanofluids is based on imparting the thermal conductivity on the nanoparticles (Asadikia et al. 2020). The higher the thermal conductivity the more effectiveness is achieved through the conduction of heat. The fluid helps in reducing the thermal resistances. It thus facilitates the efficient heat exchange between the fluid and the heat exchanger surfaces.

The nanoparticles possess a high surface and volume ratio. In comparison to larger particles, it is more efficient. This increase in the volume of the surface area makes way for more opportunities (Vinod and Philip 2022). This increases the interactions of the particles with the fluid, which leads to improving the heat transfer efficiency. It is important to notice that the introduction of nanoparticles alters the fluid dynamics. In the heat exchanger, this modification and upgradation changes the flow patterns. It thus affects the turbulence and also the convective heat transfer. According to Gupta et al. (2021), the use of nanoparticles thus optimizes the efficiency of the heat exchange process. The thermophoresis effect is the motion of particles in response to temperature gradients. It can play a role in the movement and distribution of nanoparticles. Within the nanofluid, the thermophoresis effect is capable of improving the overall convective current.

7.5 THE USE OF NANOFLUIDS IN PLATE HEAT EXCHANGERS

The plate heat exchangers work on the principles of combined thermal plates through which interconnected pipes provide heat circulation. Researchers recommend the use of stainless steel which provides a sustainable energy consumption within the walls of the plates (Figure 7.3).

The use of the nanofluids in plate heat exchangers improves the heat transfer efficiency. The nanofluids are engineered by dispersing the nanoparticles in conventional heat transfer fluids. They have a tendency to exhibit superior thermal conductivity. The characteristics improve the overall heat exchange performance in the plate exchangers and they lead to enhancing the heat absorption. The small size of the nanoparticles allows increased surface area and facilitates improved fluid–solid interactions, promoting efficient heat transfer. Consequently, the nano-enhanced plate exchangers are found useful in applications of diverse industries, such as electronics cooling and renewable energy systems, where the optimization of heat transfer processes is crucial for enhanced performance and energy efficiency (Pandya

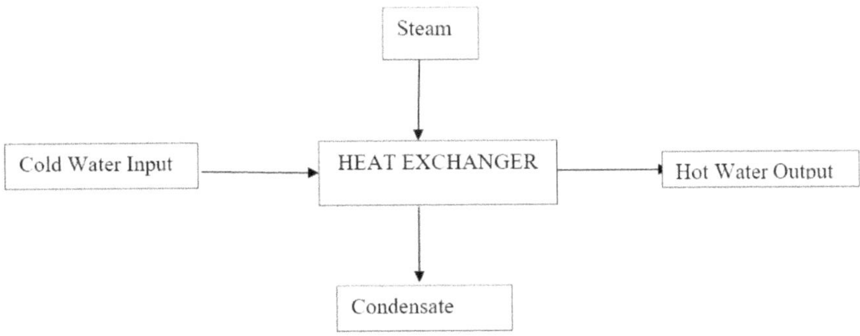

Figure 7.3 Flow principle of heat exchanger (*Source*: self-developed).

et al. 2020). By doing so, the heat is transferred through the plates which leads to a temperature exchange between the fluids without mixing them. This design influences the effective heat transfer property of the fluids.

The use of stainless steel in the plates ensures compatibility with the diverse nanofluids. The given spacing within the plates determines the flow characteristics and the drop of the pressure inside (Manikandan and Baskar 2021). These characteristics also influence the temperature and chemical conditions. The temperature, material, design, and chemical conditions act as a pivotal parameter in achieving optimal heat transfer efficiency (Menni et al. 2020). It brings in maintenance of the heat exchangers which increases reliability over its operational life.

Nanofluids have a visible impact on the performance of the plate heat exchangers. The use of them in the plates increases their potential of enhancing heat transfer. The results of heat exchange have revealed that nanofluids are predicted by the Nusselt number to be reflective of the heightened convective heat transfer efficiency. Various researches have explored the influence of the nanoparticles in the hat chambers. According to Bhattad et al. (2020), their concentrations on the Nusselt numbers have shown improvement in the Plate Heat Exchangers. As nanoparticles tend to concentrate more, the enhanced thermal conductivity becomes stronger. This also improves the rate of heat transfer. It is necessary to notice that there is a gap beyond which any further concentration may damage the entire process.

7.6 MECHANISMS FOR HEAT TRANSFER ENHANCEMENT

The increase in the particles of the nanofluid initiates thermal conductivity among each other. This boosts the nanofluids which promote a better conduction process (Guo 2020). The flow through the plates of the heat exchangers results in the enhancement of the overall heat transfer rates. As per Shi et al. (2023), convective heat management manipulates the characteristics of nanofluids. By manipulating it, they play a crucial role in enhancing the optimal utilization of the Plate Heat Exchangers performance. It is so that the particles change the fluid flow patterns. This results in turbulence which contributes to improved convective heat transfer efficiency.

The turbulence generated by the combined plates intensifies the flow of heat in the convective heat transfer. The nanoparticles within the fluid influence this turbulence. According to Mousa et al. (2021), the plate heat exchangers become playful in their role which leads to additional enhancements. The combined and interrelated structure promotes turbulence. They also provide increased surface area to the nanoparticle within these plates which improves their space for interaction. These further aid the process of conductive heat transfer by acting as a catalyst. The combined effect of the enhanced turbulence and the increased thermal conductivity also contributes to improving the heat transfer.

Nanofluids enhance conductive and convective heat transfer in heat exchangers through various mechanisms. First, the addition of nanoparticles increases thermal conductivity by enhancing surface area and inter-particle interactions, facilitating efficient heat conduction. Second, nanoparticles induce Brownian motion, promoting convective heat transfer by random particle motion. Third, thermophoresis, the particle movement in response to temperature gradients, contributes to convective heat enhancement. Additionally, nanoparticle agglomeration and sedimentation, though challenging for nanofluid stability, can influence convective currents. Understanding these mechanisms elucidates how nanofluids optimize conductive and convective heat transfer processes, improving overall heat exchanger performance (Figure 7.4).

The various experimental and theoretical studies consistently demonstrate the positive impact of nanofluids on heat transfer rates. This showcases their potential to improve the overall performance of the plate heat exchangers (Bezaatpour and Goharkhah 2020). The influence of the nanoparticles' concentration on Nusselt number has the area to improve the nanofluid formulations for specific industrial applications. The ongoing developments have facilitated the synergy between nanofluids and plate heat exchangers. This can prove to have a pivotal role in shaping the future of heat exchange technology (Table 7.3).

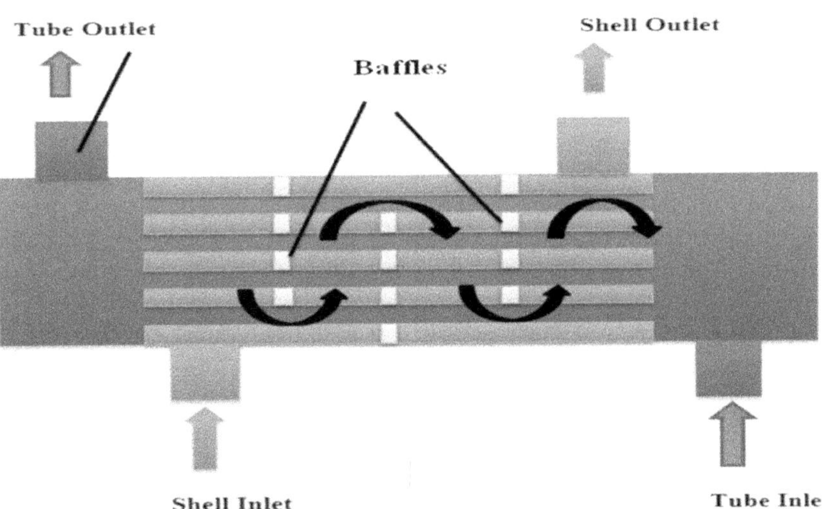

Figure 7.4 Heat transfer enhancement (*Source*: self-developed).

Table 7.3 Nanofluid heat transfer enhancement mechanisms.

Mechanism	*Description*
Thermal conductivity enhancement	Improved heat transfers due to increased thermal conductivity of nanofluids.
Brownian motion	Nanoparticles exhibit random motion, enhancing convective heat transfer.
Thermophoresis	Nanoparticles move toward regions of higher temperature, augmenting heat transfer.
Agglomeration and Sedimentation	Challenges in nanofluid stability leading to aggregation or settling of nanoparticles.

Source: Self-developed.

7.7 HEAT EXCHANGERS WITH DOUBLE TUBE

The Double-Tube heat exchangers have the characteristics of a versatile and widely used thermal management solution. Their configuration follows an arrangement of two tubes. The design allows circulation in the annular space that exists between the inner and outer tubes. The designs are very simple and reliable (Table 7.4).

In traditional heat exchangers with double tubes, the performance is highly limited by the characteristics of the flow. They succumb to exhibit effective heat transfer under the interference of the thermal conductivity (Bahmani et al. 2020). Hence, the use of nanofluids has emerged as a transformative approach. They have helped in enhancing the efficiency of the heat exchange. The nanofluids are composed of base fluids and nanoparticles. They introduce unique thermophysical properties within the components of heat chambers (Corcoles et al. 2020). They include improvement in the thermal conductivity and thus alter the convective heat transfer characteristics. According to Pourahmad et al. (2021), traditional heat exchangers are inferior to the use of nanofluids as the nanofluids demonstrate superior heat transmission capabilities. The potential of the nanofluids lies in their capacity to create a nanoscale network in the double-tube exchangers. They use the two fluid streams to circulate through the space between the tubes. According to Cao et al. (2021), by doing so, the nanoparticles facilitate rapid and efficient heat exchange, contributing to elevating the overall performance (Figure 7.5).

The Double-Tube Heat Exchangers go beyond the local designs and their improvements. They throw an unusual effect on the efficiency of the heat exchanger. The nanofluids are being better utilized through the heightened thermal conductivity which allows energy conservation and provides operational cost reduction. As per Nakhchi et al. (2021), the nanofluid-enhanced

Table 7.4 Case studies—heat exchanger performance with nanofluids

Study	*Nanofluid type*	*Base fluid*	*Nanoparticle concentration*	*Heat transfer enhancement (%)*	*Key findings*
Pourahmad et al. (2021)	Carbon nanotubes	Water	1.5%	30%	Significant improvement in heat transfer observed.
Cao et al. (2021)	Copper oxide	Ethylene glycol	3%	20%	Enhanced thermal performance with stable nanofluid.
Corcoles et al. (2020)	Aluminium oxide	Engine oil	2%	15%	Viscosity increase noted, but heat transfer improvement achieved.

Source: Self-developed.

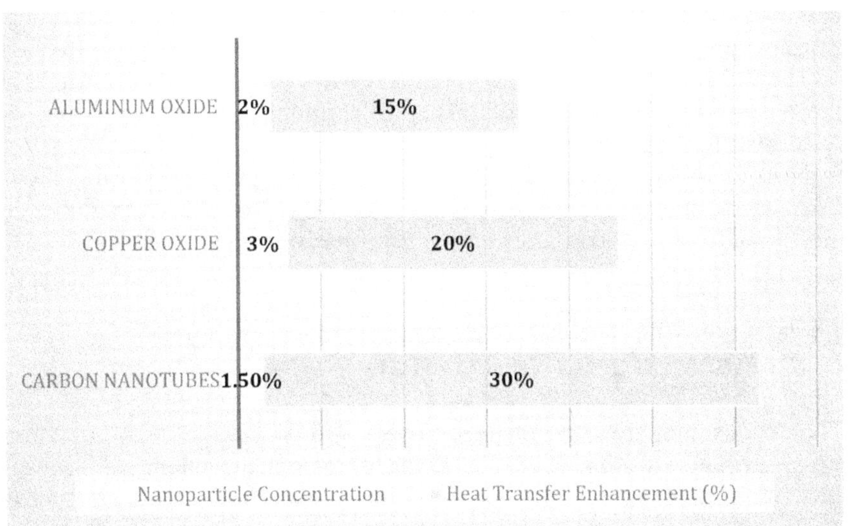

Figure 7.5 Heat exchanger performance with nanofluids (Source: Self-developed).

double-tube heat exchanger does not only facilitate increased heat transfer rates. They have shown paramount solutions over more precise temperature control. This enhanced control of the double-tube plates is vital in industries. In the industrial sectors, it is necessary to maintain the specific operating temperatures is critical. Thus, the process helps in improving product quality.

7.8 THE DESIGN AND FLOW TYPE OF DOUBLE-TUBE HEAT EXCHANGERS

A combination of the double-tube heat exchangers significantly affects the performance of the heat exchangers. The higher and lower medium in the design facilitates an optimal structure for increasing the heat flow and exchange. A double-pipe heat transfer exchanger consists of one or more pipes that are placed concentrically inside another pipe of a larger diameter. They are so structured with appropriate fittings that they direct the flow from one section to the next. One fluid flows through the inner pipe which is the tube side and the other flows through the annular space. Their flexibility and design allow them to be used in many industries. Because of their low design and maintenance costs, flexibility, and low installation costs, they have grown very popular. The arrangement of counter flow region makes lesser flow of the fluids to move in opposite directions.

The use of nanofluids in the double-tube heat exchangers offers advanced techniques in the process of heat transfer. Thus, by improving the heat transmission devices the overall performance can be made better. In terms of energy efficiency and temperature control, the design allows for considerations that include the various requirements of tube parameters. Relying on the industrial need, the flow characteristics are critical in optimizing the synergy between the nanofluids and the double-tube heat exchangers. According to the industrial requirements, there is a need to carefully select materials (Anderson et al. 2020). The various fluid dynamics bring about challenges that are associated with nanofluid stability. This contributes to unlocking the full potential of this transformative technology.

7.9 ROLE OF FLUID CHARACTERISTICS IN HEAT EXCHANGER DESIGN

The various industries stay inclined toward approaches that provide energy efficiency and sustainability. Thus, the developments taken in the heat exchange design allow for the incorporation of nanofluids. Doing so speeds up the overall performance in comparison to the traditional designs. The nanofluids which consist of base fluids exhibit distinct thermophysical properties. These play a pivotal role in heat exchange efficiency as the base fluids alter the nanofluid properties.

The different types of nanofluids have varying effects on the heat exchanger performance. The metallic nanoparticles are known for their high thermal conductivity. This makes them more effective among heat transfer enhancers. Examples of metallic nanoparticles include copper or aluminium oxide. Non-metallic nanoparticles offer additional advantages such as chemical stability to the entire operation of heat transfer. This includes elements like titanium dioxide. The properties of nanofluids allow improving the performance of heat exchangers. For instance, the addition of nanoparticles increases the thermal conductivity that facilitates improved heat transfer rates within the heat exchanger.

7.10 IMPACT OF NANOPARTICLE VOLUME PERCENTAGE ON HEAT TRANSFERS

The volume percentage of nanofluids increases through the concentration of the nanoparticles in their returns. The increase in its viscosity has a positive impact on the efficiency of the heat transfer. Emphasizing on careful consideration and experimentation is necessary. According to Hughes et al. (2021), it allows the identification of the ideal volume fraction which can help to balance the enhanced thermal conductivity with the practical real-life challenges. These challenges are associated with higher nanoparticle concentrations.

The nanoparticle's percentage of volume involves a trade-off between the improved heat transfer and its potential drawbacks. It requires a critical approach to verify and consider the specific characteristics of the nanofluid. These explore the heat exchanger design and its intended industrial application. The studies were made to understand the experiments, and computational modelling plays a crucial role in determining the optimal nanoparticle concentration.

While nanofluids offer promising improvements in heat exchanger performance, several challenges and considerations warrant attention. Challenges include potential nanoparticle agglomeration, which can affect stability and sedimentation issues. Achieving optimal nanoparticle concentrations without compromising fluid stability is a delicate balance. Moreover, the cost of nanoparticles and potential long-term effects on equipment need consideration. The increased viscosity associated with higher nanoparticle concentrations may pose challenges in some applications. Additionally, the potential environmental impact of nanoparticle disposal requires scrutiny. Striking a balance between enhanced heat transfer and overcoming these challenges is crucial for practical implementation, emphasizing the need for ongoing research and careful engineering considerations.

7.11 THE OVERALL EFFICIENCY ACHIEVED BY HEAT EXCHANGER

Heat exchangers are influenced by a multitude of factors beyond fluid characteristics, volume, percentage, and the applicator design. The other influential factors of heat exchangers are the tube material and design. According to Awais et al. (2021b), they affect the flow design by their impact and pressure on the nanoparticles. The tube material and designs of the applicator device ensure the availability of compatibility within the particles of nanofluids. This choice of material and design is further overpowered by the choice of counterflow and parallel flow. The flow arrangement affects the temperature and the overall heat transfer.

There has been the establishment of a wholesome approach in the thermal efficiency of heat exchangers. The sync is established between the varied components and needs to be carefully balanced to achieve maximum performance. As per Dagdevir and Ozceyhan (2021), the various and rigorous testing of the computational simulations and iterative designs allow for evaluating the process. These are essential to establish tune and stability of these parameters and unlock the full potential of the nanofluid heat exchangers. The optimal performance of the various heat exchanges depends on the designs and strategic developments. The performance of heat exchangers with nanofluids remains a promising frontier in the pursuit of energy-efficient and sustainable thermal systems. According to Sim et al. (2021), researchers have looked out for recommendations which can optimize the overall thermal efficiency. Considering a specific applications' requirement, there needs to be a continuous refining of the existing designs. This requires collaborative efforts between researchers, engineers, and industry experts. The implications of the research show the potential of applications in various sectors that can improve their productivity with the help of nanofluids in heat exchangers.

7.12 SUMMARY OF THIS CHAPTER

In conclusion, this chapter serves as a comprehensive exploration of the capacity and capabilities inherent in heat exchanger devices, particularly with a focus on the integration of nanofluids to drive design and performance improvements. The journey embarked upon in this exploration initiates with a nuanced understanding of the challenges associated with traditional heat exchanger technologies. This recognition of challenges becomes a catalyst, propelling the motivation to seek alternatives that are not only more efficient but also transformative.

The primary thrust of this chapter is to underscore the pivotal role played by nanofluids in elevating key parameters crucial to heat exchanger efficiency, namely the enhancement of the heat transfer coefficient, Nusselt number, and thermal conductivity. Traditional heat exchangers are scrutinized, not merely as existing entities but as a baseline to comprehend the transformative potential that nanofluids bring to the table. This understanding becomes a foundational element for subsequent discussions.

Moving forward, this chapter meticulously delves into the characteristics and properties inherent in nanofluids. This detailed examination sets the stage for a comprehensive analysis of various comparative designs, shedding light on their promising and contrasting impacts on different types of heat exchangers. Special attention is dedicated to specific applications, with a focus on how nanofluids influence plate heat exchangers. Detailed discussions unravel the intricacies of design principles and operational principles governing plate heat exchangers, substantiated by both experimental and theoretical evidence supporting the effectiveness of nanofluids in enhancing heat transfer.

Exploration extends to encompass design principles, fluid characteristics, and various flow types, exposing the intricate relationships among these elements and emphasizing the critical role played by nanoparticles in influencing heat exchanger performance. The identification of nanoparticle volume and percentage as key variables, along with their positive impact on the heat transfer rate, solidifies the significance of nanofluids in the realm of heat exchange.

In essence, this chapter stands as a crucial bridge, connecting theoretical frameworks with practical considerations. It unfolds the diverse facets and phases of nanofluid potential in conducting heat transfer within various industrial sectors. The concrete knowledge derived from this exploration not only ensures a safer and more accurate utilization of nanofluids but also enriches professional experiences in the realm of heat exchange technologies.

REFERENCES

Anderson, D., Tannehill, J.C., Pletcher, R.H., Munipalli, R. and Shankar, V., 2020. Computational *Fluid Mechanics and Heat Transfer*. CRC Press.

Asadikia, A., Mirjalily, S.A.A., Nasirizadeh, N. and Kargarsharifabad, H., 2020. Characterization of thermal and electrical properties of hybrid nanofluids prepared with multi-walled carbon nanotubes and Fe2O3 nanoparticles. *International Communications in Heat and Mass Transfer*, *117*, p. 104603.

Awais, M., Bhuiyan, A.A., Salehin, S., Ehsan, M.M., Khan, B. and Rahman, M.H., 2021a. Synthesis, heat transport mechanisms and thermophysical properties of nanofluids: A critical overview. *International Journal of Thermofluids*, *10*, p. 100086.

Awais, M., Ullah, N., Ahmad, J., Sikandar, F., Ehsan, M.M., Salehin, S. and Bhuiyan, A.A., 2021b. Heat transfer and pressure drop performance of Nanofluid: A state-of-the-art review. *International Journal of Thermofluids*, *9*, p. 100065.

Bahmani, M.H., Akbari, O.A., Zarringhalam, M., Ahmadi Sheikh Shabani, G. and Goodarzi, M., 2020.Forced convection in a double tube heat exchanger using nanofluids with constant and variable thermophysical properties. International *Journal of Numerical Methods for Heat & Fluid Flow*, *30*(6), pp. 3247–3265.

Bezaatpour, M. and Goharkhah, M., 2020. Convective heat transfer enhancement in a double pipe mini heat exchanger by magnetic field induced swirling flow. *Applied Thermal Engineering*, *167*, p. 114801.

Bhattad, A., Sarkar, J. and Ghosh, P., 2020. Heat transfer characteristics of plate heat exchanger using hybrid nanofluids: Effect of nanoparticle mixture ratio. *Heat and Mass Transfer*, *56*, pp. 2457–2472.

Cao, Y., Ayed, H., Dizaji, H.S., Hashemian, M. and Wae-hayee, M., 2021.Entropic analysis of a double helical tube heat exchanger including circular depressions on both inner and outer tube. Case *Studies in Thermal Engineering*, *26*, p. 101053.

Chai, L. and Tassou, S.A., 2020. A review of printed circuit heat exchangers for helium and supercritical CO2 Brayton cycles. Thermal *Science and Engineering Progress*, *18*, p. 100543.

Corcoles, J.I., Moya-Rico, J.D., Molina, A.E. and Almendros-Ibáñez, J.A., 2020. Numerical and experimental study of the heat transfer process in a double pipe heat exchanger with inner corrugated tubes. *International Journal of Thermal Sciences*, *158*, p. 106526.

Dagdevir, T. and Ozceyhan, V., 2021. An experimental study on heat transfer enhancement and flow characteristics of a tube with plain, perforated and dimpled twisted tape inserts. *International Journal of Thermal Sciences*, *159*, p. 106564.

Guo, Z., 2020. A review on heat transfer enhancement with nanofluids. *Journal of Enhanced Heat Transfer*, *27*(1), pp. 1–70.

Gupta, S.K., Gupta, S., Gupta, T., Raghav, A. and Singh, A., 2021. A review on recent advances and applications of nanofluids in plate heat exchanger. *Materials Today: Proceedings*, *44*, pp. 229–241.

Hughes, M.T., Fronk, B.M. and Garimella, S., 2021. Universal condensation heat transfer and pressure drop model and the role of machine learning techniques to improve predictive capabilities. *International Journal of Heat and Mass Transfer*, *179*, p. 121712.

Kaur, I. and Singh, P., 2021. State-of-the-art in heat exchanger additive manufacturing. International *Journal of Heat and Mass Transfer*, *178*, p. 121600.

Kilkovsky, B., 2020. Review of design and modeling of regenerative heat exchangers. *Energies*, *13*(3), p. 759.

Manikandan, S.P. and Baskar, R., 2021. Experimental heat transfer studies on copper nanofluids in a plate heat exchanger. *Chemical Industry and Chemical Engineering Quarterly*, *27*(1), pp. 15–20.

Menni, Y., Chamkha, A.J. and Ameur, H., 2020. Advances of nanofluids in heat exchangers-A review. *Heat Transfer*, *49*(8), pp. 4321–4349.

Mousa, M.H., Miljkovic, N. and Nawaz, K., 2021. Review of heat transfer enhancement techniques for single phase flows. *Renewable and Sustainable Energy Reviews*, *137*, p. 110566.

Nakhchi, M.E., Hatami, M. and Rahmati, M., 2021. Experimental investigation of performance improvement of double-pipe heat exchangers with novel perforated elliptic turbulators. *International Journal of Thermal Sciences*, *168*, p. 107057.

Pandya, N.S., Shah, H., Molana, M. and Tiwari, A.K., 2020. Heat transfer enhancement with nanofluids in plate heat exchangers: A comprehensive review. *European Journal of Mechanics-B/Fluids*, *81*, pp. 173–190.

Pourahmad, S., Pesteei, S.M., Ravaeei, H. and Khorasani, S., 2021. Experimental study of heat transfer and pressure drop analysis of the air/water two-phase flow in a double tube heat exchanger equipped with dual twisted tape turbulator: Simultaneous usage of active and passive methods. *Journal of Energy Storage*, *44*, p. 103408.

Said, Z., Arora, S., Farooq, S., Sundar, L.S., Li, C. and Allouhi, A., 2022a. Recent advances on improved optical, thermal, and radiative characteristics of plasmonicnanofluids: Academic insights and perspectives. *Solar Energy Materials and Solar Cells*, *236*, p. 111504.

Said, Z., Sundar, L.S., Tiwari, A.K., Ali, H.M., Sheikholeslami, M., Bellos, E. and Babar, H., 2022b. Recent advances on the fundamental physical phenomena behind stability, dynamic motion, thermophysical properties, heat transport, applications, and challenges of nanofluids. *Physics Reports*, *946*, pp. 1–94.

Sait, S.M., Mehta, P., Gürses, D. and Yildiz, A.R., 2023. Cheetah optimization algorithm for optimum design of heat exchangers. *Materials Testing*, *65*(8), pp. 1230–1236.

Sekulic, D.P. and Shah, R.K., 2023. *Fundamentals of Heat Exchanger Design*. John Wiley & Sons.

Shi, J., Du, H., Chen, Z. and Lei, S., 2023. Review of phase change heat transfer enhancement by metal foam. *Applied Thermal Engineering*, *219*, p. 119427.

Sim, J., Lee, H. and Jeong, J.H., 2021. Optimal design of variable-path heat exchanger for energy efficiency improvement of air-source heat pump system. *Applied Energy*, *290*, p. 116741.

Sivaraj, R. and Banerjee, S., 2021. Transport properties of non-Newtonian nanofluids and applications. *The European Physical Journal Special Topics*, *230*(5), pp. 1167–1171.

Smaisim, G.F., Mohammed, D.B., Abdulhadi, A.M., Uktamov, K.F., Alsultany, F.H., Izzat, S.E., Ansari, M.J., Kzar, H.H., Al-Gazally, M.E. and Kianfar, E., 2022. Nanofluids: properties and applications. *Journal of Sol-Gel Science and Technology*, *104*(1), pp. 1–35.

Sofiah, A.G.N., Samykano, M., Pandey, A.K., Kadirgama, K., Sharma, K. and Saidur, R., 2021. Immense impact from small particles: Review on stability and thermophysical properties of nanofluids. *Sustainable Energy Technologies and Assessments*, *48*, p. 101635.

Urmi, W.T., Rahman, M.M., Kadirgama, K., Ramasamy, D. and Maleque, M.A., 2021. An overview on synthesis, stability, opportunities and challenges of nanofluids. *Materials Today: Proceedings*, *41*, pp. 30–37.

Vinod, S. and Philip, J., 2022. Thermal and rheological properties of magnetic nanofluids: Recent advances and future directions. *Advances in Colloid and Interface Science*, 307, p. 102729.

Younes, H., Mao, M., Murshed, S.S., Lou, D., Hong, H. and Peterson, G.P., 2022. Nanofluids: Key parameters to enhance thermal conductivity and its applications. *Applied Thermal Engineering*, *207*, p. 118202.

Zheng, D., Wang, J., Chen, Z., Baleta, J. and Sundén, B., 2020. Performance analysis of a plate heat exchanger using various nanofluids. International *Journal of Heat and Mass Transfer*, *158*, p. 119993.

Chapter 8

Nanofluids in automotive engineering

Cooling and energy efficiency enhancement

Raja Gunasekaran, Gobinath Velu Kaliyannan, Suganeswaran Kandasamy, Vinodhini Chinnathambi, and Nithyavathy Nagarajan

8.1 INTRODUCTION

The need for fuel-efficient engines is expanding in tandem with the ever-increasing pace of technical advancement in the automobile sector. Rising thermal loads become a direct result of advancements in vehicle technology, necessitating more rapid cooling (Wang and Mujumdar 2008; Mukkamala 2017). Many different types of organizations, including production, technological devices, and automotive, have cooling as a part of their primary technological concerns. Conventional methods of increasing the radiator's rate of cooling that includes the application of microchannels, turbulators, and fins have reached their limitations (Mukkamala 2017; Alam and Kim 2018). The automobile sector has recently prioritized in reducing the weight of vehicles as a means to enhance the efficiency of fuel and related operating expenses (Bigdeli et al. 2016). This could be possible by the process of cooling in the vehicle's engine. Environmentally friendly vehicles must reduce weight by enhancing the layout and capacity of the radiator.

The transmission of heat is crucial for heating as well as cooling processes within the automobile industry. A number of automotive characteristics are significantly impacted through convection of heat, including consumption of gasoline, exhaust, and reliability of engine (Zhao et al. 2016; Gupta et al. 2017). Therefore, cutting-edge coolants are desperately needed to accomplish such a level of effective cooling. Several traditional coolants like water and lubricating substances possess minimal thermal conductivity, resulting in limited transmission of heat (Mukkamala 2017; Mannekote et al. 2018). Developing sustainable heat-transferring fluids featuring much greater thermal conductivity compared to the existing fluid is a critical requirement for organizations in light of the intensifying worldwide competitiveness. As a

 DOI: 10.1201/9781003494454-8

result, to meet the increasing demands for safe and sustainable thermal management, novel, creative, and condensed coolant & coolant technologies will be required (Gupta et al. 2018).

Many different fields are making use of nanotechnologies as a result of the rapid advancements in technological and scientific research. Some of these fields include solar power, biomedicine, cooling of electronics, and heat systems for conserving energy & heat exchangers. Currently, it seems like every substance, and equipment is becoming increasingly smaller. Recent years have seen a proliferation of "nano fluids," a novel kind of coolants made possible by the exponential growth of nanotechnology. Nanofluids represent an innovative and unprecedented conception. Choi Eastman (1995) was the first to suggest the concept of nano fluid, it is essentially a dual-phase structure with a base fluid component and nanoparticles of solid form floating in the fluid. A new class of heat-transferring fluid called nanoparticle-containing fluids has emerged to tackle increasingly difficult cooling problems. In vehicle cooling processes, it shows promise as a substitute for traditional coolant.

Heat transmission coefficients, rigidity, and enhanced thermal conductivity, together with decreased total usage and expenses, have been shown by nanofluids during the last 20 years. Multiple industries might greatly benefit from the use of nanofluids. A growing number of heat transfer systems are making use of nanofluids to maximize efficiency. By improving the rate of cooling, decreasing size, and simplifying thermal management processes, nanofluids possess tremendous potential for use in automobiles and heavy-duty vehicles (Bigdeli et al. 2016; Gupta et al. 2017, 2018; Sidik et al. 2017, Alam and Kim 2018). The increased area of surface and atomic density of nanofluids cause them to outperform traditional fluids in terms of coefficient of heat transfer and thermal conductivity (Said et al. 2015; Bigdeli et al. 2016; Gupta et al. 2017; Tawfik 2017). Radiators for automobiles may be designed to have minimal size and improved thermal efficiency by using nanofluid. It also results in reduced usage of fuel and reduced drag force.

There are several significant investigation discoveries that were published in modern times that emphasize the improved heat transfer characteristics of nanofluids. Yu et al. (2007) discovered that nanofluids improve heat transmission around 15%–40%. The weight of an automobile radiator may be decreased while maintaining its heat-transmission effectiveness because of these improved features. For the structure of the front end of an automobile, this results in a more aerodynamically advantageous characteristic. It is possible to enhance the consumption of fuel effectiveness while minimizing the coefficient of drag.

According to the findings of a number of scholars, the Brownian motion serves as the primary characteristics that contribute to the reduction of heat transmission. According to Xuan and Roetzel's hypothesis, "Brownian motion is brought about by the minimal changes in the formulation with respect to temperature and velocity" (Xuan and Roetzel 2000). The irregular mobility of nanomaterials in the fluid decreases the temperature barrier layer, thereby improving heat transmission (Maiga et al. 2005; Gupta et al. 2017). There exists a difference in velocity among the medium and nanoparticles of fluid due to their irregular mobility (Kakaç and Pramuanjaroenkij 2009; Said et al. 2018). Additionally, because these microscopic particles are tiny, they might greatly lower rust and deterioration (Das et al. 2007). Cooling agents, gasoline compounds, refrigerants lubricating oils, and suspension components are some of the diverse automobile usages of those fluids (Rashmi et al. 2014; Gupta et al. 2017).

Choi and Eastman (1995) demonstrated that it is possible to create a novel category of heat-transferring fluids through floating nanoparticles of metallic material in traditional heat-transmission fluids. Additionally, their findings indicate that generated nanofluids will probably possess greater thermal conductivity than the heat-transferring fluids which are now in use. Theoretical calculations of nanofluid thermal conduction in water containing copper nanoparticles demonstrate promising applications for these materials.

In significant research carried out by Wang et al., a pair of separate nanoparticles containing Al_2O_3 and CuO, were mixed with various fluids, including engine oil, ethylene glycol (EG), vacuum pump and water (Wang et al. 1999). According to the results, nanofluids had greater thermal conductivity compared to their corresponding base fluids and it was measured by steady-state parallel-state approach. An additional innovative investigation was conducted by Das and colleagues to examine the connection between thermal conductivities and temperatures (Das et al. 2003). A rise in heat transfer for the nanofluid containing 4 vol% Al_2O_3 and water was reported to range between 9.4% and 24.3% while the temperature was raised from 21°C to 51°C, as shown by their findings.

Nanoparticles improve the fundamental fluids' blending flow and increase their ability to transfer heat over clear fluid. Recent research accomplished with Masuda et al. (1993) found that 4.3 vol% of γ-$A1_2O_3$ nanoparticles dispersed into water may enhance its efficient thermal conductivity over 30%. Through the use of a chemical-reduction approach, Liu et al. (2006) were able to increase the thermal conduction of Cu-water nanofluid by 23.8%.

Leong et al. (2010) studied the radiator's heat-transmission properties employing copper nanofluids made from EG. Researchers found that employing nanofluids with some EG as the fundamental fluid significantly improved the rate of heat transfer in cooling system of the vehicle. According to the findings, it is possible to produce an enhanced heat transmission around 3.8% by incorporating 2% copper particles into a fundamental fluid at various Reynolds numbers for coolant and air, which are 5000 and 6000, correspondingly.

This chapter explores the transformative potential of nanofluids in the field of automobile engineering, concentrating on their usages in cooling systems for improving energy efficiency. This chapter starts with an examination of the importance of nanofluids. Following that, particular usages of nanofluids in automobile cooling systems that include thermal conductivity, thermophysical characteristics, and stabilization of nanofluids which are used to improve heat transfer and overall efficiency have been explored. Essentially, this exploration into the collaboration between nanofluids and automobile cooling systems acts as evidence to the dynamic interaction between nanotechnology and conventional engineering fields.

8.2 ENGINE COOLANT AND VEHICLE RADIATOR SYSTEM

Automobile engine systems have been steadily improving over the last several decades, because of ever-more-powerful engines. Production of heat by the block system of the engine is a significant factor that inhibits the efficiency of the engine. At the outset of an automobile's cooling mechanism, flow tubes carry coolant directly from the radiators to the block system of the engine. After that, coolant is responsible for absorbing the heat that is produced by the block system. This heat is mostly caused by friction that occurs by the pistons movement as they spin the crankshaft, which subsequently in response causes the wheels to rotate. From there, the thermostat is used to initiate the flow of coolant toward the radiator at specific temperatures. It performs the function of a valve, controlling the coolant flow toward the radiator in order to manage the temperature within the engine's system.

Once the temperature of the engine exceeds the set point, a wax that is present on the thermostat starts to melt. Then, it makes the rod to push out from the valve and finally, it leads the valve to open. Extreme temperatures can generate significant pressure throughout the flow tubes prior to hot coolant reaching the radiators. A pressure-regulating valve can relieve this high pressure. During the final step, air from the outside environment is pushed into the radiator's fan to cool the heated coolant that flows within. Ineffective heat transmission might result in the production of excessive heat in the engine, which would then cause damage to the engine's body. Consequently, it is essential to regulate the temperature at a consistent level in order to improve both the lifetime and capacity of the engine.

A radiator helps in transferring heat to the environment and thus it is also called as heat exchanger. Because of the friction, excessive heat produced in the engine can be protected by the radiators. To prevent the formation of an outer layer as well as to increase heat transmission across the surface level, radiators are constructed using a louvered fin as illustrated in Figure 8.1. Antifreeze is a substance that is used as a supplement throughout the

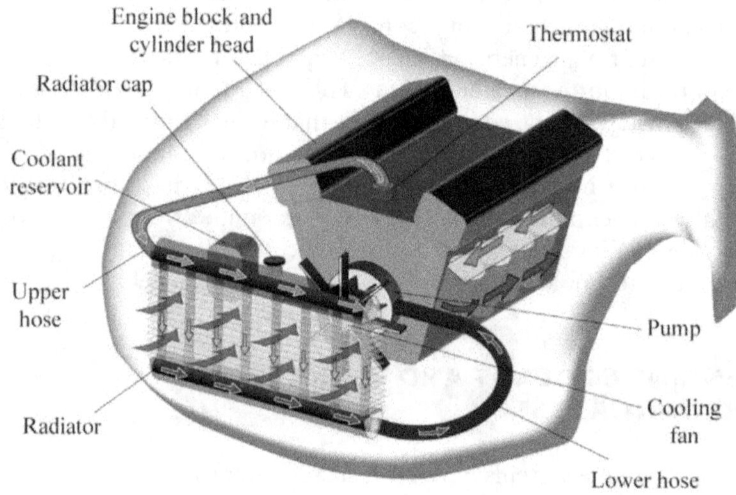

Figure 8.1 Engine and radiator system (Sidik et al. 2017).

Table 8.1 Characteristics of pure EG and water (Sidik et al. 2017)

	Ethylene Glycol	*Water*
Density (g/cm^3)	1.1132	1.0
Molar mass (g/mol)	62.07	18.02
Freezing point (°C)	−12.9	0
Boiling points (°C)	197.3	100
Viscosity (Ns/m^2)	1.61×10^2	1.002×10^3
Thermal conductivity (W m/K)	0.258	0.609

countries that experience severe weather. Its purpose is to either raise the boiling point or to reduce the melting point of the fluid. Due to its cooling characteristics, water is often used as a fundamental fluid in mixtures with glycols, particularly EG, in varying amounts according to the environment. Tables 8.1 and 8.2 demonstrate the characteristics of a water-based EG with different mixing proportions, as well as pure EG and water.

The effectiveness of automobile engines has improved dramatically during the past several years. Producers of engines are currently in a competitive struggle to provide affordable, higher performance engines to satisfy the needs of consumers. The incorporation of nanocoolant represents a novel approach to improving heat transmission in automobile radiators. The research shows that automobile radiators were the first complicated cooling systems for nanocoolants (Choi et al. 2002). Researchers Choi et al. (2002) employed a transient hot wire technique for testing the heating capacity

Table 8.2 Boiling and freezing points of water/EG vs. proportions of EG (Sidik et al. 2017)

Percentage of EG in Water	*Freezing Point (°C)*	*Boiling Point (°C)*
0	0	100
10	−4	102
20	−7	102
30	−15	104
40	−23	104
50	−34	107
60	−48	110
70	−51	116
80	−45	124
90	−29	140
100	−12	197

of nanofluids that consisted of EG and oxides of metal. According to their suggestions, the observed thermal conductivities were significantly greater compared to the expected values. The transient planar sources approach employed by Maranville et al. (2006) to evaluate the thermal conduction properties of water and EG nanofluids aligns with their results. A few years subsequently, Goldstein et al. demonstrated the incorporation of nanomaterials into water results in an excellent thermal diffusivity of nanofluid. This was accomplished through the use of the differential scanning calorimetry method" (Goldenstein et al. 2004). An automobile's radiator, for example, would benefit greatly from a particular coolant's outstanding characteristics as it allows for a rapid reaction to variations in temperature.

8.3 POTENTIAL CHARACTERISTICS OF NANOFLUIDS

Nanofluids possess unique characteristics which make them necessary for a wide range of technical uses. Among such distinct characteristics are (Xie et al. 2002; Witharana et al. 2013; Zhang et al. 2013; Moh'd A and Al-Dafaie 2014; Murshed and De Castro 2014; Peyghambarzadeh et al. 2014; Tie et al. 2014; Ali et al. 2015; Xing et al. 2015):

- Extremely rapid heat-transmission capability
- Excellent theoretical forecasts with higher thermal conductivities
- Less force required to pump fluids
- Substantial stability
- Excellent lubrication
- Reduced coefficient of friction
- Minimal blockage and destruction within the microchannels

8.3.1 Nanofluids—thermal conductivity

In nanofluid, thermal conductivity is a desirable feature for multiple purposes. A material's thermal conductivity is a measure of its heat transfer capability. Numerous studies were conducted on this area from a variety of perspectives. It was discovered by Eastman et al. (2001), indicating the thermal conductivity of EG nanofluids containing 0.3% nanoparticles of copper is improved by as much as 40% in comparison to the fundamental fluid. Possessing this characteristic is crucial for building heat-transferring technology which requires fewer resources, as the researchers pointed out.

According to Hwang et al. (2007), nanofluid thermal conductivity relies on particle volume ratio and fundamental fluid and nanoparticle thermal conductivities. A two-step procedure was used for producing nanofluids comprised of aqueous alumina (Al_2O_3), and Lee et al. (2008) evaluated the heat-transferring capabilities of these nanofluids at limited volume concentrations. According to the researchers, adding alumina fragments to water-based nanofluids causes a gradual rise in their thermal conductivity.

Compared with standard fluids, nanofluids were observed to possess better thermal conductivity. This was successfully observed by the majority of researchers. In nanofluids, thermal conductivity is affected by a number of factors, including humidity, size of particles, distribution and stabilization (Murshed et al. 2009). Figure 8.2 displays the results of comparing the nanofluids thermal conductivity with heat-transferring fluids.

According to studies, nanoparticles in suspension significantly improve the fluid's thermal conductivity and convective heat transfer coefficients

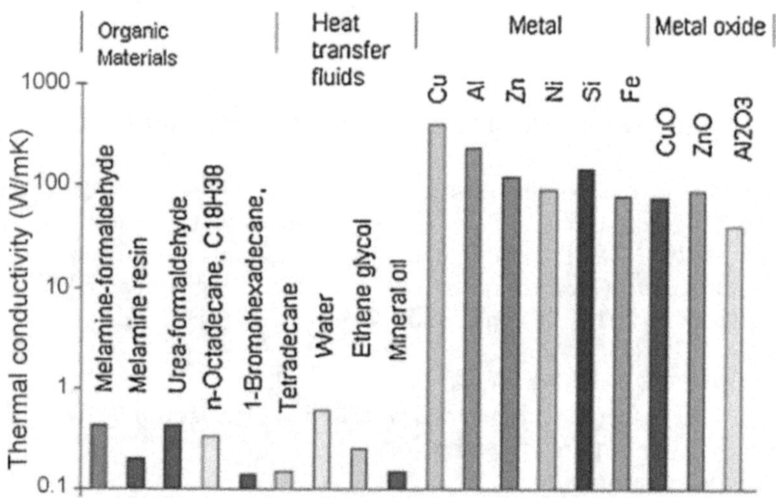

Figure 8.2 Comparison of the thermal conductivity of common liquids, polymers and solids (Wen et al. 2009).

(Xuan and Roetzel 2000; Keblinski et al. 2002). According to Choi et al. (2001) compared to oil on its own, these nanofluids had a thermal conductivity exceeding 150% higher. Current studies in tribology have shown that lubricants containing nanoparticle such as diamond, titanium dioxide and copper oxide have better load-carrying ability, wear-resistant characteristics and anti-friction capability (Xu et al. 1996; Verma et al. 2007). The cooling and lubrication properties of nanofluids make them appealing to several sectors, such as technology, transport, electricity, production and more.

8.3.2 Nanofluids—thermophysical characteristics

In comparison to the fundamental fluid of origin, nanofluids have greater advantages because they belong to a new category of fluids that possess significantly different thermophysical properties. These properties include viscosity, concentration, specific heat power, thermal conductance, accessible heat transfer, heat dispersion and heat diffusion (Eggers and Kabelac 2016; Kianfar et al. 2016; Kianfar 2020a, 2020b). These properties, including their optimal density and viscosity, are often referred to as “effective”. The rationale behind this might be for distinguishing the thermophysical properties of the basic fluid with nanofluids that is created (Kianfar 2015; Kianfar et al. 2015, 2016). Figure 3 shows the nanofluid’s thermophysical properties.

When nanoparticles are added to the fluid, four of its thermophysical properties are changed. Specific heat, heat transfer coefficient, viscosity and densities are every aspect of those properties (Hwang et al. 2007; Pantzali et al. 2009; Sato et al. 2011; Kianfar and Moghadam 2015; Hajimirzaee et al. 2022). Although there are differing opinions among intellectuals on how nanoparticles affect those features, in general, they enhance qualities

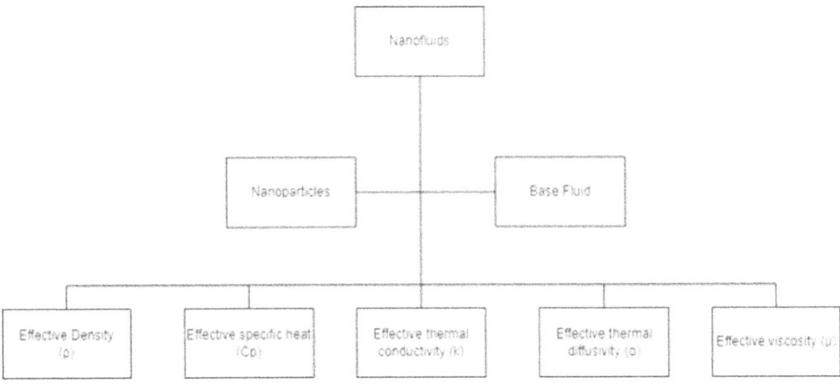

Figure 8.3 Thermophysical properties of nanofluids (Sato et al. 2011; Smaisim et al. 2022).

(as well as the exception of actual heat, which gets reduced) when added to a material (Hwang et al. 2008; Li et al. 2008; Chandrasekar et al. 2010; Yu et al. 2010; Kong et al. 2017). A number of variables, including particle percentage of volume, characteristics of nanoparticles and parameters of fundamental fluid determine the magnitude of the enhancement. Furthermore, since such attributes are dependent upon the nanoparticles' concentration in the fundamental fluid, the attributes may be altered by adjusting the quantity of nanoparticles. This is mainly due to the incorporation of nanoparticles into the fundamental fluid (Li et al. 2007; Kim et al. 2009; Madni et al. 2010; Dey et al. 2017; Abdelbasset et al. 2022; Salah Aldeen et al. 2022; Suryatna et al. 2022).

8.3.3 Nanofluids in vehicle thermal management

Thermal management in vehicles is a multi-faceted field since it influences many different aspects of the automobile, including but not limited to efficiency of the engine, energy consumption, protection, dependability, aerodynamics, convenience for drivers and passengers, pollutants, service and lifespan of the equipment. As a consequence of this, it is essential to have a thermal management in vehicles that is both productive and adaptable in order to build and operate vehicles that have been economical in terms of fuel consumption and are able to comply with continuously growing emission limits (Wambsganss 1999). In thermal management, nanofluids are considered as novel coolants as they possess much greater thermal conductivity and it was demonstrated by Choi et al. (2008). The need for heat dissipation in vehicles is rising steadily is a result of rising power generation standards. A technique aimed at reducing energy consumption represents thermal management, which is applied to the engines and assistance systems of heavy vehicles. This method improves thermal effectiveness and decreases excessive energy consumption and inefficiencies in the engines.

Oils used in engines are much less effective in transferring heat. It is possible to increase the rate of cooling of automobiles and powerful engines by adding nanomaterials to the conventional engine fluid. This development is often employed to dissipate heat from the engine using a smaller coolant arrangement. It is possible that such action will lower the drag coefficient, leading to reduced fuel usage. It is also possible to utilize a larger coolant system featuring better rates of cooling that eliminate additional heat from engines which have greater horsepower. These engines are frequently found in trucks and automobiles.

Reportedly, surface-modified nanomaterials stabilized in mineral-based oils have a positive impact on wear reduction and load-bearing ability when it is used as lubricants in automobiles. A collaborative university-industry study found that nanoparticles added to lubricating substances improve tribological characteristics, including reducing the friction between the dynamic parts in mechanics. From various researches, it has been clear that the implementation of nanoparticles or nanofluids in automobile applications can significantly enhance the rate of heat transfer (Wang and Mujumdar 2008).

8.4 STABILIZATION CHARACTERISTICS OF NANOFLUID

The stability of nanoparticles can be improved by the addition of surfactants. In particular, for applications that require extreme temperature, the functioning of surfactant remains an important consideration. A highly accurate and commonly employed approach for determining a nanofluid's stabilization is the sedimentation process, which is additionally known as the settling bed (Witharana et al. 2012; Witharana et al. 2013). This technique involves physically monitoring the level of the bed or measuring the light absorption throughout a particular duration. Applying this approach to somewhat slow-settling suspensions has the major downside of requiring an extended period of observations; a quicker option is to evaluate the sample's zeta value with temporal resolution.

Agglomerate development, produced by nanoparticles' larger surfaces as well as intense van der Waals force, was the primary reason for the low stabilization of nanofluids, making their preparation a difficult one. Stabilization of nanofluid is crucial for real-world uses due to the fact that thermophysical characteristics of fluids that are not stable vary periodically (Haghighi et al. 2013; Bianco et al. 2014). Nanoparticles must be evenly distributed throughout the fundamental fluid for stabilization. Maintaining nanofluid stability improves their electrical and thermal characteristics (Philip and Shima 2012; Ponmani et al. 2014). Agglomerated nanoparticles hold microchannels and lower the ability of nanofluid to transfer heat (Beck et al. 2010). Since nanofluid stability greatly affects their application-specific features, it is important to study this characteristic.

Due to the robust Brownian movement of nanoparticles that are embedded in base fluids, it was just discovered that nanofluids are better stable compared to suspension of particles that are in micrometer size (Hwang et al. 2008; Tantra et al. 2010). Incorporating surfactant into nanofluid preparations is one of the most effective ways to ensure that nanoparticles are distributed evenly throughout the fundamental fluids. Adding surfactant can reduce the agglomeration among the particles due to van der Waals forces (Fendler 2001; Tantra et al. 2010). In Figure 4, two samples of nanofluid are shown: the first appears optically steady, whereas the second clearly demonstrates particles suspended in the fluid. The definition of stabilized nanofluids is unclear. However, nanofluids are typically regarded as stable if they maintain their solid state over a long enough duration (Rao 2010; Sarkar et al. 2013).

Figure 8.4 Stable (left) and unstable (right) nanofluids (Solangi et al. 2015).

Hwang et al. (2007) used UV-visible spectroscopy in order to assess the nanofluid's stability. A straight-line correlation was observed in a spectrophotometer using absorbance and concentrations of particles using fullerene/oil and multiwalled nanotubes made of carbon. A Malvern ZS Nano S analyzer (DLS) was used by Wang et al. (2009) to evaluate the nanofluids stabilization. Highly charged, evenly distributed suspensions were produced. The authors Chen et al. (2008) used Ludox in conjunction using small-angle X-ray diffraction. A complex issue with X-ray scattering analysis is the interaction between X-rays in which they are dispersed elastically from different nanoparticles in the fluid. Visual inspection of the dispersed stable nanofluids containing metallic oxide substances and nanotube was used to determine their stability and it was reported by Wensel et al. (2008). Considering uniform distribution, Chiesa and Simonsen (2007) employed transmission electron microscopy in conjunction with oil and alumina to produce stable nanofluids.

8.5 NANOFLUID AS COOLANT IN ENGINE COOLING

Automobile manufacturers are always seeking methods to enhance the vehicle's efficiency via aerodynamic layout; however, one of their primary goals involves reducing the total amount of power required to withstand air resistance while driving. When traveling at incredibly rapid speeds, roughly 65% of the entire power production of a large vehicle is utilized up in the process of reducing the impact of aerodynamic drag. The primary reason behind this is because of the strategically placed massive radiators directly at the forefront of the vehicle, which helps to optimize the impact of cooling of the atmospheric air.

A decrease of as much as 10% of the size of the radiator may be achieved through the incorporation of nanofluids that possess high thermal conductivity, which was discovered by Singh et al. (2006). Reduced aerodynamic drag saves approximately 5% gasoline. Nanofluid technology helped cut down on erosion and abrasion, which in turn reduced operational damages, and improved the efficiency of equipment like air compressors, which ultimately saved over 6% of gasoline. It's possible that spending outcomes may change better in subsequent years. Researchers have created and adjusted equipment that may simulate the flow of coolant inside a radiator; researchers are now conducting tests and measurements of the amount of substance lost through distinct nanofluids with the aim to ascertain if the nanofluids deteriorate radiator components. Weight loss measures, which are dependent on the fluid speed and impacting angle, are used to evaluate the rate of deterioration of radiator component. The researchers tested nanofluids composed of ethane and tri-chloroethylene glycols at an angle of impact varying between 90° and 30° and speeds up to 9 m/s without identifying any deterioration. Copper nanofluids were found to exhibit deterioration at a speed of 9.6 m/s and an angle of impact of nearly 90°C. Based upon the computations, the rate of recession that corresponds to the operating of the vehicle is estimated as 0.065 million per year.

Initial research found that nanofluid comprised of copper particles wears faster compared to the standard fluid, presumably owing to the oxidation process of copper. Nanofluids made of aluminum oxide reduce both wear and friction more effectively than a standard fluid (Singh et al. 2006).

Selvam et al. (2017) conducted an investigation on the level of improvement that might be achieved through distributing nanoplatelets of graphene throughout a solution containing water and EG. This was done using louvered fin flattened tubes. Together, such factors—concentrations of nanoparticle approximately 0.5% of volume, rate of flow nearly 62.5 g/s, as well as the speed of surrounding air which is nearly 5 m/s— were added up to around 104% and 81% of augmentation at 35°C and 45°C, correspondingly, based on their findings. Researchers Azmi et al. (2016a) studied the heat-transmission capability exhibited by a nanofluid that contains water, EG and TiO_2 inside a round tube subjected to a turbulence flow. The resulting nanofluid exhibited a 28.9% improvement at a temperature of 70°C over the standard fluid while the titanium dioxide content was raised about 0.5%–1.5% of volume. The researchers found relationships among friction factor & Nusselt number, in addition nearly 4.9% and 3.3% of typical errors were found. The convective heat transfer coefficients of aluminum oxide and titanium dioxide that were distributed in a combination containing water and EG at a ratio of 60:40 were subsequently contrasted by Azmi et al. (2016b). After evaluating various temperatures, at 30°C, aluminum oxide had a greater coefficient of heat transfer compared with titanium dioxide nanofluids. When contrasted with a standard fluid, aluminum oxide and titanium dioxide nanofluids demonstrated an improvement in heat transmission approaching 23.8% and 24.2% over a 1% volume concentration, respectively, when subjected to a temperature of 70°C. The researchers Peyghambarzadeh et al. (2011), and Chavan and Pise (2014) investigated the possibility of dispersing one volume percent of aluminum oxide nanoparticles into the water. With this method, they were able to achieve an enhanced heat transmission of about 45% in comparison to using water alone. Notably, in a related study, Chougule and Sahu (2014) declared that their approach achieved an improvement in heat transmission that is found to be 52.03%.

By incorporating the cooling capabilities of nanofluid into the layout of the engine, upcoming vehicles may achieve optimum operating conditions, resulting in higher power production. Due to the implementation of nanofluids into the engine, the parts will become more compact and lightweight, which allows for improved fuel economy, consequently it will save customers expenditures and produce fewer pollutants, which could lead to safer surroundings.

8.5 CONCLUSION

Exploring nanofluids and their multiple uses in the automobile industry served as the focus of this chapter. Numerous research indicates nanofluids might play an important role in the next generation of engine technology. The versatility of nanofluids makes them useful in many fields, including heat transmission and automotive engineering. Being an innovative substance, nanofluids are additionally employed by the automotive industry for absorbing vibrations. Reducing the dimensions as well as the mass of the vehicle's radiator while maintaining their heat-transferring effectiveness is possible through the application of nanofluid. Nanoparticles, when mixed with engine oil, may increase the fluid's thermal conductivity, thus making them ideal for usage in automotive cooling systems. Furthermore, the inclusion of nanoparticle in automotive oils will increase the lubricant's efficiency as well as decrease the level of friction that occurs. The dispersion of nanoparticles inside a traditional coolant radiator is a further technique that may be used to effectively cool the engine's cooling systems. The coefficient of heat transfer may be enhanced by as much as 50% in comparison with traditional coolant. According to the majority of studies, cooling systems operate optimally with a minimal volume percentage of nanoparticles.

When used with commercialized cooling fluid that contains nanoparticles, internal combustion engines may have an improved efficiency of 5910%. Improving the tribological characteristics of nanoparticle dispersed lubricating substances including load-bearing capability, resistance to wear, and reducing friction may extend the lifespan and efficiency of an automobile. By incorporating a nanoparticle catalyst into conventional fossil fuels, the combustion process may be enhanced while simultaneously decreasing the release of pollutants. As a sophisticated liquid, nanofluids additionally proved to possess potential applications. The technology and science nanofluids, as well as their prospective uses, will allow for their mass production and a wide range of prospective uses.

REFERENCES

Abdelbasset, W. K., S. A. Jasim, D. O. Bokov, M. S. Oleneva, A. Islamov, A. T. Hammid, Y. F. Mustafa, G. Yasin, A. C. Alguno and E. Kianfar (2022). "Comparison and evaluation of the performance of graphene-based biosensors." *Carbon Letters* **32** (4): 927–951.

Alam, T. and M.-H. Kim (2018). "A comprehensive review on single phase heat transfer enhancement techniques in heat exchanger applications." *Renewable and Sustainable Energy Reviews* **81**: 813–839.

Ali, H. M., H. Ali, H. Liaquat, H. T. B. Maqsood and M. A. Nadir (2015). "Experimental investigation of convective heat transfer augmentation for car radiator using ZnO-water nanofluids." *Energy* **84**: 317–324.

Azmi, W., K. A. Hamid, R. Mamat, K. Sharma and M. Mohamad (2016a). "Effects of working temperature on thermo-physical properties and forced convection heat transfer of TiO2 nanofluids in water-Ethylene glycol mixture." *Applied Thermal Engineering* **106**: 1190–1199.

Azmi, W., K. A. Hamid, N. Usri, R. Mamat and M. Mohamad (2016b). "Heat transfer and friction factor of water and ethylene glycol mixture based TiO2 and Al2O3 nanofluids under turbulent flow." *International Communications in Heat and Mass Transfer* **76**: 24–32.

Beck, M. P., Y. Yuan, P. Warrier and A. S. Teja (2010). "The thermal conductivity of aqueous nanofluids containing ceria nanoparticles." *Journal of Applied Physics* **107** (6): 066101.

Bianco, V., O. Manca and S. Nardini (2014). "Performance analysis of turbulent convection heat transfer of Al2O3 water-nanofluid in circular tubes at constant wall temperature." *Energy* 77: 403–413.

Bigdeli, M. B., M. Fasano, A. Cardellini, E. Chiavazzo and P. Asinari (2016). "A review on the heat and mass transfer phenomena in nanofluid coolants with special focus on automotive applications." *Renewable and Sustainable Energy Reviews* **60**: 1615–163

Chandrasekar, M., S. Suresh and A. C. Bose (2010). "Experimental investigations and theoretical determination of thermal conductivity and viscosity of Al2O3/water nanofluid." *Experimental Thermal and Fluid Science* **34** (2): 210–216.

Chavan, D. and A. T. Pise (2014). "Performance investigation of an automotive car radiator operated with nanofluid as a coolant." *Journal of Thermal Science and Engineering Applications* **6** (2): 021010.

Chen, G., W. Yu, D. Singh, D. Cookson and J. Routbort (2008). "Application of SAXS to the study of particle-size-dependent thermal conductivity in silica nanofluids." *Journal of Nanoparticle research* **10**: 1109–1114.

Chiesa, M. and A. Simonsen (2007). "The importance of suspension stability for the hot-wire measurements of thermal conductivity of colloidal suspensions." *16th Australasian Fluid Mechanics Conference Crown Plaza, Gold Coast, Australia*, pp. 1154–1157.

Choi, C., H. Yoo and J. Oh (2008). "Preparation and heat transfer properties of nanoparticle-in-transformer oil dispersions as advanced energy-efficient coolants." *Current Applied Physics* 8 (6): 710–712.

Choi, S., W. Yu, J. R. Hull, Z. Zhang and F. Lockwood (2002). "Nanofluids for vehicle thermal management." *SAE Transactions* **111**: 38–43.

Choi, S., Z. G. Zhang, W. Yu, F. Lockwood and E. Grulke (2001). "Anomalous thermal conductivity enhancement in nanotube suspensions." *Applied Physics Letters* 79 (14): 2252–2254.

Choi, S. U. and J. A. Eastman (1995). *Enhancing Thermal Conductivity of Fluids with Nanoparticles*. Argonne National Lab, Argonne, IL.

Chougule, S. S. and S. K. Sahu (2014). "Comparative study of cooling performance of automobile radiator using Al2O3-water and carbon nanotube-water nanofluid." *Journal of Nanotechnology in Engineering and Medicine* 5 (1): 010901.

Das, S. K., S. U. Choi, W. Yu and T. Pradeep (2007). *Nanofluids: Science and Technology*. John Wiley & Sons.

Das, S. K., N. Putra, P. Thiesen and W. Roetzel (2003). "Temperature dependence of thermal conductivity enhancement for nanofluids." *J. Heat Transfer* **125** (4): 567–574.

Dey, D., P. Kumar and S. Samantaray (2017). "A review of nanofluid preparation, stability, and thermo-physical properties." *Heat Transfer-Asian Research* **46** (8): 1413–1442.

Eastman, J. A., S. Choi, S. Li, W. Yu and L. Thompson (2001). "Anomalously increased effective thermal conductivities of ethylene glycol-based nanofluids containing copper nanoparticles." *Applied Physics Letters* 78 (6): 718–720.

Eggers, J. R. and S. Kabelac (2016). "Nanofluids revisited." *Applied Thermal Engineering* **106**: 1114–1126.

Fendler, J. H. (2001). "Colloid chemical approach to nanotechnology." *Korean Journal of Chemical Engineering* **18**: 1–13.

Goldenstein, L. K., D. Radford and P. Fitzhorn (2004). "The effect of nanoparticle additions on the heat capacity of common coolants." *Vehicle Thermal Management: Heat Exchangers & Climate Control* **97**: 139.

Gupta, M., V. Singh, R. Kumar and Z. Said (2017). "A review on thermophysical properties of nanofluids and heat transfer applications." *Renewable and Sustainable Energy Reviews* **74**: 638–670.

Gupta, M., V. Singh, S. Kumar, S. Kumar, N. Dilbaghi and Z. Said (2018). "Up to date review on the synthesis and thermophysical properties of hybrid nanofluids." *Journal of Cleaner Production* **190**: 169–192.

Haghighi, E., N. Nikkam, M. Saleemi, M. Behi, S. A. Mirmohammadi, H. Poth, R. Khodabandeh, M. S. Toprak, M. Muhammed and B. Palm (2013). "Shelf stability of nanofluids and its effect on thermal conductivity and viscosity." *Measurement Science and Technology* **24** (10): 105301.

Hajimirzaee, S., A. Soleimani Mehr and E. Kianfar (2022). "Modified ZSM-5 zeolite for conversion of LPG to aromatics." *Polycyclic Aromatic Compounds* **42** (5): 2334–2347.

Hwang, Y.-j. , J. Lee, C. Lee, Y. Jung, S. Cheong, C. Lee, B. Ku and S. Jang (2007). "Stability and thermal conductivity characteristics of nanofluids." *Thermochimica Acta* **455** (1-2): 70–74.

Hwang, Y., J.-K. Lee, J.-K. Lee, Y.-M. Jeong, S.-i. Cheong, Y.-C. Ahn and S. H. Kim (2008). "Production and dispersion stability of nanoparticles in nanofluids." *Powder Technology* **186** (2): 145–153.

Kakaç, S. and A. Pramuanjaroenkij (2009). "Review of convective heat transfer enhancement with nanofluids." *International Journal of Heat and Mass Transfer* **52** (13-14): 3187–3196.

Keblinski, P., S. Nayak, P. Zapol and P. Ajayan (2002). "Charge distribution and stability of charged carbon nanotubes." *Physical Review Letters* **89** (25): 255503.

Kianfar, E. (2015). "Production and identification of vanadium oxide nanotubes." *Indian Journal of Science and Technology* **8** (S9): 455–464.

Kianfar, E. (2020a). "Enhanced light olefins production via methanol dehydration over promoted SAPO-34." *Advances in Chemistry Research* **63**: 1–17.

Kianfar, E. (2020b). "Gas hydrate: applications, structure, formation, separation processes, Thermodynamics." *Advances in Chemistry Research* **62**: 1–24.

Kianfar, F. and S. R. M. Moghadam (2015). "Synthesis of spiro pyran by using silica-bonded N-propyldiethylenetriamine as recyclable basic catalyst." *Indian Journal of Science and Technology.*

Kianfar, F., S. R. M. Moghadam and E. Kianfar (2015). "Energy optimization of ilam gas refinery unit 100 by using HYSYS refinery software." *Indian Journal of Science and Technology* **8** (S9): 431–436.

Kianfar, M., F. Kianfar and E. Kianfar (2016). "The effect of nano-composites on the mechanic and morphological characteristics of NBR/PA6 blends." *American Journal of Oil and Chemical Technologies* **4** (1): 29–44.

Kim, H. J., I. C. Bang and J. Onoe (2009). "Characteristic stability of bare Au-water nanofluids fabricated by pulsed laser ablation in liquids." *Optics and Lasers in Engineering* **47** (5): 532–538.

Kong, L., J. Sun and Y. Bao (2017). "Preparation, characterization and tribological mechanism of nanofluids." *Rsc Advances* **7** (21): 12599–12609.

Lee, J.-H., K. S. Hwang, S. P. Jang, B. H. Lee, J. H. Kim, S. U. Choi and C. J. Choi (2008). "Effective viscosities and thermal conductivities of aqueous nanofluids containing low volume concentrations of Al2O3 nanoparticles." *International Journal of Heat and Mass Transfer* **51** (11-12): 2651–2656.

Leong, K. Y., R. Saidur, S. Kazi and A. Mamun (2010). "Performance investigation of an automotive car radiator operated with nanofluid-based coolants (nanofluid as a coolant in a radiator)." *Applied Thermal Engineering* **30** (17-18): 2685–2692.

Li, X., D. Zhu and X. Wang (2007). "Evaluation on dispersion behavior of the aqueous copper nano-suspensions." *Journal of Colloid and Interface Science* **310** (2): 456–463.

Li, X., D. Zhu, X. Wang, N. Wang, J. Gao and H. Li (2008). "Thermal conductivity enhancement dependent pH and chemical surfactant for Cu-H2O nanofluids." *Thermochimica Acta* **469** (1-2): 98–103.

Liu, M.-S., M. C.-C. Lin, C. Tsai and C.-C. Wang (2006). "Enhancement of thermal conductivity with Cu for nanofluids using chemical reduction method." *International Journal of Heat and Mass Transfer* **49** (17-18): 3028–3033.

Madni, I., C.-Y. Hwang, S.-D. Park, Y.-H. Choa and H.-T. Kim (2010). "Mixed surfactant system for stable suspension of multiwalled carbon nanotubes." *Colloids and Surfaces A: Physicochemical and Engineering Aspects* **358** (1-3): 101–107.

Maiga, S. E. B., S. J. Palm, C. T. Nguyen, G. Roy and N. Galanis (2005). "Heat transfer enhancement by using nanofluids in forced convection flows." *International Journal of Heat and Fluid Flow* **26** (4): 530–546.

Mannekote, J. K., S. V. Kailas, K. Venkatesh and N. Kathyayini (2018). "Environmentally friendly functional fluids from renewable and sustainable sources-A review." *Renewable and Sustainable Energy Reviews* **81**: 1787–1801.

Maranville, C., H. Ohtani, D. Sawall, J. Remillard and J. Ginder (2006). *Thermal Conductivity Measurements in Nanofluids Via the Transient Planar Source method*. SAE Technical Paper.

Masuda, H., A. Ebata and K. Teramae (1993). "Alteration of thermal conductivity and viscosity of liquid by dispersing ultra-fine particles. Dispersion of Al2O3, SiO2 and TiO2 ultra-fine particles." *Netsu Bussei* 7 (4): 227–233. doi: 10.2963/jjtp.7.227.

Moh'd A, A.-N. and A. M. A. Al-Dafaie (2014). "Using nanofluids in enhancing the performance of a novel two-layer solar pond." *Energy* **68**: 318–326.

Mukkamala, Y. (2017). "Contemporary trends in thermo-hydraulic testing and modeling of automotive radiators deploying nano-coolants and aerodynamically efficient air-side fins." *Renewable and Sustainable Energy Reviews* **76**: 1208–1229.

Murshed, S., K. Leong and C. Yang (2009). "A combined model for the effective thermal conductivity of nanofluids." *Applied Thermal Engineering* **29** (11-12): 2477–2483.

Murshed, S. S. and C. N. De Castro (2014). "Superior thermal features of carbon nanotubes-based nanofluids: A review." *Renewable and Sustainable Energy Reviews* **37**: 155–167.

Pantzali, M., A. Mouza and S. Paras (2009). "Investigating the efficacy of nanofluids as coolants in plate heat exchangers (PHE)." *Chemical Engineering Science* **64** (14): 3290–3300.

Peyghambarzadeh, S., S. Hashemabadi, A. Chabi and M. Salimi (2014). "Performance of water based CuO and Al_2O_3 nanofluids in a Cu-Be alloy heat sink with rectangular microchannels." *Energy Conversion and Management* **86**: 28–38.

Peyghambarzadeh, S., S. Hashemabadi, S. Hoseini and M. S. Jamnani (2011). "Experimental study of heat transfer enhancement using water/ethylene glycol based nanofluids as a new coolant for car radiators." *International Communications in Heat and Mass Transfer* **38** (9): 1283–1290.

Peyghambarzadeh, S., S. Hashemabadi, M. S. Jamnani and S. Hoseini (2011). "Improving the cooling performance of automobile radiator with Al_2O_3/water nanofluid." *Applied Thermal Engineering* **31** (10): 1833–1838.

Philip, J. and P. D. Shima (2012). "Thermal properties of nanofluids." *Advances in Colloid and Interface Science* **183**: 30–45.

Ponmani, S., J. K. M. William, R. Samuel, R. Nagarajan and J. S. Sangwai (2014). "Formation and characterization of thermal and electrical properties of CuO and ZnO nanofluids in xanthan gum." *Colloids and Surfaces A: Physicochemical and Engineering Aspects* **443**: 37–43.

Rao, Y. (2010). "Nanofluids: stability, phase diagram, rheology and applications." *Particuology* **8** (6): 549–555.

Rashmi, W., A. Ismail, M. Khalid, A. Anuar and T. Yusaf (2014). "Investigating corrosion effects and heat transfer enhancement in smaller size radiators using CNT-nanofluids." *Journal of Materials Science* **49**: 4544–4551.

Said, Z., A. Allagui, M. A. Abdelkareem, H. Alawadhi and K. Elsaid (2018). "Acid-functionalized carbon nanofibers for high stability, thermoelectrical and electrochemical properties of nanofluids." *Journal of Colloid and Interface Science* **520**: 50–57.

Said, Z., M. Sajid, R. Saidur, G. Mahdiraji and N. Rahim (2015). "Evaluating the optical properties of TiO2 nanofluid for a direct absorption solar collector." *Numerical Heat Transfer, Part A: Applications* **67** (9): 1010–1027.

Salah Aldeen, O. D. A., M. Z. Mahmoud, H. S. Majdi, D. A. Mutlak, K. Fakhriddinovich Uktamov and E. Kianfar (2022). "Investigation of effective parameters Ce and Zr in the synthesis of H-ZSM-5 and SAPO-34 on the production of light olefins from naphtha." *Advances in Materials Science and Engineering* **2023**: 1–22.

Sarkar, S., S. Ganguly, A. Dalal, P. Saha and S. Chakraborty (2013). "Mixed convective flow stability of nanofluids past a square cylinder by dynamic mode decomposition." *International Journal of Heat and Fluid Flow* **44**: 624–634.

Sato, M., Y. Abe, Y. Urita, R. Di Paola, A. Cecere and R. Savino (2011). "Thermal performance of self-rewetting fluid heat pipe containing dilute solutions of polymer-capped silver nanoparticles synthesized by microwave-polyol process." *International Journal of Transport Phenomena* **12**: 339–345.

Selvam, C., R. S. Raja, D. M. Lal and S. Harish (2017). "Overall heat transfer coefficient improvement of an automobile radiator with graphene based suspensions." *International Journal of Heat and Mass Transfer* **115**: 580–588.

Sidik, N. A. C., M. N. A. W. M. Yazid and R. Mamat (2017). "Recent advancement of nanofluids in engine cooling system." *Renewable and Sustainable Energy Reviews* **75**: 137–144.

Singh, D., J. Toutbort and G. Chen (2006). "Heavy vehicle systems optimization merit review and peer evaluation." *Annual Report, Argonne National Laboratory* **23**: 405–411.

Smaisim, G. F., D. B. Mohammed, A. M. Abdulhadi, K. F. Uktamov, F. H. Alsultany, S. E. Izzat, M. J. Ansari, H. H. Kzar, M. E. Al-Gazally and E. Kianfar (2022). "Nanofluids: properties and applications." *Journal of Sol-Gel Science and Technology* **104** (1): 1–35.

Solangi, K., S. Kazi, M. Luhur, A. Badarudin, A. Amiri, R. Sadri, M. Zubir, S. Gharehkhani and a. H. Teng (2015). "A comprehensive review of thermophysical properties and convective heat transfer to nanofluids." *Energy* **89**: 1065–1086.

Suryatna, A., I. Raya, L. Thangavelu, F. R. Alhachami, M. M. Kadhim, U. S. Altimari, Z. H. Mahmoud, Y. F. Mustafa and E. Kianfar (2022). "A review of high-energy density lithium-air battery technology: investigating the effect of oxides and nanocatalysts." *Journal of Chemistry* **2022**: 1–32.

Tantra, R., P. Schulze and P. Quincey (2010). "Effect of nanoparticle concentration on zeta-potential measurement results and reproducibility." *Particuology* **8** (3): 279–285.

Tawfik, M. M. (2017). "Experimental studies of nanofluid thermal conductivity enhancement and applications: A review." *Renewable and Sustainable Energy Reviews* **75**: 1239–1253.

Tie, P., Q. Li and Y. Xuan (2014). "Heat transfer performance of Cu-water nanofluids in the jet arrays impingement cooling system." *International journal of thermal sciences* **77**: 199–205.

Verma, P., P. Chaturvedi, J. B. Rawat, M. Kumar, S. Pal, M. Bal, D. Rawal, Harsh, H. P. Vyas and P. Ghosal (2007). "Elimination of current non-uniformity in carbon nanotube field emitters." *Journal of Materials Science: Materials in Electronics* **18**: 677–680.

Wambsganss, M. W. (1999). "Thermal management concepts for higher-efficiency heavy vehicles." *SAE Transactions* **108**: 41–47.

Wang, X.-J. , X. Li and S. Yang (2009). "Influence of pH and SDBS on the stability and thermal conductivity of nanofluids." *Energy & Fuels* **23** (5): 2684–2689.

Wang, X.-Q. and A. S. Mujumdar (2008). "A review on nanofluids-part II: experiments and applications." *Brazilian Journal of Chemical Engineering* **25**: 631–648.

Wang, X., X. Xu and S. U. Choi (1999). "Thermal conductivity of nanoparticle-fluid mixture." *Journal of Thermophysics and Heat Transfer* **13** (4): 474–480.

Wen, D., G. Lin, S. Vafaei and K. Zhang (2009). "Review of nanofluids for heat transfer applications." *Particuology* 7 (2): 141–150.

Wensel, J., B. Wright, D. Thomas, W. Douglas, B. Mannhalter, W. Cross, H. Hong, J. Kellar, P. Smith and W. Roy (2008). "Enhanced thermal conductivity by aggregation in heat transfer nanofluids containing metal oxide nanoparticles and carbon nanotubes." *Applied Physics Letters* **92** (2): 023110.

Witharana, S., C. Hodges, D. Xu, X. Lai and Y. Ding (2012). "Aggregation and settling in aqueous polydisperse alumina nanoparticle suspensions." *Journal of Nanoparticle Research* **14**: 1–11.

Witharana, S., I. Palabiyik, Z. Musina and Y. Ding (2013). "Stability of glycol nanofluids-the theory and experiment." *Powder technology* **239**: 72–77.

Xie, H., J. Wang, T. Xi, Y. Liu, F. Ai and Q. Wu (2002). "Thermal conductivity enhancement of suspensions containing nanosized alumina particles." *Journal of Applied Physics* **91** (7): 4568–4572.

Xing, M., J. Yu and R. Wang (2015). "Experimental study on the thermal conductivity enhancement of water based nanofluids using different types of carbon nanotubes." *International Journal of Heat and Mass Transfer* **88**: 609–616.

Xu, J., J. Zhang, Y. Du, X. Zhang and Y. Li (1996). "Ultrasonic velocity and attenuation in nano-structured Zn materials." *Materials Letters* **29** (1-3): 131–134.

Xuan, Y. and W. Roetzel (2000). "Conceptions for heat transfer correlation of nanofluids." *International Journal of heat and Mass transfer* **43** (19): 3701–3707.

Yu, W., D. M. France, S. U. Choi and J. L. Routbort (2007). *Review and Assessment of Nanofluid Technology for Transportation and Other Applications*. Argonne National Lab, Argonne, IL.

Yu, W., H. Xie, L. Chen and Y. Li (2010). "Enhancement of thermal conductivity of kerosene-based Fe3O4 nanofluids prepared via phase-transfer method." *Colloids and Surfaces A: Physicochemical and Engineering Aspects* **355** (1–3): 109–113.

Zhang, H., S. Shao, H. Xu and C. Tian (2013). "Heat transfer and flow features of Al_2O_3-water nanofluids flowing through a circular microchannel: Experimental results and correlations." *Applied Thermal Engineering* **61** (2): 86–92.

Zhao, N., S. Li and J. Yang (2016). "A review on nanofluids: Data-driven modeling of thermalphysical properties and the application in automotive radiator." *Renewable and Sustainable Energy Reviews* **66**: 596–616.

Chapter 9

Flow of Reiner–Rivlin nanofluid between two rotating disks

Reshu Gupta

9.1 INTRODUCTION

The study of steady flow caused by rotating disks has become one of the most popular topics in academia because of its many technical uses including computer storage, rotating machinery, lubrication, and more. Karman's [1] ground-breaking research on the Newtonian fluid flow on an endless rotating disk involved converting the Navier-Stokes equations into a set of ordinary differential equations (ODEs) using classical similarity transformations. Numerous scientists have spent a lot of time investigating the flow issues on rotating disks by considering many industrial and technological issues. Khan et al. [2] featured the flow of Ree-Eyring fluid between two rotating disks with a magnetic field. Lin and Ghaffari [3] analyzed the flow and the heat transfer of power-law fluid flowing between stretched and rotating disks. They solved the obtained ODEs by using Newton–Raphson and Runge–Kutta methods. Turkyilmazoglu [4] also developed the fluid flow by a disk, which is rotating as well as moving in upward and downward directions. Ejaz and Mustafa [5] compared different models of different viscosity on a rotating disk. Rout et al. [6] studied the flow of a nanofluid in their model. They considered copper and silver as nanoparticles. They applied the Adomian decomposition method to get the solution. Also, compared the results with the literature. Khan et al. [7] discussed the flow of hybrid nanofluids between stretchable and rotating disks. They formulated the system of nonlinear equations by the ND-solve method. Agarwal and Mishra [8] presented the model of disks. They discussed the forced flow as well as the heat transfer of non-Newtonian fluid. They compared the results of Homotopy Perturbation Method (HPM) and the numerical method. Some of the latest research works on the different models of disks have been published by Tabassum and Mustafa [9], Shehzad et al. [10], Naqvi et al. [11], Das and Sarkar [12], Sabu et al. [13], Rashid and Mustafa [14], Gupta and Agarwal [15–19] and Kumar and Sharma [20].

A particle that has less than 100 nm diameter is known as a nanoparticle and a fluid formed by these nanoparticles is known as a nanofluid. In other

DOI: 10.1201/9781003494454-9

words, nanoparticles are suspended in the nanofluid to increase the thermal conductivity of the fluid. First, Choi [21] used the word nanofluid in his research. Kumar et al. [22] explored the problem of dusty nanofluid. They solved by RKF45 method with the help of the shooting method. Khan et al. [23] investigated thermal properties through a disk. They used nanofluid as a source fluid and applied the Reiner-Rivlin (RK) method to solve differential equations. Hayat et al. [24] examined the flow of silver nanoparticles suspended in water and flowing in the disk. Reddy et al. [25] presented the profile of the temperature and flow of copper water and silver water nanofluid in a disk with porosity. Agarwal [26,27] explored micropolar fluid characteristics in the model of disks.

Reiner [28] and Rivlin [29] discussed the non-Newtonian fluid model, which has many uses in different areas. Agarwal [30] studied Reiner–Rivlin fluid flow through stretchable disks and applied an HPM technique in her study. Sadiq and Hayat [31] address irreversibility in Reiner-Rivlin (RR) nanofluid in stretchable as well as rotating disks. Gupta and Agarwal [32] and Gupta and Wakif [33] studied the flow of nanofluid between two parallel disks.

The prime objective of the research is to investigate the Reiner–Rivlin nanofluid between two rotating disks numerically. In this model, different cases for both disks have been discussed. graphene oxide (GO)-water nanofluid is considered and the variations of many parameters are discussed and pictured graphically.

9.2 FORMULATION AND SOLUTION OF THE PROBLEM

As suggested by Reiner [28] and Rivlin [29], the equations of Reiner–Rivlin fluid are

$$\tau_{ij} = -p\delta_{ij} + \mu_{nf} e_{ij} + 2\mu_c e_{ik} e_{kj} \tag{9.1}$$

where

$$\delta_{ij} = \begin{cases} 1 & i = j \\ 0 & i \neq j \end{cases} \tag{9.2}$$

A system of cylindrical coordinates is chosen to solve this model. The Reiner–Rivlin nanofluid is flowing between two coaxially rotating disks. Graphene GO is considered a nanoparticle and water is a base fluid. u, v, and w are velocities in the direction of cylindrical coordinates. The lower disk rests at the plane $z = 0$ and is rotating with the angular velocity $s\Omega$. The upper disk is resting at the plane $z = d$ and is rotating with different angular velocity Ω as shown in Figure 9.1.

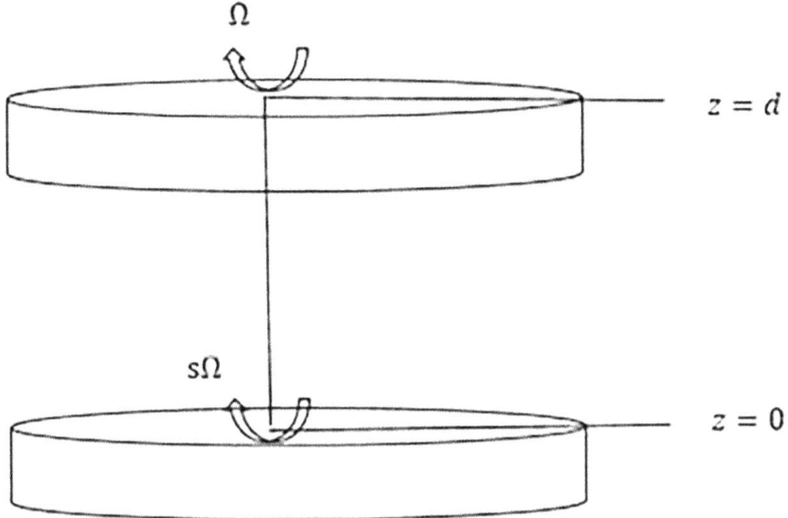

Figure 9.1 Model picture.

The governing equations are

$$\frac{\partial u}{\partial r}+\frac{1}{r}\frac{\partial v}{\partial \theta}+\frac{u}{r}+\frac{\partial w}{\partial z}=0 \tag{9.3}$$

$$\rho_{nf}\left(u\frac{\partial u}{\partial r}+\frac{v}{r}\frac{\partial u}{\partial \theta}-\frac{v^2}{r}+w\frac{\partial u}{\partial z}\right)=\frac{\partial \tau_{rr}}{\partial r}+\frac{1}{r}\frac{\partial \tau_{r\theta}}{\partial \theta}+\frac{\partial \tau_{rz}}{\partial z}+\frac{\tau_{rr}-\tau_{\theta\theta}}{r} \tag{9.4}$$

$$\rho_{nf}\left(u\frac{\partial v}{\partial r}+\frac{v}{r}\frac{\partial v}{\partial \theta}+\frac{uv}{r}+w\frac{\partial v}{\partial z}\right)=\frac{\partial \tau_{r\theta}}{\partial r}+\frac{1}{r}\frac{\partial \tau_{\theta\theta}}{\partial \theta}+\frac{\partial \tau_{\theta z}}{\partial z}+\frac{2\tau_{r\theta}}{r} \tag{9.5}$$

$$\rho_{nf}\left(u\frac{\partial w}{\partial r}+w\frac{\partial w}{\partial z}\right)=\frac{\partial \tau_{rz}}{\partial r}+\frac{\partial \tau_{zz}}{\partial z}+\frac{\tau_{rz}}{r} \tag{9.6}$$

The non-dimensional form of velocity components, which helps to convert nonlinear partial differential equations into a set of nonlinear ODEs and is suggested by Bhatnagar [34] is

$$u=-\frac{1}{2}r\Omega H'(\xi),\ v=r\Omega G(\xi),\ w=d\Omega H(\xi) \tag{9.7}$$

with boundary conditions

$$\begin{aligned} &\text{At}\, z=0:\ u=0,\ v=rs\Omega,\ w=0,\\ &\text{At}\, z=d:\ u=0,\ v=r\Omega,\ w=0 \end{aligned} \tag{9.8}$$

where $\xi\left[=\frac{z}{d}\right]$ is a non-dimensional parameter. The values of ρ_{nf} and μ_{nf} as introduced by Zangooee et al. [35] are

$$\mu_{nf}=\frac{\mu_f}{(1-\phi)^{2.5}},\ \nu_{nf}=\frac{\mu_{nf}}{\rho_{nf}},\ \rho_{nf}=(1-\phi)\rho_f+\phi\rho_s \tag{9.9}$$

Solving momentum Eqs. (9.4)–(9.6) with the help of the similarity transformation (9.7) and Eq. (9.9) and then eliminating the pressure factor, we get

$$\begin{aligned} &H^{iv}-R\left(1-\phi+\phi\frac{\rho_s}{\rho_f}\right)^{2.5}(HH'''+4GG')\\ &+\frac{1}{2}LR(1-\phi)^{2.5}(H'H^{iv}+12G'G''+2H''H''')=0 \end{aligned} \tag{9.10}$$

$$\begin{aligned} &G''-R\left(1-\phi+\phi\frac{\rho_s}{\rho_f}\right)^{2.5}(HG'-H'G)\\ &+\frac{1}{2}LR(1-\phi)^{2.5}(H'G''-H''G')=0 \end{aligned} \tag{9.11}$$

where $R\left[=\frac{\rho_f\Omega d^2}{\mu_f}\right]$ be the Reynolds number, and $L\left[=\frac{\mu_c}{\rho_f d^2}\right]$ be the non-Newtonian parameter. ρ_s and ρ_f are the densities of GO and water, respectively, and ϕ is nanoparticle volume fraction.

The transformed boundary conditions are

$$\begin{aligned} &\text{At}\,\xi=0: G=s,\ H=0,\ H'=0,\\ &\text{At}\,\xi=1: G=1,\ H=0,\ H'=0 \end{aligned} \tag{9.12}$$

Now solve Eqs. (9.10) and (9.11) with Eq. (9.12) numerically with the help of MATLAB software. We get values of profiles for different parameters.

9.3 RESULT AND DISCUSSION

In this section, the response of the different parameters is analyzed and presented graphically. We considered Graphene oxide Go as nanoparticles and water as a base fluid. So, the values of $\rho_s=1{,}800\,\text{kg/m}^3$ and $\rho_f=997.1\ \text{kg/m}^3$.

The impact of the Reynolds number by fixing other variables for $s = 2$ and -1, $L = 5$, $\phi = 0.1$ for all profiles is displayed in Figures 9.2a, 9.3a, 9.5a, 9.6a, 9.7a, and 9.8a. The variation of profiles for the different values of non-Newtonian parameter by keeping other variables fixed like $s = 2$ and -1, $R = 0.1$, $\phi = 0.1$ are presented in Figures 9.2b, 9.3b, 9.5b, 9.6b, 9.7b, and Figure 9.8b. The effect of ϕ at $s = 2$ and -1, $L = 5$, $R = 0.1$ for all profiles is shown in Figures 9.2c, 9.3c, 9.5d, 9.6d, 9.7c and 9.8c.

9.3.1 Discussion on the transverse velocity $G(\xi)$

The behavior of the transverse velocity is presented in Figures 9.2–9.4. Figure 9.2 depicts the nature of the velocity when both disks rotate with different angular velocities in the same direction. Figure 9.3 presents the

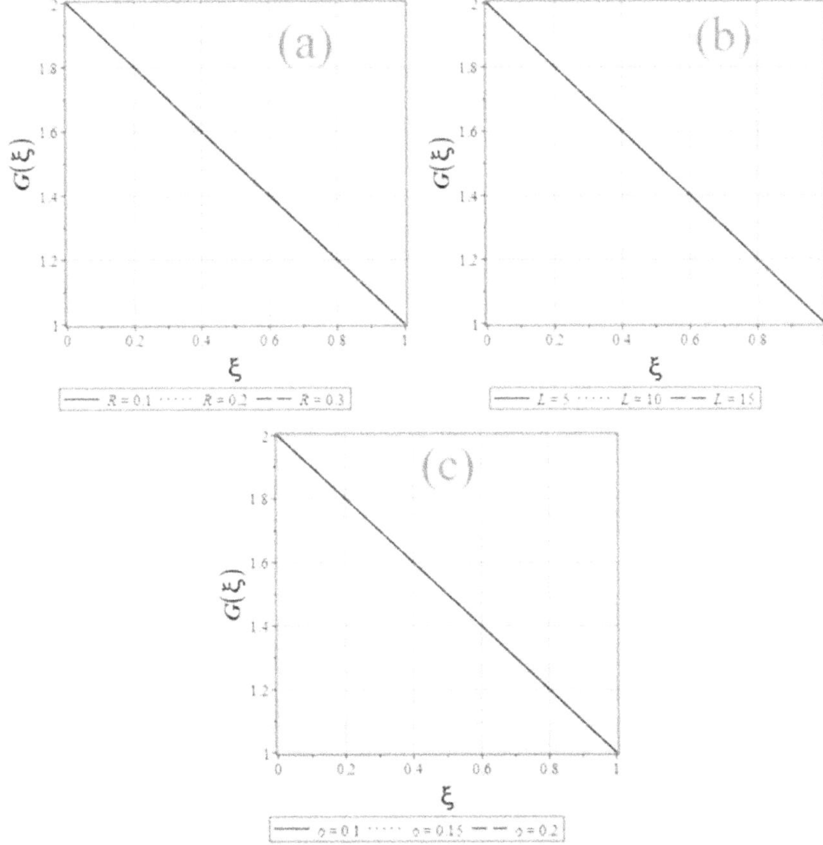

Figure 9.2 (a) Effect of R on transverse velocity when s is positive. (b) Effect of L on transverse velocity when s is positive. (c) Effect of ϕ on transverse velocity when s is positive.

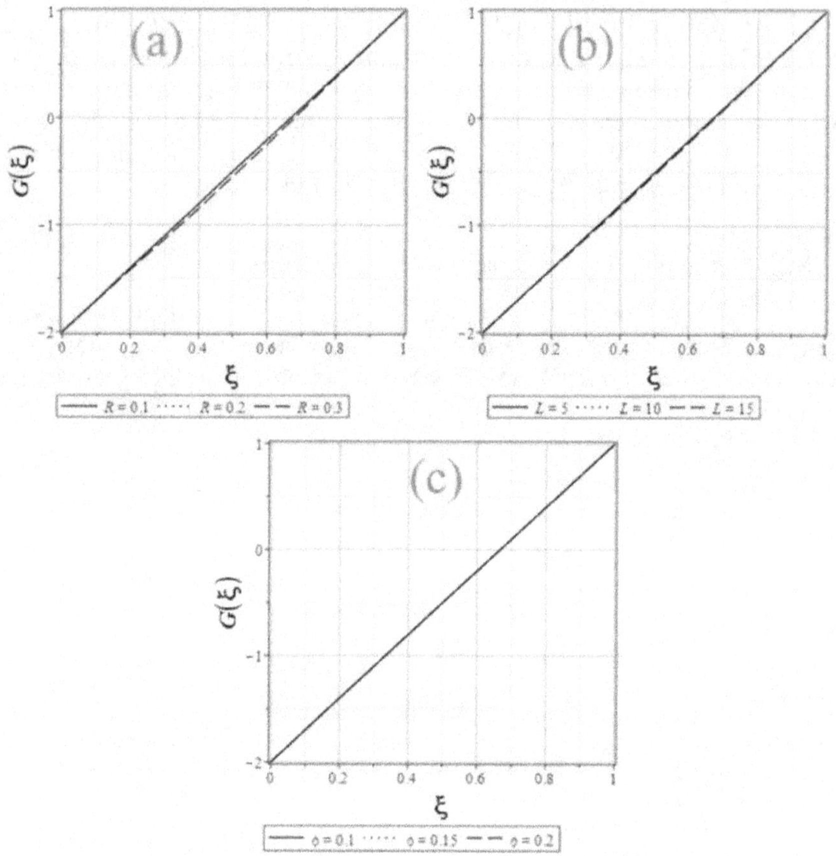

Figure 9.3 (a) Effect of R on transverse velocity when s is negative. (b) Effect of L on transverse velocity when s is negative. (c) Effect of ϕ on transverse velocity when s is negative.

velocity profile when both disks rotate with different angular velocities but in the opposite direction. It can be elucidated from these figures that $G(\xi)$ has linear behavior in all cases. The increased and decreased nature of transverse velocity depends on the direction of the rotating disk. The effect of the other parameters on the velocity is almost null.

9.3.2 Discussion on the axial velocity $H(\xi)$

The effect of various parameters like the Reynolds number (R), rotation parameter (s), non-Newtonian parameter (L), and the nanoparticle volume fraction (ϕ) on the axial velocity is presented in Figures 9.5 and 9.6.

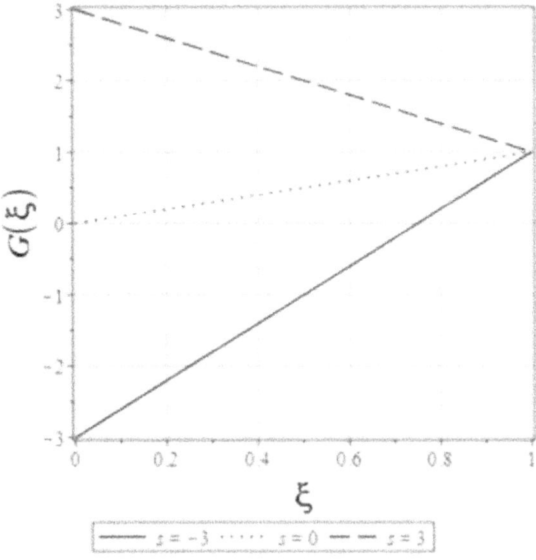

Figure 9.4 Transverse velocity when s is negative, positive, and zero.

The case, when both disks are in the same direction, is framed in Figure 9.5 and the other case, when disks are in the opposite direction, is framed in Figure 9.6. In all figures, the axial velocity is negative in the entire gap length. One more observation can be made from these figures that in both cases the nature of the axial velocity for the other parameters is the same. The behavior of the velocity is more negative on increasing values of R, s, and L. In the case of rising values of ϕ, velocity nature is less negative.

9.3.3 Discussion on the radial velocity $H'(\xi)$

The variation of the radial velocity for different parameters is presented in Figures 9.7 and 9.8. Figure 9.7 displays when s is positive while Figure 9.8 is for negative s. In both cases and all subfigures of Figures 9.7 and 9.8, the radial velocity is negative near the lower disk and positive near the upper disk. The velocity falls with increased values of R, s, and L in the lower half region and raised with increased values in the upper half region. Figures 9.7d and 9.8d displayed the less negative nature in the first half and the more positive nature of $H'(\xi)$ in the second half for ϕ.

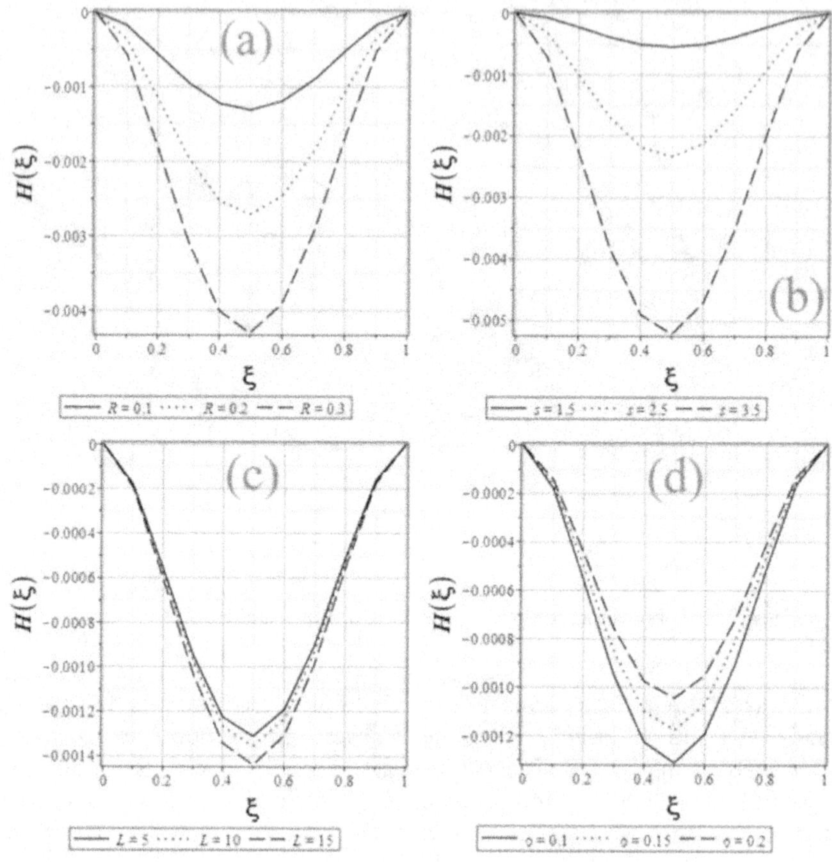

Figure 9.5 (a) Effect of R on axial velocity when s is positive. (b) Effect of s on axial velocity when s is positive. (c) Effect of L on axial velocity when s is positive. (d) Effect of ϕ on axial velocity when s is positive.

9.4 CONCLUSION

A numerical method is successfully applied for the model of the flow of the Reiner–Rivlin nanofluid between two rotating disks with the help of Maple software. Two different cases have been discussed here. The first is when both disks rotate with different angular velocities but in the same direction, and the second one is when both disks rotate with different angular velocities but in opposite directions. The variations of many parameters have been discussed and pictured successfully. It has been concluded that all the parameters do not affect the transverse velocity and it has linear nature. The effect of R, s, and L on the axial and the radial velocity is almost the same.

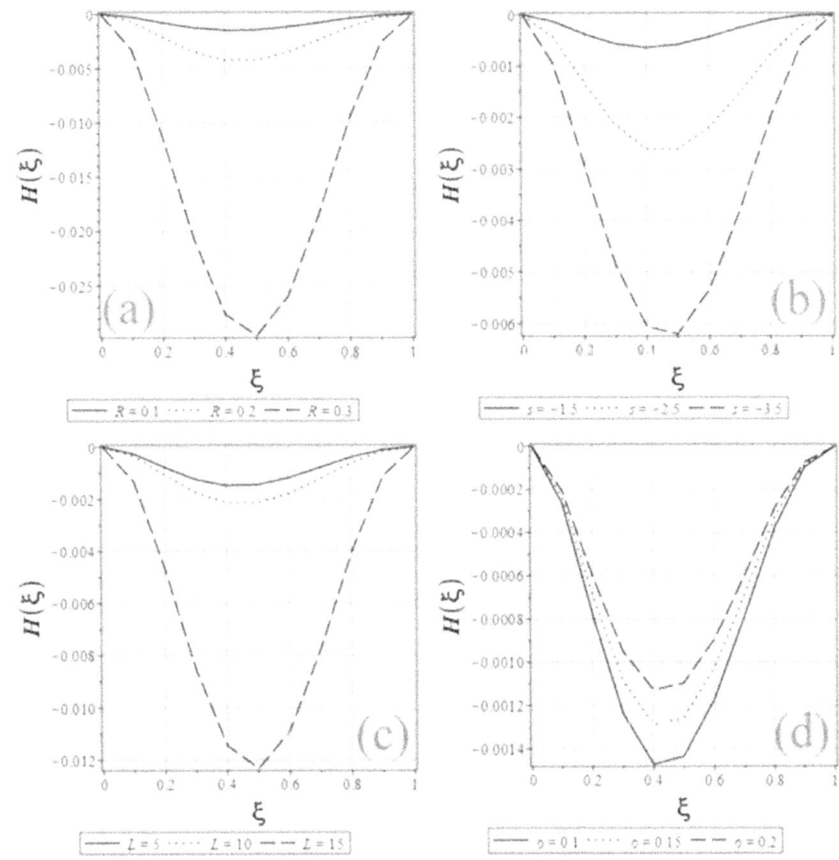

Figure 9.6 (a) Effect of R on axial velocity when s is negative. (b) Effect of s on axial velocity when s is negative. (c) Effect of L on axial velocity when s is negative. (d) Effect of ϕ on axial velocity when s is negative.

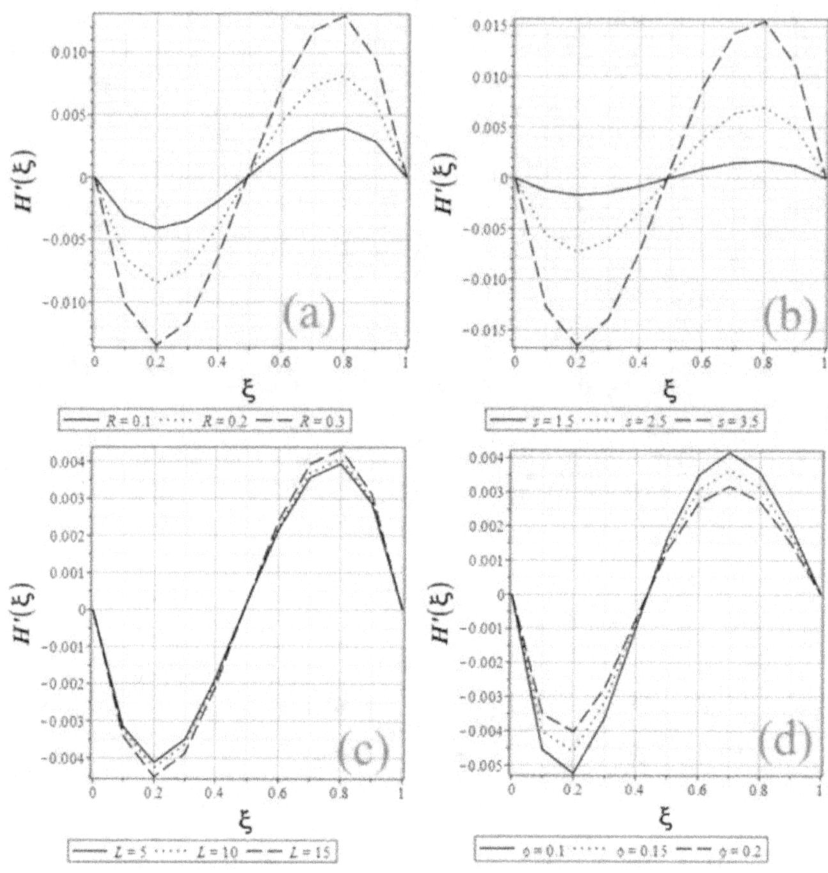

Figure 9.7 (a) Effect of R on radial velocity when s is positive. (b) Effect of s on radial velocity when s is positive. (c) Effect of L on radial velocity when s is positive. (d) Effect of ϕ on radial velocity when s is positive.

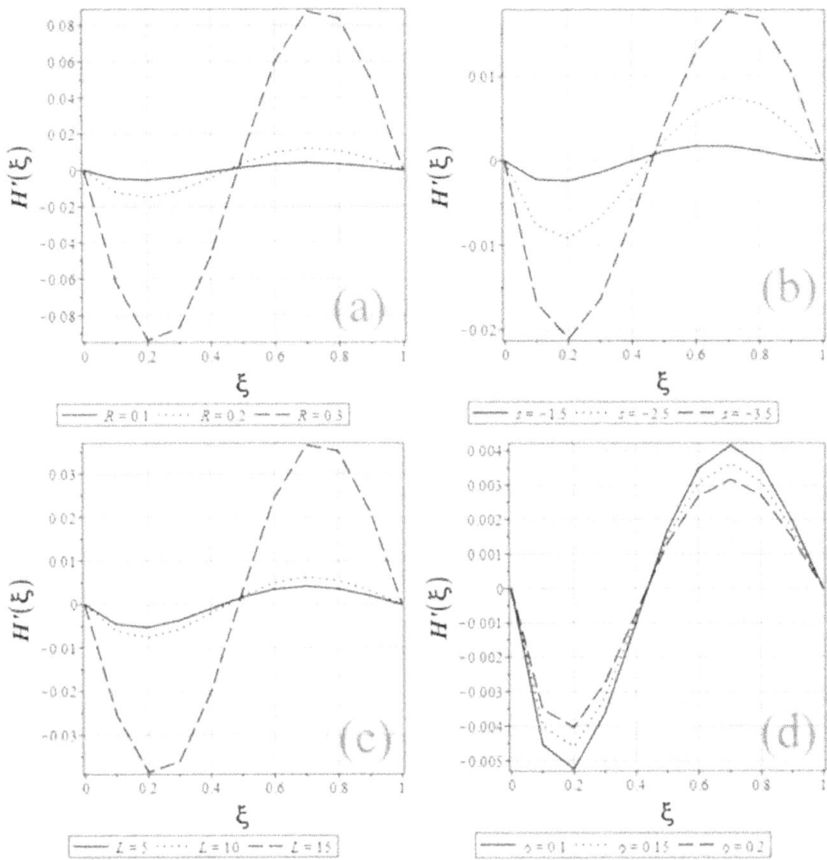

Figure 9.8 (a) Effect of *R* on radial velocity when s is negative. (b) Effect of s on radial velocity when s is negative. (c) Effect of *L* on radial velocity when s is negative. (d) Effect of ϕ on radial velocity when s is negative.

REFERENCES

1. Von Kármán, T. (1921). Uber laminare und turbulente Reibung. *Journal of Applied Mathematics and Mechanics*, *1*, 233–252.
2. Khan, M. I., Khan, S. A., Hayat, T., Javed, M. F., & Waqas, M. (2019). Entropy generation in radiative flow of Ree-Eyring fluid due to due rotating disks. *International Journal of Numerical Methods for Heat & Fluid Flow*. doi:10.1108/HFF-11-2018-0642
3. Lin, P., & Ghaffari, A. (2021). Steady flow and heat transfer of the power-law fluid between two stretchable rotating disks with non-uniform heat source/sink. *Journal of Thermal Analysis and Calorimetry*, *146*(4), 1735–1749.

4. Turkyilmazoglu, M. (2018). Fluid flow and heat transfer over a rotating and vertically moving disk. *Physics of Fluids*, *30*(6), 063605.
5. Ejaz, I., & Mustafa, M. (2022). A comparative study of different viscosity models for unsteady flow over a decelerating rotating disk with variable physical properties. *International Communications in Heat and Mass Transfer*, *135*, 106155.
6. Rout, B. C., Mishra, S. R., & Nayak, B. (2019). Semi analytical solution of axisymmetric flows of Cu- and Ag-water nanofluids between two rotating disks. *Heat Transfer-Asian Research*, *48*(3), 957–981.
7. Khan, M. I., Alzahrani, F., & Hobiny, A. (2020). Mathematical modeling and heat transfer in nanofluid flow of Newtonian material between two rotating disks. *Applied Nanoscience*, 1–12. doi:10.1007/s13204-020-01586-6
8. Agarwal, R., & Mishra, P. K. (2021). Analytical solution of the MHD forced flow and heat transfer of a non-newtonian visco-inelastic fluid between two infinite rotating disks. *Materials Today: Proceedings*, *46*, 10153–10163.
9. Tabassum, M., & Mustafa, M. (2018). A numerical treatment for partial slip flow and heat transfer of non-Newtonian Reiner–Rivlin fluid due to rotating disk. *International Journal of Heat and Mass Transfer*, *123*, 979–987.
10. Shehzad, S. A., Mabood, F., Rauf, A., Izadi, M., & Abbasi, F. M. (2021). Rheological features of non-Newtonian nanofluids flows induced by stretchable rotating disk. *Physica Scripta*, *96*(3), 035210.
11. Naqvi, S. M. R. S., Kim, H. M., Muhammad, T., Mallawi, F., & Ullah, M. Z. (2020). Numerical study for slip flow of Reiner–Rivlin nanofluid due to a rotating disk. *International Communications in Heat and Mass Transfer*, *116*, 104643.
12. Das, A., & Sarkar, S. (2020). Flow analysis of Reiner–Rivlin fluid between two stretchable rotating disks. In: Chakraverty, S. & Biswas, P. (eds) *Recent Trends in Wave Mechanics and Vibrations* (pp. 61–70). Springer, Singapore.
13. Sabu, A. S., Mackolil, J., Mahanthesh, B., & Mathew, A. (2021). Reiner–Rivlin nanomaterial heat transfer over a rotating disk with distinct heat source and multiple slip effects. *Applied Mathematics and Mechanics*, *42*(10), 1495–1510.
14. Rashid, M. U., & Mustafa, M. (2021). A study of heat transfer and entropy generation in von Kármán flow of Reiner–Rivlin fluid due to a stretchable disk. *Ain Shams Engineering Journal*, *12*(1), 875–883.
15. Agarwal, R. (2022). Squeezing MHD flow along with heat transfer between parallel plates by using the differential transform method. *Journal of Nano-and Electronic Physics*, *14*(6), 06010 (5 pp).
16. Gupta, R. (2022). Homotopy perturbation method for the MHD second-order fluid flow through a channel with permeable sides. *Harbin Gongye Daxue Xuebao/Journal of Harbin Institute of Technology*, *54*(12), 38–45.
17. Gupta, R. (2022). Comparative study of micropolar fluid flow between two disks. *Harbin Gongye Daxue Xuebao/Journal of Harbin Institute of Technology*, *54*(12), 61–76.
18. Gupta, R., Selvam, J., Vajravelu, A., & Nagapan, S. (2023). Analysis of a squeezing flow of a casson nanofluid between two parallel disks in the presence of a variable magnetic field. *Symmetry*, *15*(1), 120.

19. Gupta, R. (2023). Flow of a second-order fluid due to disk rotation. In: Mukesh Kumar Awasthi, Maitri Verma, & Mangey Ram (eds) *Advances in Mathematical and Computational Modeling of Engineering Systems* (pp. 315–333). CRC Press.
20. Kumar, S., & Sharma, K. (2022). Impacts of Stefan blowing on Reiner–Rivlin fluid flow over moving rotating disk with chemical reaction. *Arabian Journal for Science and Engineering*, *48*(3), 1–10.
21. Choi, S. U., & Eastman, J. A. (1995). *Enhancing Thermal Conductivity of Fluids with Nanoparticles* (No. ANL/MSD/CP-84938; CONF-951135-29). Argonne National Lab, Argonne, IL.
22. Naveen Kumar, R., Mallikarjuna, H. B., Tigalappa, N., Punith Gowda, R. J., & Umrao Sarwe, D. (2022). Carbon nanotubes suspended dusty nanofluid flow over stretching porous rotating disk with non-uniform heat source/sink. *International Journal for Computational Methods in Engineering Science and Mechanics*, *23*(2), 119–128.
23. Khan, U., Ahmed, N., Mohyud-Din, S. T., Alharbi, S. O., & Khan, I. (2022). Thermal improvement in magnetized nanofluid for multiple shapes nanoparticles over radiative rotating disk. *Alexandria Engineering Journal*, *61*(3), 2318–2329.
24. Hayat, T., Rashid, M., Imtiaz, M., & Alsaedi, A. (2017). Nanofluid flow due to rotating disk with variable thickness and homogeneous-heterogeneous reactions. *International Journal of Heat and Mass Transfer*, *113*, 96–105.
25. Reddy, P. S., Sreedevi, P., & Chamkha, A. J. (2017). MHD boundary layer flow, heat and mass transfer analysis over a rotating disk through porous medium saturated by Cu-water and Ag-water nanofluid with chemical reaction. *Powder Technology*, *307*, 46–55.
26. Agarwal, R. (2020). Analytical Study of micropolar fluid flow between two porous disks. *PalArch's Journal of Archaeology of Egypt/Egyptology*, *17*(12), 903–924.
27. Agarwal, R. (2021). Heat and mass transfer in electrically conducting micropolar fluid flow between two stretchable disks. *Materials Today: Proceedings*, *46*, 10227–10238
28. M. Reiner (1945). A mathematical theory of dilatancy. *Journal of the American Mathematical Society*, *67*, 350–362.
29. R.S. Rivlin (1948). The hydrodynamics of non-Newtonian fluids. *Proceedings of the Royal Society of London. Series A*, *193*, 260–281.
30. Agarwal, R. (2022). An analytical study of non-Newtonian visco-inelastic fluid flow between two stretchable rotating disks. *Palestine Journal of Mathematics*, *11*, 184–201.
31. Sadiq, M. A., & Hayat, T. (2022). Entropy optimized flow of Reiner–Rivlin nanofluid with chemical reaction subject to stretchable rotating disk. *Alexandria Engineering Journal*, *61*(5), 3501–3510.
32. Gupta, R., & Agrawal, D. (2023). Flow analysis of a micropolar nanofluid between two parallel disks in the presence of a magnetic field. *Journal of Nanofluids*, *12*(5), 1320–1326.
33. Gupta, R., & Wakif, A. (2023). Computing neural network to analyze heat and mass transfer in the flow of nanofluid between two disks. *Numerical Heat Transfer, Part A: Applications*. doi:10.1080/10407782.2023.2292197

34. Bhatnagar, R. K. (1981). Flow between coaxial rotating disks: with and without externally applied magnetic field. *International Journal of Mathematics and Mathematical Sciences,* 4, 181–200.
35. Zangooee, M. R., Hosseinzadeh, K., & Ganji, D. D. (2019). Hydrothermal analysis of MHD nanofluid (TiO2-GO) flow between two radiative stretchable rotating disks using AGM. *Case Studies in Thermal Engineering, 14,* 100460.

Chapter 10

Flow and thermal behavior of Ellis hybrid nanofluid

Neelav Sarma and Ashish Paul

10.1 INTRODUCTION

In the recent 20 years, significant advancements have occurred in the exploration of nanofluids, spurred by their essential significance in diverse engineering implementations. These include currency production, electric machine cooling, nanomaterial fabrication, optical devices, transportation, as well as motorcycle and car engine efficiency. Nanoparticles, typically ranging from 1 to 100 nm, are the key components within these fluids. By dispersing these nanoparticles into base liquids such as H_2O, $C_2H_6O_2$, kerosene, and certain oils, nanofluids are formed. These base fluids generally exhibit low thermal conductivity, a limitation that can be effectively addressed by the incorporation of nanoparticles. Understanding the heat transfer properties of nanofluids has garnered significant scientific interest due to their extensive applications in solar and wind energy generation, as well as microchip manufacturing. The core aim of the creation of nanofluids is to improve the heat transfer abilities of base fluids that naturally possess low thermal conductivity. The inception of nanofluids, originally proposed by Choi and Eastman (1995), emphasized numerous potential benefits in this regard. Subsequent research by Jang and Choi (2004) demonstrated that the thermal conductivity of numerous liquids could increase by over 20% upon the introduction of nanoparticles. Sheikholeslami et al. (2014) investigated the dynamics of nanofluid movement and the heat transfer properties within a rotating setup consisting of two parallel plates oriented horizontally. Hatami et al. (2014) analyzed the Jeffery-Hamel nanofluid flow influenced by magnetohydrodynamic (MHD) in non-converging walls, utilizing various distinct analytical techniques. Salahuddin et al. (2022) investigated how the flow properties of nanofluid in a permeable channel are influenced by the presence of a magnetic field and the generation of heat. Mahmood et al. (2023) explored the impact of thermal generation/absorption and mass removal on MHD stagnation point flow over a nonlinearly extending/contracting surface composed of a tri-hybrid nanofluid. Paul et al. (2023) studied the Casson-Maxwell hybrid nanofluid flow across a cylinder.

DOI: 10.1201/9781003494454-10

Non-Newtonian fluids display unusual behavior while flowing over a stretched surface, confounding the standard viscosity patterns of conventional fluids. Their deformation response varies greatly, altering boundary layer dynamics and flow properties. Understanding this phenomenon is critical in a variety of domains where these fluids are widespread, such as industrial operations and biological applications. Because of the intricacy of their behavior on stretched surfaces, standard fluid dynamics models are challenged, demanding specialized methodologies for reliable predictions. Exploring the complex interaction of surface stretching and non-Newtonian fluid flow opens up new options for innovative study and technological improvement. Hameed and Nadeem (2007) investigated a pair of specific solutions concerning the dynamic motion of a magnetized, electrically conductive, second-order incompressible fluid. Ellahi and Hameed (2012) examined the impact of partial slip that is nonlinear on channel wall surfaces concerning the constant flow and heat exchange of an incompressible third-grade fluid consistent with thermodynamics. Sharma and Shaw (2022) examined the transfer of heat and mass in a 2D MHD flow involving the Casson and Williamson movements, which account for nonlinear radiation, dissipation due to viscosity, and the effects of thermo-diffusion and Dufour. Ashraf et al. (2023) explored the impact of the Cattaneo-Christov heat flux on the flow of a Jeffrey nanofluid past a cylindrical surface undergoing stretching. The Ellis fluid flow is a non-Newtonian fluid behavior defined by viscoelastic qualities that display both viscous and elastic responses under stress. Rashid et al. (2023) explored the characteristics of natural convection flow and heat transfer in a crown enclosure with non-Newtonian properties. Jalili et al. (2023) used the semi-analytical approach of the HAN method to analyze the non-transient, forced flow of a Reiner–Rivlin viscoelastic fluid with MHD characteristics, occurring between two parallel plates. This flow, named after John F. Ellis, exhibits complicated behavior in which the material's response to deformation relies on the rate of strain. The Ellis fluids frequently exhibit shear-thinning or shear-thickening tendencies, impacting a variety of industrial applications such as polymer processing, food production, and biomedical engineering due to their distinctive rheological properties. Matsuhisa and Bird (1965) investigated the behavior of the non-Newtonian Ellis fluid in laminar flow, examining both analytical and computational methods for resolution. Ali et al. (2015) analyzed the propulsion of an Ellis non-Newtonian fluid through a flat conduit using peristaltic motion. Ali et al. (2019) investigated the mathematical formulation of the peristaltic pumping of an Ellis fluid using electro-osmosis in a two-fluid system within a cylindrical tube. Ahmed et al. (2022) delved into pioneering bioconvective treatments focused on controlling the movement of Ellis nanofluids, which don't adhere to traditional Newtonian physics, over a cone in rotation. Awan et al. (2023) conducted an investigation involving a numerical examination of thermal conduction within the flow of an Ellis hybrid nanofluid encompassing a stretched cylinder. Abbasi et al. (2023) examined the analysis of heat and mass transfer in the peristaltic motion of chemically reactive Ellis fluid within an asymmetrical conduit.

Incorporating thermal radiation into MHD flow considerably modifies fluid dynamics, influencing heat transfer properties and flow behavior. Grasping the effect of heat radiation on MHD flow is critical for understanding its impact on a variety of physical phenomena and providing critical insights into engineering applications and natural processes involving conducting fluids in magnetic fields, as cited by Raptis et al. (2004), Abbas and Hayat (2008). Bilal et al. (2018) delved into the computational analysis of magnetohydrodynamics and thermal emission in the context of Williamson nanofluid motion around an elongated cylinder experiencing variable thermal conductivity. Prasad et al. (2012) studied the influence of thermal radiation on the transfer of heat and mass in a magnetohydrodynamic context from a horizontal cylinder within a regime of variable porosity. Farooq et al. (2023) explored the computational methods used for the Cattaneo-Christov model, investigating how it was employed in the study of Williamson nanofluid dynamics. The study further extended its analysis to include bioconvection and thermal radiation effects in a vertical, slim cylinder configuration. The effect of viscous dissipation on MHD flow is an important subject of research in fluid dynamics. Understanding the impact of this dissipation phenomenon is critical for understanding the changes it causes in flow characteristics, energy transfer, and thermal behavior in MHD systems. Showkat et al. (2023) investigated the computational analysis of the flow, considering the influence of viscous dissipation of a pseudoplastic fluid inspired by buoyancy around a vertical cylinder. Awais et al. (2024) delved into the impact of viscous dissipation and activation energy on the flow of an MHD Eyring-Powell fluid under the influence of Darcy-Forchheimer effects and fluid properties that vary. Rikitu (2023) investigated the flow characteristics of Eyring-Powell nanofluid on a porous cylinder experiencing stretching while considering the influence of viscous dissipation and magnetic field.

The investigation undertaken herein delves comprehensively into the analysis of flow and thermal behavior, encompassing heat transfer rates and skin friction, within the context of TiO_2-Cu/Water Ellis hybrid nanofluid flow across a horizontally stretching cylinder. This study holds significance, as there is a notable absence of comprehensive investigations into the flow and thermal characteristics of Ellis hybrid nanofluid utilizing TiO_2 and Cu nanoparticles. Furthermore, to date, no comparative analysis has been conducted examining the absolute skin friction and thermal transfer rates between Ellis nanofluid and Ellis hybrid nanofluid. This study holds significant practical implications, particularly in fields involving heat transfer applications and engineering designs. By examining the impact of multiple factors such as magnetic fields, radiation, and viscous dissipation on the nanofluid flow, this research contributes essential insights instrumental in enhancing the efficiency and optimization of thermal systems and heat transfer processes in various industrial and technological applications.

10.2 MATHEMATICAL MODELING

Consider a flow involving the Ellis hybrid nanofluid comprising TiO_2-Cu/H_2O, characterized by incompressibility, stability, and two-dimensionality, incorporating magnetic field, viscous dissipation, and thermal radiation effects around a horizontal stretching cylinder. The velocity components u' and v' correspond to motion along the horizontal (x'-axis) and radial (r'-axis) directions of the cylinder. This cylinder's length, represented by l', is subject to stretching via the velocity $u'_w(x') = u'_0 \frac{x'}{l'}$, where u'_0 remains constant. A radial magnetic field B_0 is imposed. The physical setup for this problem is depicted in Figure 10.1.

The equations dictating continuity, momentum, and energy in the boundary layer approximation are articulated as follows (Sarma and Paul 2023):

$$\frac{\partial(r'u')}{\partial x'} + \frac{\partial(r'v')}{\partial r'} = 0 \tag{10.1}$$

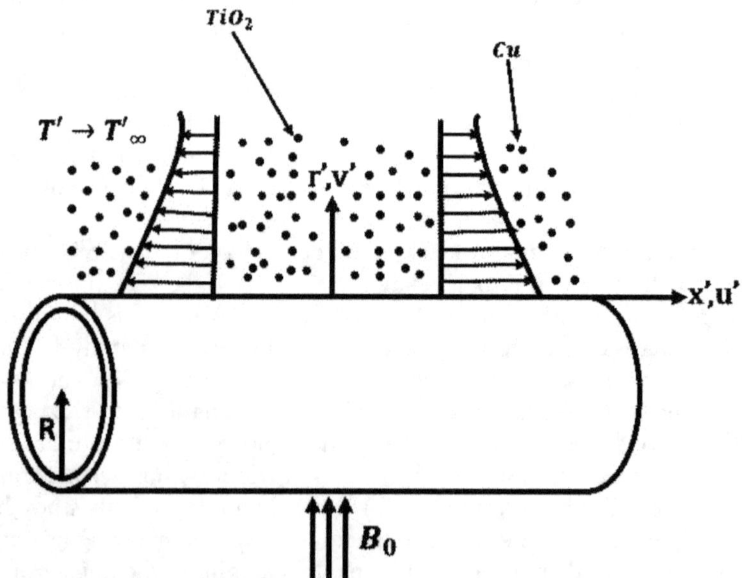

Figure 10.1 Physical setup of the problem.

$$u'\frac{\partial u'}{\partial x'}+v'\frac{\partial u'}{\partial r'}=\frac{1}{r'\rho_{hnf}}\frac{\partial}{\partial r'}\left(\frac{r'\mu_{hnf}}{1+\left(\frac{1}{\sqrt{2}r_0^2}\frac{\partial u'}{\partial x'}\right)^{\alpha-1}}\frac{\partial u'}{\partial r'}\right)-\frac{\sigma_{hnf}}{\rho_{hnf}}B_0^2u' \tag{10.2}$$

$$\begin{aligned} u'\frac{\partial T'}{\partial x'}+v'\frac{\partial T'}{\partial r'}=&\frac{K_{hnf}}{(\rho C_p)_{hnf}}\left(\frac{\partial^2T'}{\partial r'^2}+\frac{1}{r'^2}\frac{\partial T'}{\partial r'}\right)\\ &+\frac{\mu_{hnf}}{\rho_{hnf}}\left(\frac{\partial u'}{\partial r'}\right)^2+\frac{16\sigma^*T_\infty'^3}{3k^*}\frac{\partial^2T'}{\partial r'^2}\end{aligned} \tag{10.3}$$

With boundary conditions

$$\left.\begin{aligned} &u'=u_0'\frac{x'}{l'},v'=0,T'=T_w'(x')\ \text{at}\ r'=r_0'\\ &u'=0,\ T'\to T_\infty'\ \text{at}\ r'\to\infty\end{aligned}\right\} \tag{10.4}$$

When contemplating the selection of suitable factors for similarity alterations, the examination encompasses the subsequent elements:

$$\eta=\frac{r'^2-r_0'^2}{2r_0'^2}\sqrt{\frac{u_0'}{v_fl'}};\Psi=\sqrt{\frac{u_0'v_f}{l'}}x'r_0'\,h(\eta);\ \chi=\frac{T'-T_\infty'(x')}{T_w'(x')-T_0'} \tag{10.5}$$

where $u'=\frac{1}{r'}\frac{\partial\Psi}{\partial r'};v'=-\frac{1}{r'}\frac{\partial\Psi}{\partial x'}$

Using (10.5) in (10.1)–(10.4), we have

$$\begin{aligned}&\frac{\mu_{hnf}}{\mu_f}\frac{\rho_f}{\rho_{hnf}}\left[\alpha(1+2\zeta\eta)\left(1+(2-\alpha)(\Lambda_eh'')^{\alpha-1}\right)h'''+\zeta\left(1+(2-\alpha)(\Lambda_eh'')^{\alpha-1}\right)h''\right]\\ &+\left((1+\Lambda_eh'')^{\alpha-1}\right)^2(hh''-h'^2)-\frac{\sigma_{hnf}}{\sigma_f}\frac{\rho_f}{\rho_{hnf}}\left((1+\Lambda_eh'')^{\alpha-1}\right)^2M_ph'=0\end{aligned} \tag{10.6}$$

$$\left(\frac{K_{hnf}}{K_f}+R\right)(1+2\zeta\eta)\chi''+\left(2\frac{K_{hnf}}{K_f}+R\right)\zeta\chi'+\frac{(\rho C_p)_{hnf}}{(\rho C_p)_f}\Pr(h\chi'-2h'\chi)$$

$$+\frac{\mu_{hnf}}{\mu_f}\frac{(\rho C_p)_{hnf}}{(\rho C_p)_f}Ec\ h''^2=0 \tag{10.7}$$

With boundary conditions

$$\left.\begin{array}{c} h(0)=0,\ h'(0)=1,\ \chi(0)=1 \\ h'(\infty)\to 0, \chi(\infty)\to 0 \end{array}\right\} \tag{10.8}$$

Our curiosity extends to assessing the impact of dimensionless factors on the following two engineering-utilized physical quantities mentioned subsequently (Tables 10.1 and 10.2):

Table 10.1 Thermophysical correlation of hybrid nanofluid (Takabi and Salehi 2014)

Properties	*Hybrid nanofluid*
Density (ρ)	$\rho_{hnf} = (1-\varphi_{Cu})\left[(1-\varphi_{TiO_2})\rho_f + \varphi_{TiO_2}\rho_{TiO_2}\right] + \varphi_{Cu}\rho_{Cu}$
Heat capacity (ρC_p)	$(\rho C_p)_{hnf} = (1-\varphi_{Cu})\left[(1-\varphi_{TiO_2})(\rho C_p)_f + \varphi_{TiO_2}(\rho C_p)_{TiO_2}\right] + \varphi_{Cu}(\rho C_p)_{Cu}$
Thermal conductivity (K)	$\frac{K_{hnf}}{K_f} = \left[\frac{\left(\frac{\varphi_{TiO_2}K_{TiO_2} + \varphi_{Cu}K_{Cu}}{\varphi_{hnf}}\right) + 2K_f + 2(\varphi_{TiO_2}K_{TiO_2} + \varphi_{Cu}K_{Cu}) - 2\varphi_{hnf}K_f}{\left(\frac{\varphi_{TiO_2}K_{TiO_2} + \varphi_{Cu}K_{Cu}}{\varphi_{hnf}}\right) + 2K_f - (\varphi_{TiO_2}K_{TiO_2} + \varphi_{Cu}K_{Cu}) + \varphi_{hnf}K_f}\right]$ where $\varphi_{hnf} = \varphi_{TiO_2} + \varphi_{Cu}$
Dynamic viscosity (μ)	$\frac{\mu_{hnf}}{\mu_f} = \frac{1}{(1-\varphi_{TiO_2})^{2.5}(1-\varphi_{Cu})^{2.5}}$
Electrical conductivity (σ)	$\frac{\sigma_{hnf}}{\sigma_f} = \left[\frac{\left(\frac{\varphi_{TiO_2}\sigma_{TiO_2} + \varphi_{Cu}\sigma_{Cu}}{\varphi_{hnf}}\right) + 2\sigma_f + 2(\varphi_{TiO_2}\sigma_{TiO_2} + \varphi_{Cu}\sigma_{Cu}) - 2\varphi_{hnf}\sigma_f}{\left(\frac{\varphi_{TiO_2}\sigma_{TiO_2} + \varphi_{Cu}\sigma_{Cu}}{\varphi_{hnf}}\right) + 2\sigma_f - (\varphi_{TiO_2}\sigma_{TiO_2} + \varphi_{Cu}\sigma_{Cu}) + \varphi_{hnf}\sigma_f}\right]$ where $\varphi_{hnf} = \varphi_{TiO_2} + \varphi_{Cu}$

Table 10.2 Thermophysical properties of H_2O, TiO_2, Cu (Gohar et al. 2022; Patil and Kulkarni 2021; Zeeshan et al. 2022)

	H_2O	*TiO_2*	*Cu*
ρ (kg/m^3)	997.1	4,250	8,933
C_p (J/kg/K)	4,179	686.2	385
K (W/m/K)	0.613	40	400
σ (S/m)	0.05	3.5×10^6	59.6×10^6

10.2.1 Skin friction coefficient

$$\mathrm{Re}_x^{1/2} C_h = \frac{\mu_{hnf}}{\mu_f}\left(\frac{h''(0)}{\left(1+\Lambda_e h''(0)\right)^{\alpha-1}}\right) \tag{10.9}$$

10.2.2 Nusselt number

$$\mathrm{Re}_x^{-1/2} \mathrm{Nu}_x = -\left(\frac{K_{hnf}}{K_f}+\frac{4}{3}R\right)\chi'(0) \tag{10}$$

10.3 RESULTS AND DISCUSSIONS

Equations (10.6)–(10.8) are solved through numerical methods employing MATLAB's "bvp4c" function. The attained local Nusselt number results were cross-referenced with the prior publication by Ramzan et al. (2021) in Table 10.3 for validation. The comparison indicated a substantial concurrence between the findings. Our primary focus involves examining how different factors impact the flow of fluids and the temperature within a field. We explore the influence of various parameters, including the Ellis fluid parameter, curvature parameter, magnetic parameter, Prandtl number, and Eckert number. Dimensionless parameters' numerical values are selected as $0.2 \le \Lambda_e \le 0.8$, $0.1 \le \zeta \le 0.5$, $0.2 \le M_p \le 3$, $0.2 \le R \le 1$, $6 \le \mathrm{Pr} \le 7.5$, $0.2 \le \mathrm{Ec} \le 2.2$.

Figure 10.2 illustrates how the velocity profile intensifies as the Ellis fluid parameter (Λ_e) rises. The shear-thinning characteristic of Ellis fluids results in enhanced fluid parameter values corresponding to lower viscosity, allowing for higher flow rates. In contrast, Figure 10.3 shows a decline in the temperature profile as Λ_e increases. As the Ellis fluid parameters increase, the viscosity decreases, resulting in less energy dissipation and lower temperature profiles owing to fewer internal frictional heating effects inside the fluid flow.

The augmentation exhibited in both velocity and temperature profiles as the curvature parameter (ζ) rises, as seen in Figures 10.4 and 10.5, is

Table 10.3 Validation of Nusselt number with Ramzan et al. (2021)

	Nusselt number	
Pr	*Ramzan et al. (2021)*	*Present work*
0.72	1.088632	1.0889
1	1.333333	1.3333
10	4.796346	4.7969

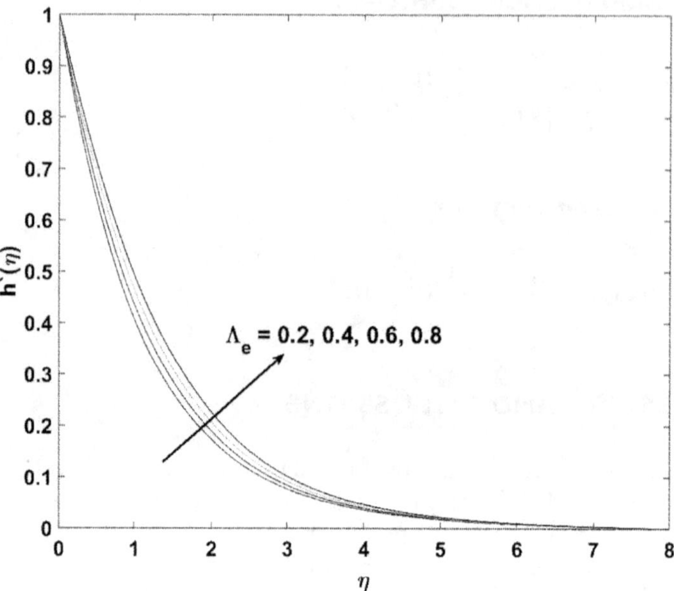

Figure 10.2 Variation of $h'(\eta)$ with Λ_e.

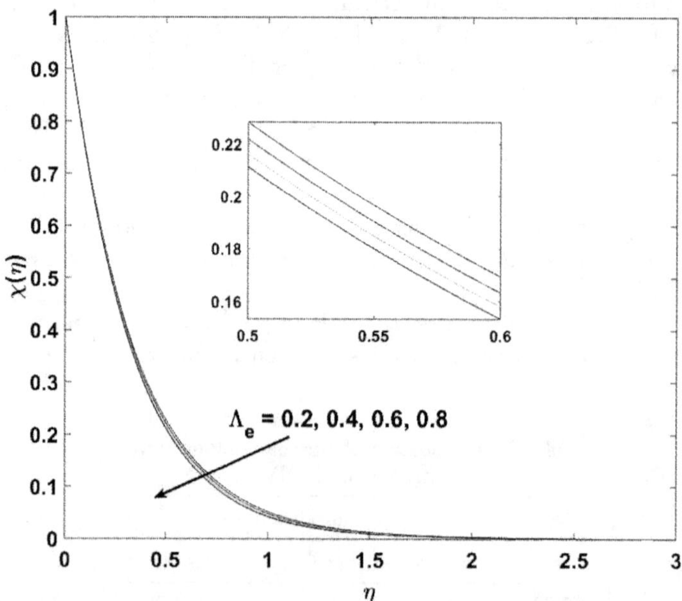

Figure 10.3 Variation of $\chi(\eta)$ with Λ_e.

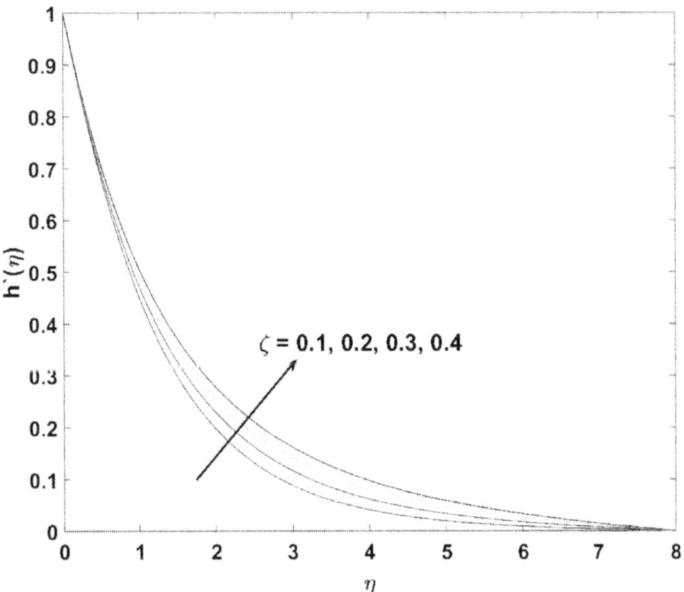

Figure 10.4 Variation of $h'(\eta)$ with ζ.

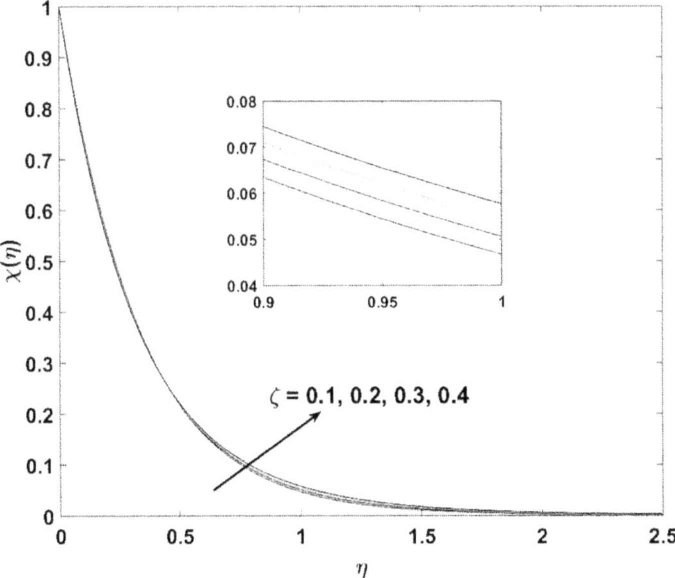

Figure 10.5 Variation of $\chi(\eta)$ with ζ.

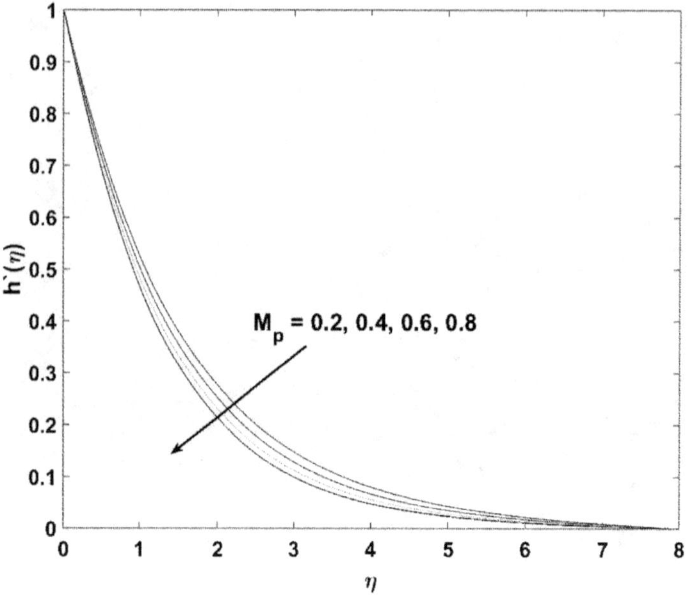

Figure 10.6 Variation of $h'(\eta)$ with M_p.

due to the changing flow dynamics caused by curvature fluctuations. The flow faces higher confinement as the curvature parameter grows, triggering increased fluid acceleration and consequently raising the velocity profile. Concurrently, this curvature-induced confinement limits convective heat dissipation, strengthening temperature gradients, and, as a result, raising the temperature profile.

The drop in velocity profile in Figure 10.6 may be accounted for by the magnetic field's impact on the flow, which induces resistance and alters momentum transfer, resulting in flow waning. In contrast, the rise in temperature profile in Figure 10.7 corresponds to magnetic parameter (M_p) escalation due to increased energy dissipation from magnetic field interactions, increasing heat exchange within the fluid.

The change observed in the temperature pattern shown in Figure 10.8 occurs as the Prandtl number (Pr) increases because of modifications in the fluid's characteristics related to the transfer of heat and momentum, specifically in terms of its thermal diffusivity and momentum transfer properties. A higher Prandtl number indicates that momentum diffusivity is greater than thermal diffusivity. As a result of the more effective transmission of momentum relative to heat, the temperature profile decreases.

According to Figure 10.9, enhancements in the Eckert number (Ec) correspond with improvements in the temperature arrangement. When the

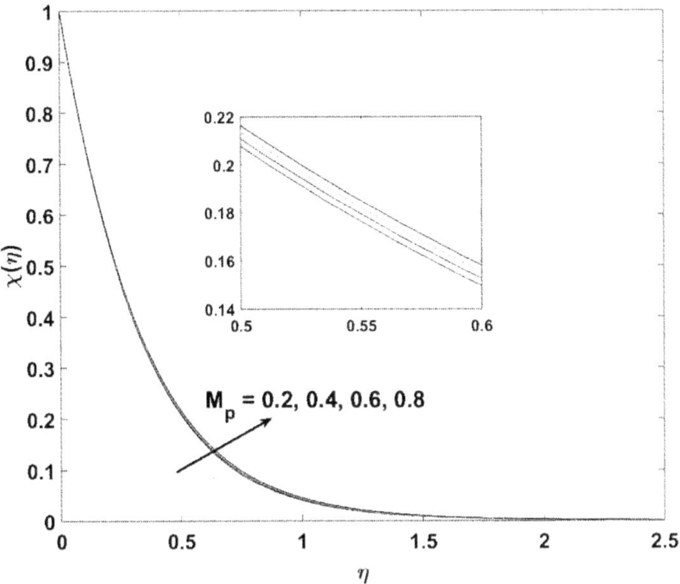

Figure 10.7 Variation of $\chi(\eta)$ with M_p.

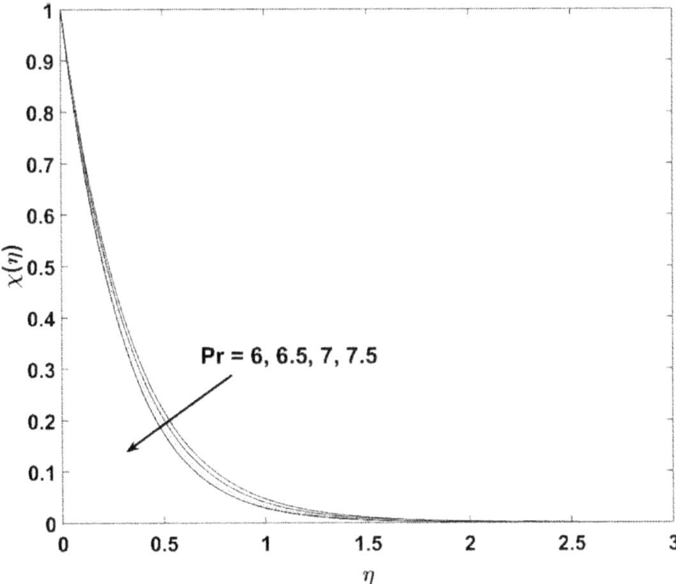

Figure 10.8 Variation of $\chi(\eta)$ with Pr.

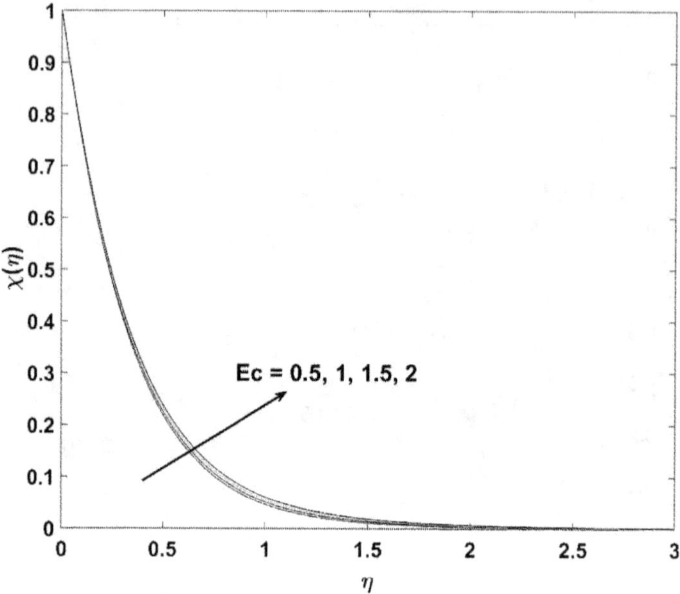

Figure 10.9 Variation of $\chi(\eta)$ with $\Lambda_{e.}$

Eckert number rises, there is a concurrent increase in the ratio between kinetic energy and enthalpy. This points toward a greater portion of kinetic energy being converted into thermal energy. Such heightened conversion signifies an increased dissipation of kinetic energy, consequently leading to the elevation of the temperature arrangement.

The observed trend in skin friction and Nusselt number variations related to the Ellis fluid parameter (Λ_e), magnetic parameter (M_p), curvature parameter (ζ), Prandtl number (Pr), radiation parameter (R), and Eckert number (Ec), as shown in Table 10.4, The increase in skin friction with increasing Λ_e can be due to increased viscosity and changed rheological behavior, which amplifies flow resistance. In contrast, a decrease in skin friction concerning ζ shows a reduction in flow resistance owing to geometric considerations. In terms of M_p, increasing it enhances the interaction between the fluid and the magnetic field, increasing skin friction by changing momentum transfer and developing additional opposition within the flow. Because of increased viscosity and changed rheological characteristics, increasing Λ_e increases convective heat transfer. Rising ζ and Pr both accelerate heat transport by changing flow patterns and thermal conductivity. Furthermore, increasing the radiation parameter (R) improves heat transmission via radiation effects. Increased M_p and Ec, on the other hand, interrupt convective heat transmission by changing flow momentum and releasing more kinetic energy, lowering the Nusselt number.

Table 10.4 Impact of Λ_e, ζ, M_p, Pr, Ec, R on skin friction and Nusselt number

Λ_e	ζ	M_p	Pr	Ec	R	*Skin friction*	*Nusselt number*
0.2	0.1	1	6	0.5	0.5	−1.6192	7.3623
0.4						−1.7346	7.4374
0.6						−1.8559	7.5015
0.5	0.1	1	6	0.5	0.5	−1.7945	7.4707
	0.2					−1.7473	7.6499
	0.3					−1.7063	7.8254
0.5	0.1	1	6	0.5	0.5	−1.7945	7.4707
		2				−2.2663	7.3602
		3				−2.2688	7.2717
0.5	0.5	1	6.5	0.5	0.5	−1.7945	7.7966
			7				8.1100
			7.5				8.4122
0.5	0.5	1	6	1	0.5	−1.7495	7.3836
				1.5			7.2966
				2			7.2095
0.5	0.5	1	6	0.5	0.2	−1.7495	5.6567
					0.4		6.8845
					0.6		8.0783

Table 10.5 Comparison of Ellis-hybridnanofluid and Ellis-nanofluid

Ellis nanofluid				*Ellis hybrid nanofluid*				*Change in %*	
φ_{TiO_2}	φ_{Cu}	*Skin friction*	*Nusselt number*	φ_{TiO_2}	φ_{Cu}	*Skin friction*	*Nusselt number*	*Skin friction*	*Nusselt number*
0	0.01	−1.1956	6.3029	0.1	0.01	−1.5662	7.0989	30.99	12.62
	0.02	−1.2454	6.3837		0.02	−1.6220	7.1911	30.23	12.64
	0.03	−1.2956	6.4651		0.03	−1.6786	7.2838	29.56	12.66
	0.04	−1.3464	6.5470		0.04	−1.7361	7.3770	28.94	12.67
	0.05	−1.3979	6.6295		0.05	−1.7945	7.4707	28.37	12.68

As per the information provided in Table 10.5, augmenting the proportion of Cu nanoparticles within the volume leads to enhancements in both the absolute skin friction and the Nusselt number while maintaining φ_{TiO_2} as zero. Additionally, when the solid volume fraction of TiO_2 is set at 0.1, the rate of heat transfer increases by more than 12.6%. These findings indicate a significantly higher thermal transport rate for the Ellis hybrid nanofluid compared to the regular nanofluid. Furthermore, there's an observed increase of about 29.6% in absolute skin friction for the Ellis hybrid nanofluid.

10.4 CONCLUSION

The examination explores the flow and heat transfer of the Ellis hybrid nanofluid, which encompasses a horizontal stretching cylinder under the influence of magnetic fields, thermal radiation, and viscous dissipation within the boundary layer. To comprehend the ramifications of various physical parameters on the flow scenario, pertinent dimensionless quantities are analyzed. Their impact on velocity and thermal distributions, Nusselt number, and skin friction coefficient are thoroughly investigated. This numerical investigation unveils the following noteworthy discoveries essential to this study:

- The velocity profile reduces as the Ellis fluid parameter and curvature parameter increase, yet it escalates with the magnetic fluid parameter.
- The temperature distribution increases with the curvature, magnetic, and radiation parameters, along with the Eckert number. However, it decreases with an elevated Ellis fluid parameter.
- The skin friction increases with an elevation in both the Ellis fluid parameter and magnetic parameter, while it decreases with a rise in the curvature parameter.
- The rates of heat transfer exhibit an upward trajectory with escalating Ellis fluid and Prandtl numbers, while conversely diminishing with elevated magnetic parameters and Prandtl numbers.
- The Ellis hybrid nanofluid demonstrates an enhancement of more than 29.6% in its absolute skin friction coefficient when juxtaposed against the Ellis nanofluid.
- The enhancement in heat transfer efficiency for the Ellis hybrid nanofluid compared to the Ellis nanofluid registers an augmentation of approximately 12.6%.

REFERENCES

Abbas, Z., & Hayat, T. (2008). Radiation effects on MHD flow in a porous space. *International Journal of Heat and Mass Transfer*, 51(5–6), 1024–1033. https://doi.org/10.1016/j.ijheatmasstransfer.2007.05.031

Abbasi, A., Khan, S. U., Farooq, W., Mughal, F. M., Khan, M. I., Prasannakumara, B. C., El-Wakad, M.T., Guedri, K.,& Galal, A. M. (2023). Peristaltic flow of chemically reactive Ellis fluid through an asymmetric channel: Heat and mass transfer analysis. *Ain Shams Engineering Journal*, 14(1), 101832. https://doi.org/10.1016/j.asej.2022.101832

Ahmed, S. E., Arafa, A. A., & Hussein, S. A. (2022). MHD Ellis nanofluids flow around rotating cone in the presence of motile oxytactic microorganisms. *International Communications in Heat and Mass Transfer*, 134, 106056. https://doi.org/10.1016/j.icheatmasstransfer.2022.106056

Ali, N., Abbasi, A., & Ahmad, I. (2015). Channel flow of Ellis fluid due to peristalsis. *AIP Advances*, 5(9). https://doi.org/10.1063/1.4932042

Ali, N., Hussain, S., Ullah, K., & Bég, O. A. (2019). Mathematical modelling of two-fluid electro-osmotic peristaltic pumping of an Ellis fluid in an axisymmetric tube. *The European Physical Journal Plus*, 134(4), 141. https://doi.org/10.1140/epjp/i2019-12488-2

Ashraf, M. B., Tanveer, A., & Ulhaq, S. (2023). Effects of Cattaneo-Christov heat flux on MHD Jeffery nano fluid flow past a stretching cylinder. *Journal of Magnetism and Magnetic Materials*, 565, 170154. https://doi.org/10.1016/j.jmmm.2022.170154

Awais, M., Salahuddin, T., & Muhammad, S. (2024). Effects of viscous dissipation and activation energy for the MHD Eyring-powell fluid flow with Darcy Forchheimer and variable fluid properties. *Ain Shams Engineering Journal*, 15(2), 102422. https://doi.org/10.1016/j.asej.2023.102422

Awan, A. U., Ali, B., Shah, S. A. A., Oreijah, M., Guedri, K., & Eldin, S. M. (2023). Numerical analysis of heat transfer in Ellis hybrid nanofluid flow subject to a stretching cylinder. *Case Studies in Thermal Engineering*, 49, 103222. https://doi.org/10.1016/j.csite.2023.103222

Bilal, M., Sagheer, M., & Hussain, S. (2018). Numerical study of magnetohydrodynamics and thermal radiation on Williamson nanofluid flow over a stretching cylinder with variable thermal conductivity. *Alexandria Engineering Journal*, 57(4), 3281–3289. https://doi.org/10.1016/j.aej.2017.12.006

Choi, S. U., & Eastman, J. A. (1995). *Enhancing Thermal Conductivity of Fluids with Nanoparticles* (No. ANL/MSD/CP-84938; CONF-951135-29). Argonne National Lab, Argonne, IL.

Ellahi, R., & Hameed, M. (2012). Numerical analysis of steady non-Newtonian flows with heat transfer analysis, MHD and nonlinear slip effects. *International Journal of Numerical Methods for Heat & Fluid Flow*, 22(1), 24–38. https://doi.org/10.1108/09615531211188775

Farooq, U., Waqas, H., Makki, R., Ali, M. R., Alhushaybari, A., Muhammad, T., & Imran, M. (2023). Computation of Cattaneo-Christov heat and mass flux model in Williamson nanofluid flow with bioconvection and thermal radiation through a vertical slender cylinder. *Case Studies in Thermal Engineering*, 42, 102736. https://doi.org/10.1016/j.csite.2023.102736

Gohar, Khan, T. S., Sene, N., Mouldi, A., & Brahmia, A. (2022). Heat and mass transfer of the Darcy-Forchheimer Casson hybrid nanofluid flow due to an extending curved surface. *Journal of Nanomaterials*, 2022, 1–12. https://doi.org/10.1155/2022/3979168

Hameed, M., & Nadeem, S. (2007). Unsteady MHD flow of a non-Newtonian fluid on a porous plate. *Journal of Mathematical Analysis and Applications*, 325(1), 724–733. https://doi.org/10.1016/j.jmaa.2006.02.002

Hatami, M., Sheikholeslami, M., Hosseini, M., & Ganji, D. D. (2014). Analytical investigation of MHD nanofluid flow in non-parallel walls. *Journal of Molecular Liquids*, 194, 251–259. https://doi.org/10.1016/j.molliq.2014.03.002

Jalili, P., Azar, A. A., Jalili, B., & Ganji, D. D. (2023). The HAN method for a thermal analysis of forced non-Newtonian MHD Reiner-Rivlin viscoelastic fluid motion between two disks. *Heliyon*, 9, e17535. https://doi.org/10.1016/j.heliyon.2023.e17535

Jang, S. P., & Choi, S. U. (2004). Role of Brownian motion in the enhanced thermal conductivity of nanofluids. *Applied Physics Letters*, 84(21), 4316–4318. https://doi.org/10.1063/1.1756684

Mahmood, Z., Eldin, S. M., Rafique, K., & Khan, U. (2023). Numerical analysis of MHD tri-hybrid nanofluid over a nonlinear stretching/shrinking sheet with heat generation/absorption and slip conditions. *Alexandria Engineering Journal*, 76, 799–819. https://doi.org/10.1016/j.aej.2023.06.081

Matsuhisa, S., & Bird, R. B. (1965). Analytical and numerical solutions for laminar flow of the non-Newtonian Ellis fluid. *AIChE Journal*, 11(4), 588–595. https://doi.org/10.1002/aic.690110407

Patil, P. M., & Kulkarni, M. (2021). Analysis of MHD mixed convection in a Ag-TiO2 hybrid nanofluid flow past a slender cylinder. *Chinese Journal of Physics*, 73, 406–419. https://doi.org/10.1016/j.cjph.2021.07.030

Paul, A., Sarma, N., & Patgiri, B. (2023). Thermal and mass transfer analysis of Casson-Maxwell hybrid nanofluids through an unsteady horizontal cylinder with variable thermal conductivity and Arrhenius activation energy. *Numerical Heat Transfer, Part A: Applications*, 1–26. https://doi.org/10.1080/10407782.2023.2297000

Prasad, V. R., Vasu, B., Bég, O. A., & Parshad, D. R. (2012). Thermal radiation effects on magnetohydrodynamic heat and mass transfer from a horizontal cylinder in a variable porosity regime. *Journal of Porous Media*, 15(3). https://doi.org/10.1615/JPorMedia.v15.i3.50

Ramzan, M., Shaheen, N., Chung, J. D., Kadry, S., Chu, Y. M., & Howari, F. (2021). Impact of Newtonian heating and Fourier and Fick's laws on a magnetohydrodynamic dusty Casson nanofluid flow with variable heat source/sink over a stretching cylinder. *Scientific Reports*, 11(1), 2357. https://doi.org/10.1038/s41598-021-81747-x

Raptis, A., Perdikis, C., & Takhar, H. S. (2004). Effect of thermal radiation on MHD flow. *Applied Mathematics and Computation*, 153(3), 645–649. https://doi.org/10.1016/S0096-3003(03)00657-X

Rashid, U., Shahzad, H., Lu, D., Wang, X., & Majeed, A. H. (2023). Non-Newtonian MHD double diffusive natural convection flow and heat transfer in a crown enclosure. *Case Studies in Thermal Engineering*, 41, 102541. https://doi.org/10.1016/j.csite.2022.102541

Rikitu, E. H. (2023). Flow dynamics of eyring-powell nanofluid on porous stretching cylinder under magnetic field and viscous dissipation effects. *Advances in Mathematical Physics*, 2023. https://doi.org/10.1155/2023/9996048

Salahuddin, T., Khan, M. H. U., Khan, M., Al Alwan, B., & Amari, A. (2022). An analysis on the flow behavior of MHD nanofluid with heat generation. *Fuel*, 311, 122548. https://doi.org/10.1016/j.fuel.2021.122548

Sarma, N., & Paul, A. (2023). Thermophoresis and Brownian motion influenced bioconvective cylindrical shaped Ag-CuO/H2O Ellis Hybrid nanofluid flow along a radiative stretched tube with inclined magnetic field. *BioNanoScience*, 1–27. https://doi.org/10.1007/s12668-023-01280-1

Sharma, R. P., & Shaw, S. (2022). MHD Non-Newtonian fluid flow past a stretching sheet under the influence of non-linear radiation and viscous dissipation. *Journal of Applied and Computational Mechanics*, 8(3), 949–961. https://doi.org/10.22055/jacm.2021.34993.2533

Sheikholeslami, M., Abelman, S., & Ganji, D. D. (2014). Numerical simulation of MHD nanofluid flow and heat transfer considering viscous dissipation. *International Journal of Heat and Mass Transfer*, 79, 212–222. https://doi.org/10.1016/j.ijheatmasstransfer.2014.08.004

Showkat, I., Mushtaq, A., & Mustafa, M. (2023). Numerical exploration of buoyancy inspired flow of pseudoplastic fluid along a vertical cylinder with viscous dissipation effects. *Alexandria Engineering Journal*, 74, 415–425. https://doi.org/10.1016/j.aej.2023.05.039

Takabi, B., & Salehi, S. (2014). Augmentation of the heat transfer performance of a sinusoidal corrugated enclosure by employing hybrid nanofluid. *Advances in Mechanical Engineering*, 6, 147059. https://doi.org/10.1155/2014/147059

Zeeshan, Khan, I., Weera, W., & Mohamed, A. (2022). Heat transfer analysis of Cu and Al2O3 dispersed in ethylene glycol as a base fluid over a stretchable permeable sheet of MHD thin-film flow. *Scientific Reports*, 12(1), 8878. https://doi.org/10.1038/s41598-022-12671-x

Chapter 11

Flow of Newtonian and non-Newtonian hybrid nanofluid in cylindrical geometry

Jintu Mani Nath, Tusar Kanti Das, and Bamdeb Dey

11.1 INTRODUCTION

Granular sedimentation of nano-sized (1–100 nm) metallic materials, nanotube-shaped carbon oxides, and grapheme nano-flakes within oil, ethanol, and water creates tiny fluids. These tiny fluids' new properties render them useful in medicinal products, nanotechnology, hybrid-powered algorithms, energy-efficient radiators, home refrigerators, and other thermal transmission technologies. According to Hayat et al. (2017), a dense nanofluid flows in three dimensions across a perforated stretching sheet, taking into account the heating ratio and fractional slip phenomena. Using a Cattaneo-Christov thermal flux approach, Reddy et al. (2021) investigated the motion of composite dusty fluids. Arooj et al. (2021) covered the topic of peristaltically generated circulation in helical channels and its potential uses in pharmacology and mechanical biomedicine. Due to its relevance in technology and production, inflow issues are a topic of ever-increasing interest in today's academics. A wide variety of materials and processes rely on non-Newtonian fluid motions caused by prolonged interfaces; they include plastique, paper, foodstuffs, fuel, chemical liquids, and many more. Because of their ubiquitous presence in the actual world, non-Newtonian solutions have attracted a lot of academic interest (Dey et al., 2019; Choudhury et al., 2018; Reddy et al., 2021). Whenever a fluid's density is not pervasive, we say that it is not a Classical fluid; instead, we say that it flows differently. A non-Newtonian fluid's viscosity may vary in response to changes in shear rate, temperature, pressure, and stress. There are primarily three categories of non-Newtonian fluids, such as distributed, ratey, and continuous. For rate-type fluids, the most basic subcategory that may foretell the relaxation of stress follows the Maxwell hypothesis. Using temperature-sensitive dynamic characteristics and the exponential thermo-solutal dispersion impact, Akolade et al. (2021) examined the unstable flow of Maxwell fluids. Mushtaq et al. (2016) examined the separation zone flowing of a Maxwell fluid in an oscillating structure including an unambiguous chemical interaction and stimulation energy. Research by Farooq et al. (2019) examined the field of MHD of a

 DOI: 10.1201/9781003494454-11

non-Newtonian Maxwell fluid including nanostructures on an exterior that is expanding at an accelerated pace. Casson fluid is yet another fluid with a non-Newtonian composition type that shows yielding stress. When describing fluids, the Casson paradigm accounts for an elastic tension that is not captured by Newtonian frameworks. Another property that Casson fluid may display is nonlinear behavior, which means it may encompass simultaneous viscous and elastic characteristics. Because of this property, it may be used to represent fluids, including predicted polymeric solutions, that exhibit behavior that is similar to a mixed fluid and a substance beneath various circumstances. Many industries, including processing food, manufacturing, tapping, and biotechnology, rely on Casson fluid, which is among the most widely used non-Newtonian fluids (Paul et al., 2022; Moatimid et al., 2022; Farooq et al., 2023). To back up the improved heating process, hybrid nanotechnology is the most remarkable kind of nanofluid. The characteristics of an initial fluid combined with the breakdown of two distinct nanomaterials characterize this hybrid category. In experiments, hybrid nanofluid heating analysis shows significantly improved thermal efficiency. Due to their distinctive thermal effect, hybrid structures are used in solar panels, temperature control systems, vehicular radiators, batteries, insulation coatings, heating items, etc. Yang et al. (2023) examine hybrid nanofluid heating owing to varied form characteristics. Famakinwaet al. (2022) investigated heat-induced impacts on varying viscosity unstable insoluble compressing flow transporting water composite nanoparticles across two parallel edges. Using the effects of condensation and radiant heat, Rehman and Khan (2023) analyze the steady-state, impermeable, viscous, two-dimensional, pair-stress flow through the boundary layer of composite nanofluids. Paul et al. (2023) conducted a computational investigation of the dynamically segregated flow of a $CuAl_2O_3$ nanofluid hybrid generated from water across a vertically mounted, radially extending cylinder in an opaque medium. Krishna et al. (2021) investigated the radiant unstable MHD circulation involving an insoluble sticky conductive unconventional Casson's nanofluid mix over a vast steadily intensified upward transitioning permeable surface in spinning structure slip rate. Research on the thermophoretic dispersion of particles and the compression motion of Casson composite nanofluid among identical plates containing thermal sources or sinks was carried out by Jyothi et al. (2021).

The transmission of heat is employed in practically all technical domains, including chemical science, biomedical sciences, thermal discharge science, astrophysics, and mechanical oceanography. Thus, heat transmission across diverse fluid motions is being studied extensively in the scientific literature (Reddy et al., 2021; Paul et al., 2023; Khashi et al., 2020; Scott et al., 2022; Dey et al., 2022). Typically, the Eckert coefficient is used to describe the impact of the dissipation of viscous energy on physical streams and, more specifically, manufacturing processes using spontaneous radiative movements. One way to think about the value of Eckert is as the force that drives behind the flow-to-heat-transition entropy proportion, which is a measure of the dissipation of viscosity. To enhance the heating process of water-soluble materials, Dero et al. (2022) investigate mixed nanofluid expanding and reducing surface contact occurs with viscous dissipation. Famakinwa et al. (2022) assessed the consequences of mechanical radiated and viscous dispersion on the time-varying intangible compressing flow of a CuO-Sl_2O_3/water hybrid nanofluid across two adjacent surfaces with varying viscosities. Choudhury et al. (2018) highlighted the effects of radiant heat on the unsteady magneto-conventional circulation for an insoluble, conductive visco-elastic fluid over an opaque medium and past semi-infinite upward permeable sheets in the context of a magnetic exterior and time-varying extraction performing normal to the movement. The influence of the chemical processes and magnetism on the inherent laminar motion of a combined nanofluid through a spherical vessel with viscous dissipation was investigated by Venkateswarlu et al. (2022). Choudhury et al. (2017) conducted an empirical evaluation of the consequence of mass and energy transmission in a visco-elastic fluid upon a conductive MHD boundary layer's flow through an upright pervasive surface. Their intent has been to determine the impact of the magnetism, pliability, and skin friction within the motion of the fluid. Taking into consideration the dispersal of solid components, a mathematical representation was recently suggested by Yimin and Qiang (2000). This hypothetical scenario is intended to characterize the thermal transfer efficiency of a tiny liquid that is circulating through a conduit. Various models for improving the heat transfer flow of non-Newtonian fluids, Casson fluids, and hybrid nanofluids have been identified in the aforementioned literature. The fluid property of Darcy Forchheimer and arsenious impacts is crucial, yet those models don't account for it.

The idea of activation energy is most often linked with chemical processes, although it may be applied to many other occurrences by drawing analogies, one of which is fluid movement. It is necessary to conquer the initial opposition in order to initiate circulation in several fluid structures. This is particularly true when dealing with thick solutions such as fuel or incandescent compounds. The activation energy of the chemical process is analogous to the first source of energy needed to start the process. Similar to how a substance's breakdown continues more quickly when the activation energy barrier is overcome when the motion is started, it could go on effortlessly owing to less impedance. Accordingly, Alfalehet al. (2023) investigated the influence of activation energy and chemical interaction on the dissipative Darcy-Forchheimer compressed stream of Casson's fluid along an axial conduit in an unstable medium with maximum hydrodynamic pressure. Madhukesh et al. (2022) investigated the impact of a heating source or sink on the movement of nanofluids via a tube, wedge, or sheets employing a water-based initial fluid and tiny particles rigidity of aluminum alloys

(AA7072 and AA7075). The simulation also took into account the activation energy and the presence of permeable substances. Guirao et al. (2022) conducted examination of the motion of Maxwell-Sutterby substance passing through a slanted stretched surface in the presence of varying thermal insulation, an increasing heat source/sink, magneto-hydrodynamics (MHD), and activation energy.

The significance of investigating the flow of hybrid nanofluids around solid geometries is underscored by the comprehensive analysis of existing literature, which also serves to stimulate further research in this field. Furthermore, the literature reveals a notable gap in addressing the flow complexities associated with hybrid nanofluids passing through cylindrical bodies, indicating an area ripe for exploration and investigation. Based on the survey aforementioned, the author's purpose is to examine the various flow characteristics exhibited by Newtonian and non-Newtonian hybrid nanofluids featuring thermophysical attributes of magnesium and magnetite nanoparticles. The flow model is further enriched by the inclusion of viscous dissipation, activation energy, and Darcy-Forchheimer effects. To solve the given issue, the present investigation also made use of the bvp4c function in the MATLAB systems program which ensures accurate results even when dealing with complex and disorganized compositions (Dey et al., 2022; Hakeem et al., 2021; Jeelani et al., 2023). Magnesium and magnetite nanoparticle-containing Newtonian and non-Newtonian hybrid nanofluids have advantageous flow characteristics that can be applied to a variety of mechanical, health care, and industrial processes. These benefits include improved thermal transfer, upgraded lubrication, tailored rheological properties, improved fluid stability, and optimized visibility. In the realm of aerospace technology, the advancements stemming from this study have the potential to significantly augment thermal management systems deployed in spacecraft and aircraft. By facilitating efficient heat transfer and precise temperature regulation, particularly within critical components like propulsion systems and avionics, these developments promise to enhance overall performance and reliability. Similarly, within the domain of commercial sanitation practices, the integration of this technology stands to revolutionize heat exchange mechanisms and sterilization processes. This enhancement could result in more effective and comprehensive sanitation of equipment and surfaces across various industries such as food processing, pharmaceuticals, and healthcare, thereby bolstering hygiene standards and mitigating the risk of contamination. Moreover, in the context of refrigerant circulation systems, the utilization of hybrid nanofluids alongside considerations for factors such as viscous dissipation and activation energy presents a promising avenue for optimizing heat exchange processes. This optimization could lead to heightened efficiency in refrigeration units, translating into improved cooling performance and substantial energy savings across diverse applications including residential air conditioning, industrial refrigeration, and cold storage facilities.

11.2 MATHEMATICAL FORMULATION

In this investigation, we employ the boundary layer approach and analyze the phenomenon of steady laminar flow to explore the two-dimensional flow of Mg-Fe_3O_4/Blood hybrid nanofluid over an exponentially stretching cylinder. The study encompasses the consideration of viscous dissipation and the Darcy-Forchheimer impact, along with the synergic effect. The flow scenario is modeled within the cylindrical polar coordinate system (z, x), where z denotes the axial direction and x represents the radial direction. Illustrated in Figure 11.1, the exponentially stretching surface undergoes motion with a velocity given$p_p = 2dc\exp(z/d)$, where c and d denote the stretching rate and base radius of the cylinder, respectively. Furthermore, T_p and wall concentration (C_p) are characterized as $T_p = T_\infty + A^* \exp(z/d)$ and $C_p = C_\infty + B^* \exp(z/d)$, where A^*, B^*, T_∞, and C_∞ represent non-negative constants, ambient temperature, and concentration, respectively.

By employing all the assumptions, the governing equation can be constituted as follows (Paul et al., 2022):

Continuity equation is as follows:

$$\frac{\partial v}{\partial x} + \frac{v}{x} + \frac{\partial p}{\partial z} = 0 \tag{11.1}$$

Momentum equation is as follows:

$$v\frac{\partial p}{\partial x} + p\frac{\partial p}{\partial z} = \frac{\mu_{hnf}}{\rho_{hnf}}\left(1 + \frac{1}{\alpha}\right)\left[\left(\frac{\partial^2 p}{\partial x^2} + \frac{1}{r}\frac{\partial p}{\partial x}\right)\right] - \frac{\sigma_{hnf}}{\rho_{hnf}} B_0^2 p - \frac{\mu_{hnf}}{\rho_{hnf}}\frac{1}{k_p} p - \frac{c_b}{\sqrt{k_p}} p^2 \tag{11.2}$$

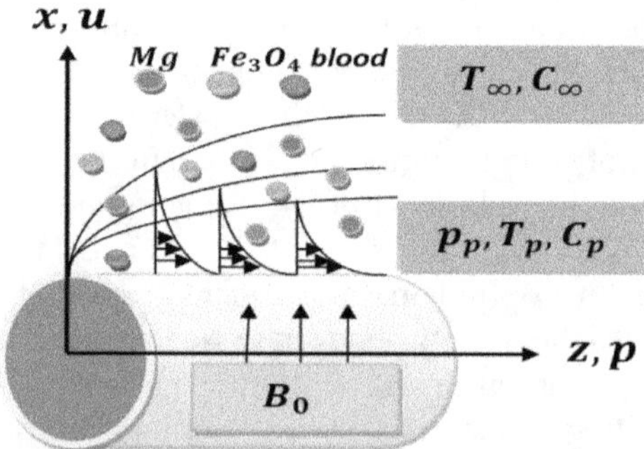

Figure 11.1 Schematic diagram of the fluid flow.

Energy equation is as follows:

$$v\frac{\partial T}{\partial x}+p\frac{\partial T}{\partial z}=k_{bnf}\frac{1}{x}\frac{\partial}{\partial x}\left(x\frac{\partial T}{\partial x}\right)+\frac{\mu_{bnf}}{\left(\rho C_p\right)_{bnf}}\left(1+\frac{1}{\alpha}\right)\left(\frac{\partial p}{\partial x}\right)^2 \tag{11.3}$$

Concentration equation is as follows:

$$v\frac{\partial C}{\partial x}+p\frac{\partial C}{\partial z}=D\left(\frac{1}{x}\frac{\partial C}{\partial x}+\frac{\partial^2 C}{\partial x^2}\right)-k_r^2\left(C-C_\infty\right)\left(\frac{T}{T_\infty}\right)^N\exp\left(\frac{E_a}{K_1 T}\right) \tag{11.4}$$

Subject to the boundary limits as

$$p(d,z)=p_p,\ u(d,\ z)=0,\ T(d,z)=T_p(z),\ C(d,z)=C_p(z)\text{ at }x=d$$

$$w(x,z)\to 0,\ T(x,z)\to 0,\ C(x,z)\to 0\text{ at }x\to\infty \tag{11.5}$$

In the context of (v,p), denoting the velocity components aligned with the (x,z) direction, the concentration equation incorporates the activation Arrhenius parameter denoted as $k_r^2\left(T/T_\infty\right)^N\exp\left(E_d/K_1T\right)$, wherein the Boltzmann constant K_1 is a crucial factor, and the condition is $-1<N<1$ observed.

The transformation variables employed in this context are outlined as follows, as referenced in Paul et al. (2022).

$$\xi=\left(\frac{x}{d}\right)^2,\ p=p_p g'(\xi),\ v=-\frac{1}{2}p_p\frac{g(\xi}{\sqrt{\xi}},$$

$$\theta=\frac{T-T_\infty}{T_p-T_\infty},\ \phi=\frac{C-C_\infty}{C_p-C_\infty}$$

Employing the dimensionless similarity parameters indicated earlier, the manifested coupled ordinary differential equations (ODEs) can be expressed in a subsequent manner:

$$\begin{aligned}g'^2-gg''=&\left(1+\frac{1}{\alpha}\right)\frac{1}{\text{Re}}\frac{\mu_{bnf}}{\mu_f}\frac{\rho_f}{\rho_{bnf}}\left[\left(\xi g'''+g''\right)\right]\\&-\frac{\sigma_{bnf}}{\sigma_f}\frac{\rho_f}{\rho_{bnf}}\text{M}.g'-\frac{\mu_{bnf}}{\mu_f}\frac{\rho_f}{\rho_{bnf}}\frac{\lambda}{\text{Re}}g'-Fr\ g'^2\end{aligned} \tag{11.6}$$

$$\begin{aligned}Pr\ Re\left(g'\theta-g\theta'\right)=&\frac{\left(\rho C_p\right)_f}{\left(\rho C_p\right)_{bnf}}\left(\frac{k_{bnf}}{k_f}\right)\left(\xi\theta''+\theta'\right)\\&+\left(1+\alpha^{-1}\right)\frac{\mu_{bnf}}{\mu_f}\frac{\left(\rho C_p\right)_f}{\left(\rho C_p\right)_{bnf}}\xi\cdot Pr\times Ec\cdot g''^2\end{aligned} \tag{11.7}$$

$$\left(g'\phi - g\phi'\right) = \frac{1}{Re}\frac{1}{Sc}\left(\xi\phi'' + \phi'\right) - C_b\left(1+\beta\theta\right)^N \exp\left(-\frac{E}{1+\beta\theta}\right)\phi \tag{11.8}$$

The transformed boundary limitations are

$$\begin{aligned} g'^2 - gg'' = \left(1+\frac{1}{\alpha}\right)\frac{1}{Re}\frac{\mu_{hnf}}{\mu_f}\frac{\rho_f}{\rho_{hnf}}\left[\left(\xi g''' + g''\right)\right] - \frac{\sigma_{hnf}}{\sigma_f}\frac{\rho_f}{\rho_{hnf}}M.g' \\ -\frac{\mu_{hnf}}{\mu_f}\frac{\rho_f}{\rho_{hnf}}\frac{\lambda}{Re}g' - Fr.g'^2 \end{aligned} \tag{11.9}$$

$$\begin{aligned} Pr \cdot Re\left(g'\theta - g\theta'\right) = \frac{\left(\rho C_p\right)_f}{\left(\rho C_p\right)_{hnf}}\left(\frac{k_{hnf}}{k_f}\right)\left(\xi\theta'' + \theta'\right) \\ +\left(1+\alpha^{-1}\right)\frac{\mu_{hnf}}{\mu_f}\frac{\left(\rho C_p\right)_f}{\left(\rho C_p\right)_{hnf}}\xi \cdot Pr \cdot Ec \cdot g''^2 \end{aligned} \tag{11.10}$$

$$\left(g'\phi - g\phi'\right) = \frac{1}{Re}\frac{1}{Sc}\left(\xi\phi'' + \phi'\right) - C_b\left(1+\beta\theta\right)^N \exp\left(-\frac{E}{1+\beta\theta}\right)\phi \tag{11.11}$$

Here,

$$\mathrm{Pr} = \frac{\upsilon_f}{\alpha_f},\ \mathrm{Re} = \frac{dp_p}{4\upsilon_f},\ H = \frac{\sigma_f B_0^2 d}{\rho_f p_p},\ \alpha_f = \frac{K_f}{\left(\rho C_p\right)_f},\ C_b = \frac{K_r^2 d}{p_p},\ \mathrm{E} = \frac{E_a}{K_1 T_\infty},$$

$$\beta = \frac{T_p - T_\infty}{T_\infty},\ \mathrm{Sc} = \frac{\nu_f}{D_f},\ \mathrm{Fr} = \frac{c_b d}{\sqrt{k_p}},\ \lambda = \frac{d^2}{4k_p},\ \mathrm{Ec} = \frac{p_p^2}{(C_p)_f\left(T_p - T_\infty\right)}$$

The relevant physical parameters within the fluid dynamics context encompass the drag force coefficient, Nusselt number, and Sherwood number. These variables function as discerning indicators for the tangential stress, thermal conduction rate, and mass diffusion rate at the surface, respectively.

Skin friction coefficient

$$\tau_p = \mu_{hnf}\left(1+\frac{1}{\alpha}\right)\left(\frac{\partial p}{\partial x}\right)$$

Nusselt number

$$Nu_p = \frac{d\exp\left(\frac{z}{d}\right)q_w}{k_f\left(T_p - T_\infty\right)}$$

where $q_p = -k_{hnf}\left(\frac{\partial T}{\partial x}\right)_{x=d}$

Table 11.1 The rheological attributes of the physical parameter (Alam et al., 2023)

Properties	*Mg*	*Fe$_3$O$_4$*	*Blood*
ρ (kg/m^3)	1,740	5,180	1,050
c_p (J/kg K)	1,050	670	3.9×10^3
k (W/ mK)	156	9.7	0.5
σ (S/m)	2.3×10^7	0.74 × 106	0.8

Sherwood number

$$Sh_p = \frac{d exp\left(\frac{z}{d}\right) q_m}{D\left(C_p - C_\infty\right)}$$

where $q_m = D\left(\frac{\partial C}{\partial x}\right)_{r=d}$

By applying the similarity transformation variables, the reformulated expressions for the transformed Skin friction coefficient, Nusselt number, and Sherwood number can be articulated as follows (Table 11.1):

$$C_f Re_p^{\frac{1}{z}} = \frac{\mu_{hnf}}{\mu_f}\left(1 + \frac{1}{\alpha}\right) g''(1)$$

$$Nu_p Re_p^{-\frac{1}{z}} = -\left(\frac{K_{hnf}}{K_f}\right)\theta'(1)$$

$$Sh_p Re_p^{-\frac{1}{z}} = -\phi'(1)$$

11.2.1 Methodology

Initiating with the refinement phase, the nondimensionalized higher-order ordinary differential Eqs. (6)–(8) are subjected to a transformation into an intermediary first-order boundary value problem, accompanied by the pertinent associated boundary impediments (9–10). Utilizing MATLAB's esteemed bvp4c algorithm in conjunction with a shooting technique, the numerical resolution of the aforementioned nondimensional ODEs is attained. Even when tackling intricate and chaotic formulations, the bvp4c decoder ensures precision in the obtained results by adeptly addressing boundary value challenges. Harnessing its robust mesh refinement capability, the computational overhead is mitigated, thereby enhancing

Table 11.2 Comparable findings with (Wang, 1988) and (Ishak et al., 2008)

Re	*Wang (1988)*	*Ishak et al. (2008)*	*Present study*
0.5	−0.88220	−0.8827	−0.8824
1	−1.17776	−1.1781	−1.1778
2	−1.59390	−1.5941	−1.5939
5	−2.41745	−2.4175	−2.4174
10	−3.34445	−3.3445	−3.3445

computational efficiency. Nevertheless, to effectively employ the bvp4c solver, an initial approximation that conforms to its spatial constraints becomes imperative. Unquestionably, the foremost choice for estimating the numerical solutions of such nondimensional complex ODEs remains the bvp4c method, which sets its convergence criterion at an impressive 10^{-6}, surpassing benchmarks established by alternative methodologies.

11.2.2 Validation

The methodology employed in this study has been delineated to be both accurate and precise, as exemplified in Table 11.2. By omitting the concentration and thermal boundary layer equations and selectively neglecting specific physical parameters from the momentum equation, the current research outcomes are juxtaposed with those reported by Wang (1988) and Ishak et al. (2008). Consequently, the reliability of the approach is underscored, as evidenced by the congruence between the computed results of the present investigation and the antecedent findings.

11.3 RESULT AND DISCUSSION

The current investigation delves into the scrutiny of various flow characteristics exhibited by a Newtonian hybrid nanofluid and a non-Newtonian hybrid nanofluid, wherein the thermophysical attributes of Magnesium (Mg) and Magnetite (Fe_3O_4) nanoparticles are integral. This analysis takes into account the influence of the Lorentz force in conjunction with the Darcy-Forchheimer impact. The flow model is further enriched by the inclusion of viscous dissipation and the synergic effect (activation energy). The spherical-shaped nanoparticles (6–50 nm) are utilized in this flow algorithm, and the volume concentration of the Mg and Fe_3O_4 nanoparticles are considered as 0.09 and 0.06, respectively. The examination encompasses a comprehensive portrayal of the graphical representation

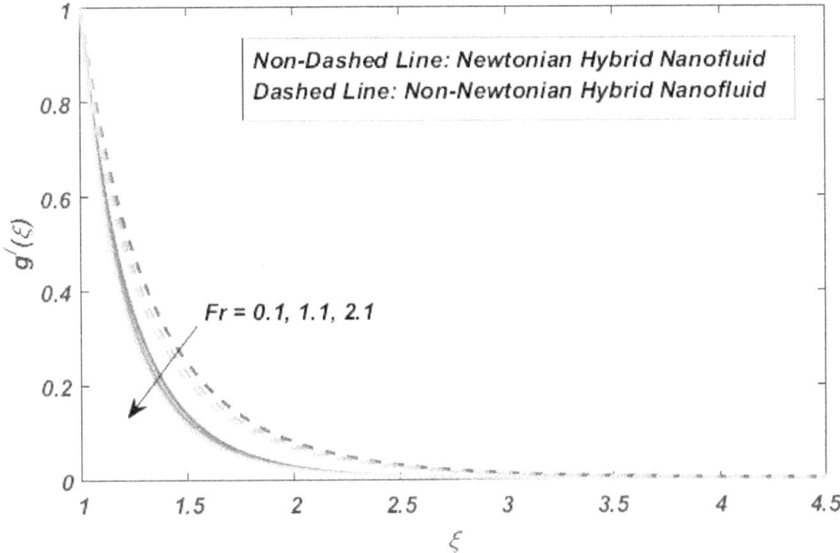

Figure 11.2 Illustration of *Fr* on velocity.

of the impact of dimensionless factors, specifically *Sc* (Schmidt number), *Fr* (Darcy-Forchheimer term), β (Temperature ratio term), *E* (Activation energy), and *Ec* (Eckert Number), on the flow, thermal, and concentration distribution curves. Figures 11.2–11.9 serve as visual aids in elucidating these effects. Notably, the dashed lines correspond to the Newtonian hybrid nanofluid flow, while the non-dashed lines represent the non-Newtonian hybrid nanofluid flow.

The vitality of the Darcy-Forchheimer inertia ingredient, *Fr*, in both Newtonian and non-Newtonian hybrid nanofluid flow is highlighted in Figures 11.2–11.4. Interestingly, fluid velocity shrinks as *Fr* enhances due to the inertia component correlating with both the medium's porosity and drag coefficient. As a result, boosting the drag coefficient augments the porosity and drag coefficient of the medium. Therefore, the fluid's resistive force goes up, which leads to a diminution in velocity with the growing Forchheimer number. As demonstrated in Figures 11.3 and 11.4, accordingly, this resistive force also assists in raising the concentration and heat profiles.

Figure 11.5 demonstrates the enhancing pattern of activation energy E on the concentration profile. The modified Arrhenius coefficient declines while E enlarges. This in turn stimulates the generative chemical process, and as a result, the concentration of nanoparticles spikes.

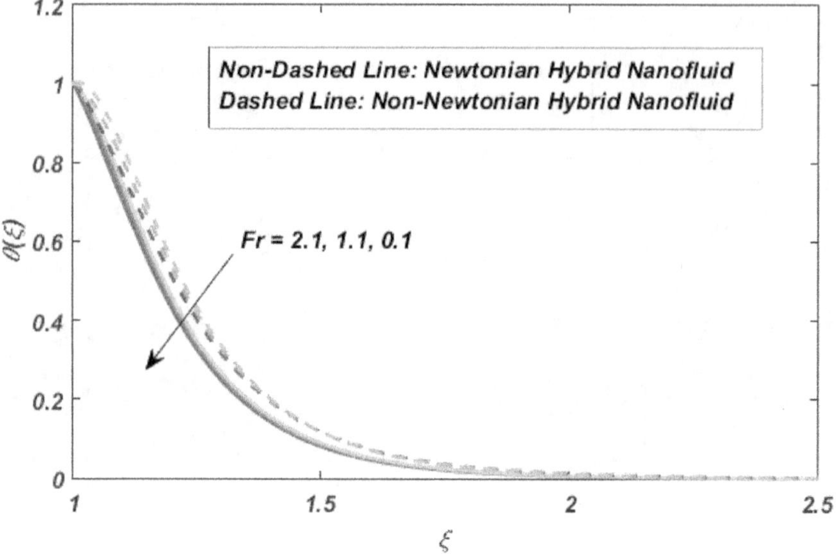

Figure 11.3 Illustration of *Fr* on temperature.

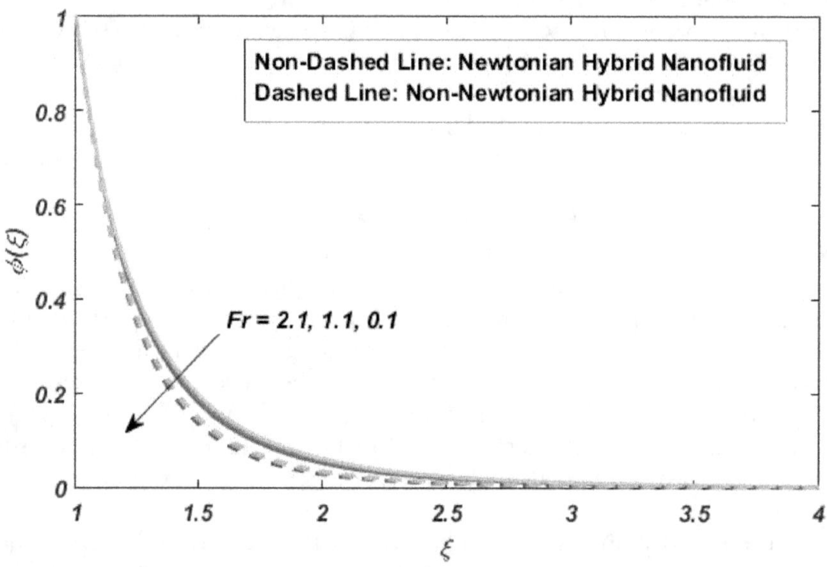

Figure 11.4 Illustration of *Fr* on concentration.

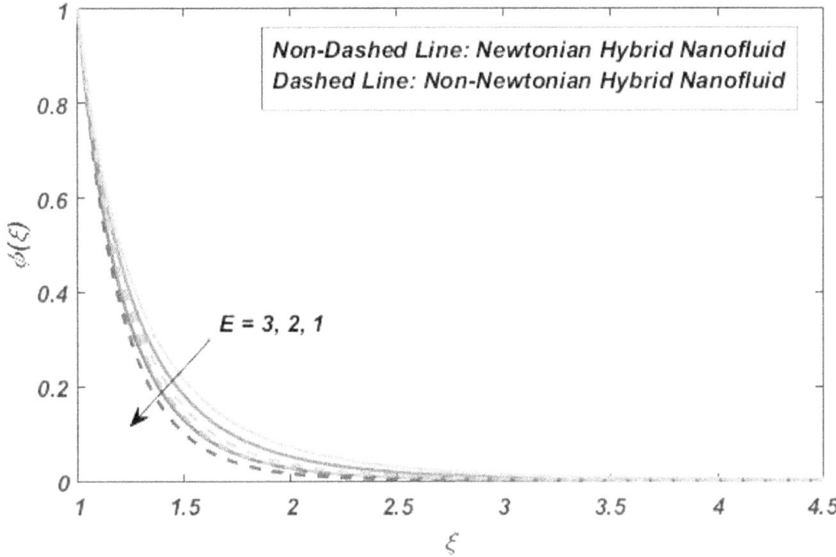

Figure 11.5 Illustration of E on concentration.

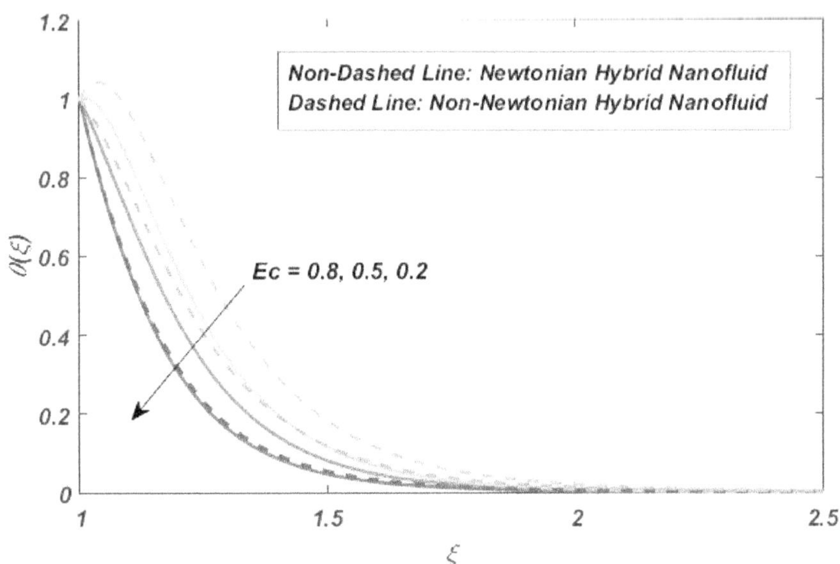

Figure 11.6 Illustration of Ec on temperature.

The implications of Eckert number, Ec, on the profile of heat, is displayed in Figure 11.6. The temperature gradient gets better as Ec ascents. The Eckert number Ec is the connection involving flow heat enthalpy differential and kinetic energy. Consequently, the kinetic energy increased by elevating the Eckert number. The "average kinetic energy" of an assembly is frequently referred to as its "temperature." Accordingly, the fluid's temperature rises. Also, an opposite pattern is detected for enhancing Eckert Number with the nanofluids concentration in Figure 11.7.

As the Schmidt number (Sc) expands, Figure 11.8 depicts that the concentration magnitudes significantly reduce. The Schmidt number is utilized for quantifying the momentum-to-mass diffusivity ratio. With a rising Schmidt number, the rate of transferring mass accelerates and the concentration patterns diminish. Also, in Figure 11.9, the correlation between the Temperature ratio factor β with the concentration distribution is pictorially reported. It has been demonstrated that escalating the magnitude of β reduces the concentration profile for both Newtonian and non-Newtonian hybrid nanofluid.

The absolute shear stress rate demonstrates an approximate 51% elevation in non-Newtonian hybrid nanofluid concerning parameters such as E, Ec, Sc, β, and Fr, as detailed in Table 11.3. As the Arrhenius energy

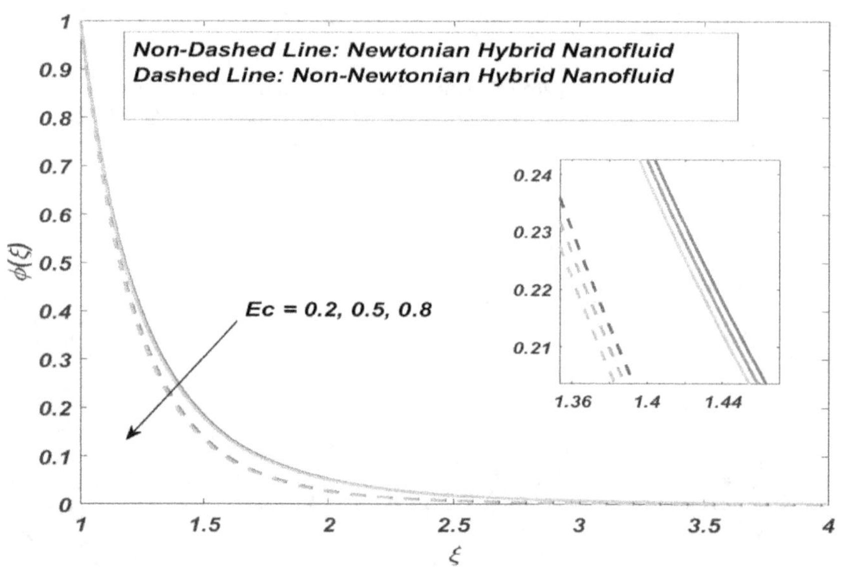

Figure 11.7 Illustration of Ec on concentration.

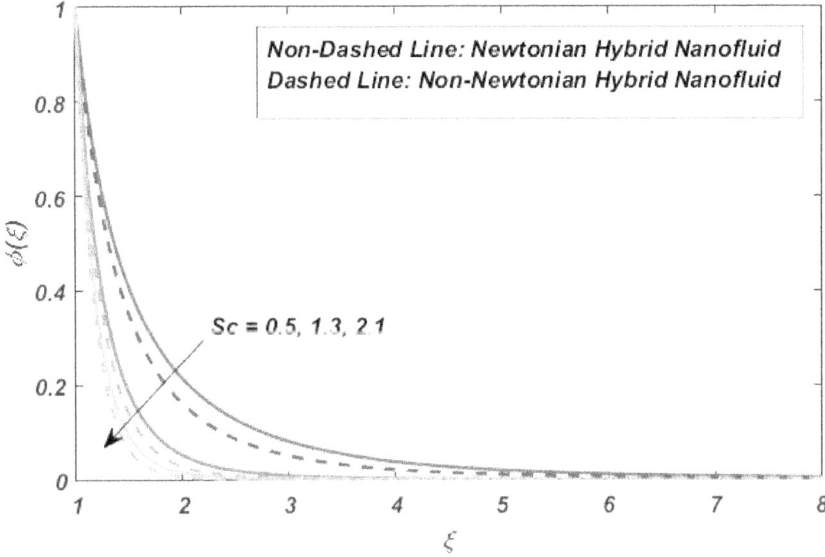

Figure 11.8 Illustration of *Sc* on concentration.

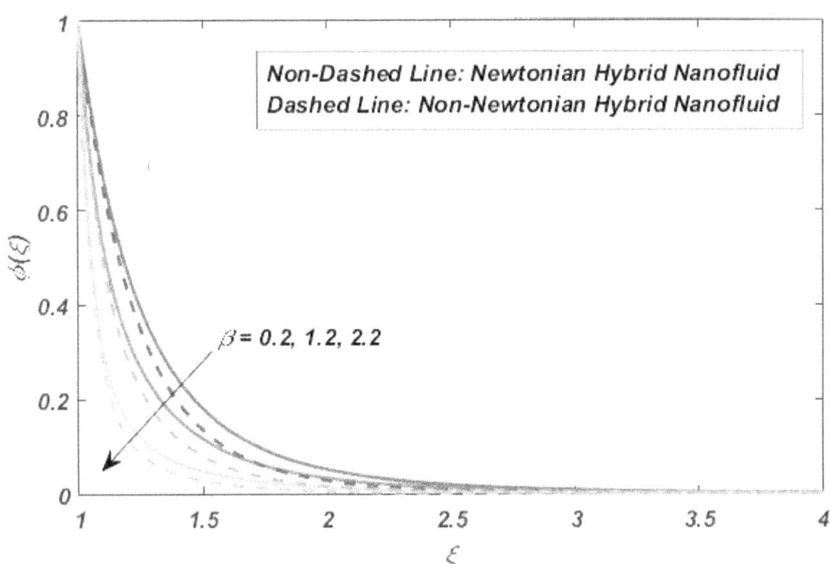

Figure 11.9 Illustration of β on concentration.

Table 11.3 Computed outcomes for numerous physical factors

		Newtonian hybrid nanofluid			*Non-Newtonian hybrid nanofluid*		
Term	*Value*	*Skin friction*	*Nusselt number*	*Sherwood number*	*Skin friction*	*Nusselt number*	*Sherwood number*
E	1	−6.7368	3.5456	5.0007	−9.9970	2.0945	5.2315
	2	−6.7368	3.5456	4.2380	−9.9970	2.0945	4.5266
	3	−6.7368	3.5456	3.8620	−9.9970	2.0945	4.1926
Ec	0.2	−6.7368	8.0825	4.1937	−9.9970	7.8162	4.4682
	0.5	−6.7368	3.5456	4.2380	−9.9970	2.0945	4.4566
	0.8	−6.7368	−0.9913	4.2849	−9.9970	−3.6273	4.5900
Sc	0.5	−6.7368	3.5456	2.3888	−9.9970	2.0945	2.6214
	1.3	−6.7368	3.5456	4.2380	−9.9970	2.0945	4.5266
	2.1	−6.7368	3.5456	5.5980	−9.9970	2.0945	5.8980
β	0.2	−6.7368	3.5456	4.2380	−9.9970	2.0945	4.5266
	1.2	−6.7368	3.5456	7.2380	−9.9970	2.0945	8.2649
	2.2	−6.7368	3.5456	15.3175	−9.9970	2.0945	15.9552
Fr	0.1	−6.7368	3.5456	4.2380	−9.9970	2.0945	4.5266
	1.1	−7.6416	2.2888	4.1722	−11.3085	0.4960	4.4756
	2.1	−8.4509	1.1382	4.1170	−12.4823	−0.9788	4.4321

value escalates, a discernible reduction in mass transport near the plate is observed. The mass transfer rate is notably elevated by approximately 4.5% in non-Newtonian hybrid nanofluid when contrasted with its Newtonian counterpart. Conversely, an escalation in the Eckert number correlates with a decrease in the Nusselt number, although this trend is inverted for the Sherwood number. The heat transfer rate is markedly more than 65% higher in Newtonian hybrid nanofluid compared to non-Newtonian hybrid nanofluid. However, the mass transport rate is approximately 5% higher in the latter.

Moreover, an augmentation in the Sherwood number is evident with increasing magnitudes of Sc and β for the fluid. The thermal transport rate shows a pronounced increase of 69% in Newtonian hybrid nanofluid concerning Sc and β, while the mass transfer rate is higher by approximately 6% and 14% in non-Newtonian hybrid nanofluid for Sc and β, respectively. Notably, an increase in the Forchheimer factor results in a concurrent rise in skin friction coefficient, accompanied by a simultaneous decrease in both Nusselt and Sherwood numbers. The thermal efficiency rate for Fr exhibits a substantial increase of 69% in Newtonian hybrid nanofluid, whereas the mass transport rate is elevated by 6% in non-Newtonian hybrid nanofluid.

11.4 CONCLUSION

The present exploration encompasses the investigation of contrasting flow characteristics of Newtonian hybrid nanofluid and non-Newtonian hybrid nanofluid featuring the thermophysical properties of Magnesium (Mg) and magnetite (Fe_3O_4) nanoparticles considering the impact of Lorentz force alongside the Darcy-Forchheimer impact. The flow model also incorporates the viscus dissipation and synergic (activation energy) effect. The outcomes which are obtained by the BVP4C approach are as follows:

- The thermal distribution profile is appeared to be mounting up with the surging value of *Ec* and *Fr*. The same trend has been noted in mass distribution profile for *Fr* and *E*.
- Thermal transport rate shows a pronounced increase of 69% in Newtonian hybrid nanofluid concerning *Sc*, *Fr*, and β than non-Newtonian hybrid nanofluid.
- The mass distribution profile decline with enhancing value of *Ec*, *Sc*, and β for both Newtonian and non-hybrid nanofluid.
- The mass transfer rate is notably elevated by approximately 4.5% in non-Newtonian hybrid nanofluid when contrasted with its Newtonian counterpart.
- Beyond the cylindrical surface, the mass distribution profile exhibits a noteworthy elevation in Newtonian hybrid nanofluids compared to non-Newtonian counterparts across all factors. A similar reverse trend is observed in the thermal distribution profile.
- The absolute shear stress rate is enhancing for escalating Darcy-Forchheimer term.

Nomenclature

Ec	Eckert number
Fr	Darcy- Forchheimer
β	Thermal ratio factor
E	Activation term
α	Casson term
Pr	Prandtl number
Re	Reynolds number
B_0, k_p	Magnetic and porous factor
C_h	Chemical reaction term
Sc	Schmidt number

REFERENCES

Akolade, M. T., Idowu, A. S., Olabode, J. O., & Titiloye, E. O. (2021). Unsteady flow analysis of Maxwell fluid with temperature dependent variable properties and quadratic thermo-solutal convection influence. *Partial Differential Equations in Applied Mathematics*, 4, 100078.

Alam, J., Murtaza, M. G., Tzirtzilakis, E. E., & Ferdows, M. (2023). A parametric simulation of MHD flow and heat transfer of blood-Fe3O4 over an exponentially stretching cylinder. *BioNanoScience*, 13(3), 891–899.

Alfaleh, A., Ali, F., Khan, M. I., Li, S., Puneeth, V., Raghunath, K., & Zaib, A. (2023). Effects of activation energy and chemical reaction on unsteady MHD dissipative Darcy-Forchheimer squeezed flow of Casson fluid over horizontal channel. *Scientific Reports*, 13(1), 2666.

Arooj, A., Javed, M., Imran, N., Sohail, M., & Yao, S. W. (2021). Pharmacological and engineering biomedical applications of peristaltically induced flow in a curved channel. *Alexandria Engineering Journal*, 60(6), 4995–5008.

B. Dey and R. Choudhury, Slip effects on heat and mass transfer in MHD visco-elastic fluid flow through a porous channel. In: *Emerging Technologies in Data Mining and Information Security: Proceedings of IEMIS 2018*, Vol. 1, pp. 553–564, Springer Singapore, 2019.

Choudhury R., Dey B. and Das B. (2018) Hydromagnetic oscillatory slip flow of a visco- elastic fluid through a porous channel. *Chemical Engineering Transactions*, 71, 961–966.

Chodhury, R., & Dey, B. (2017). Flow features of a conducting visco-elastic fluid past a vertical permeable plate. *Global Journal of Pure and Applied Mathematics*, 13(9), 5687–5702.

Choudhury, R., & Dey, B. (2018). Unsteady thermal radiation effects on MHD convective slip flow of visco-elastic fluid past a porous plate embedded in porous medium. *International Journal of Applied Mathematics and Statistics*, 57(2), 215–226.

Dero, S., Smida, K., Lund, L. A., Ghachem, K., Khan, S. U., Maatki, C., & Kolsi, L. (2022). Thermal stability of hybrid nanofluid with viscous dissipation and suction/injection applications: Dual branch framework. *Journal of the Indian Chemical Society*, 99(6), 100506.

Dey, B., Kalita, B., & Choudhury, R. (2022). Radiation and chemical reaction effects on unsteady viscoelastic fluid flow through porous medium. *Frontiers in Heat and Mass Transfer*, 18, 1–8

Dey, B., Nath, J. M., Das, T. K., & Kalita, D. (2022). Simulation of transmission of heat on viscous fluid flow with varying temperatures over a flat plate. *JP Journal of Heat and Mass Transfer*, 30, 1–18.

Farooq, U., Waqas, H., Alhazmi, S. E., Alhushaybari, A., Imran, M., Sadat, R., ... & Ali, M. R. (2023). Numerical treatment of casson nanofluid bioconvectional flow with heat transfer due to stretching cylinder/plate: Variable physical properties. *Arabian Journal of Chemistry*, 16(4), 104589.

Famakinwa, O. A., Koriko, O. K., & Adegbie, K. S. (2022). Effects of viscous dissipation and thermal radiation on time dependent incompressible squeezing flow of CuO– Al_2O_3/water hybrid nanofluid between two parallel plates with variable viscosity. *Journal of Computational Mathematics and Data Science*, 5, 100062.

Famakinwa, O. A., Koriko, O. K., & Adegbie, K. S. (2022). Effects of viscous dissipation and thermal radiation on time dependent incompressible squeezing flow of CuO- Al_2O_3/water hybrid nanofluid between two parallel plates with variable viscosity. *Journal of Computational Mathematics and Data Science*, 5, 100062.

Farooq, U., Hussain, S., Lu, D., Munir, S., Ramzan, M., & Suleman, M. (2019). MHD flow of Maxwell fluid with nanomaterials due to an exponentially stretching surface. *Scientific Reports*, 9(1), 7312.

Guirao, J. L. G., Sabir, Z., Sajid, T., & Tanveer, S. (2020). Impact of activation energy and temperature-dependent heat source/sink on Maxwell-Sutterby fluid. *Mathematical Problems in Engineering*, 2020, 1–15.

Hakeem, A. A., Kirusakthika, S., Ganga, B., Ijaz Khan, M., Nayak, M. K., Muhammad, T., & Khan, S. U. (2021). Transverse magnetic effects of hybrid nanofluid flow over a vertical rotating cone with Newtonian/non-Newtonian base fluids. *Waves in Random and Complex Media*, 1–18. doi:10.1080/17455030.2021.1983236

Hayat, T., Khan, M. I., Waqas, M., Alsaedi, A., & Khan, M. I. (2017). Radiative flow of micropolar nanofluid accounting thermophoresis and Brownian moment. *International Journal of Hydrogen Energy*, 42(26), 16821–16833.

Ishak, A., Nazar, R., & Pop, I. (2008). Uniform suction/blowing effect on flow and heat transfer due to a stretching cylinder. *Applied Mathematical Modelling*, 32(10), 2059–2066.

Jeelani, M. B., & Abbas, A. (2023). Al_2O_3-Cu\Ethylene glycol-based magnetohydrodynamic non-Newtonian Maxwell hybrid nanofluid flow with suction effects in a porous space: Energy saving by solar radiation. *Symmetry*, 15(9), 1794.

Jyothi, A. M., Madhukesh, J. K., Prasannakumara, B. C., Ramesh, G. K., & Varun Kumar, R. S. (2021). Squeezing flow of Casson hybrid nanofluid between parallel plates with a heat source or sink and thermophoretic particle deposition. *Heat Transfer*, 50(7), 7139–7156.

Krishna, M. V., Ahammad, N. A., & Chamkha, A. J. (2021). Radiative MHD flow of Casson hybrid nanofluid over an infinite exponentially accelerated vertical porous surface. *Case Studies in Thermal Engineering*, 27, 101229.

Khashi'ie, N. S., Arifin, N. M., Nazar, R., Hafidzuddin, E. H., Wahi, N., & Pop, I. (2020). Magnetohydrodynamics (MHD) axisymmetric flow and heat transfer of a hybrid nanofluid past a radially permeable stretching/shrinking sheet with Joule heating. *Chinese Journal of Physics*, 64, 251–263.

Li, Q., & Xuan, Y. (2000). Heat transfer enhancement of nanofluids. *International Journal of heat and fluid flow*, 21(1), 58–64.

Madhukesh, J. K., Prasannakumara, B. C., Raghunatha, K. R., Rekha, M. B., & Sarris, I. E. (2022). Activation energy impact on flow of AA7072-AA7075/Water-Based hybrid nanofluid through a cone, wedge and plate. *Micromachines*, 13(2), 302.

Moatimid, G. M., Mohamed, M. A., & Elagamy, K. (2022). A Casson nanofluid flow within the conical gap between rotating surfaces of a cone and a horizontal disc. *Scientific Reports*, 12(1), 11275.

Mushtaq, A., Mustafa, M., & Shafique, Z. (2016). Boundary layer flow of Maxwell fluid in rotating frame with binary chemical reaction and activation energy. *Results in Physics*, 6, 627–633.

Paul, A., Das, T. K., & Nath, J. M. (2022). Numerical investigation on the thermal transportation of MHD Cu/Al_2O_3-H_2O Casson-hybrid-nanofluid flow across an exponentially stretching cylinder incorporating heat source. *Physica Scripta*, 97(8), 085701.

Paul, A., Mani Nath, J., & Kanti Das, T. (2023). An investigation of the MHD Cu-Al_2O_3/H_2O hybrid-nanofluid in a porous medium across a vertically stretching cylinder incorporating thermal stratification impact. *Journal of Thermal Engineering*, 9(3).

Reddy, M. G., & Kumar, K. G. (2021). Cattaneo-Christov heat flux feature on carbon nanotubes filled with micropolar liquid over a melting surface: A stream line study. *International Communications in Heat and Mass Transfer*, 122, 105142.

Reddy, M. G., Rani, M. S., Praveen, M. M., & Kumar, K. G. (2021). Comparative study of different non-Newtonian fluid over an elaborated sheet in the view of dual stratified flow and ohmic heat. *Chemical Physics Letters*, 784, 139096.

Rehman, A., & Khan, I. (2023). Mixed convection flow of hybrid nanofluids with viscous dissipation and dynamic viscosity. *BioNanoScience*, 1–9. doi:10.1007/s12668-023-01271-2

Scott, T. O., Ewim, D. R., & Eloka-Eboka, A. C. (2022). Hybrid nanofluids flow and heat transfer in cavities: A technological review. *International Journal of Low-Carbon Technologies*. doi:10.1093/ijlct/ctac093.

Venkateswarlu, A., Suneetha, S., Babu, M. J., Kumar, J. G., Raju, C. S. K., & Al-Mdallal, Q. (2022). Significance of magnetic field and chemical reaction on the natural convective flow of hybrid nanofluid by a sphere with viscous dissipation: A statistical approach. *Nonlinear Engineering*, 10(1), 563–573.

Wang, C. Y. (1988). Fluid flow due to a stretching cylinder. *The Physics of fluids*, 31(3), 466–468.

Yang, H., Hayat, U., Shaiq, S., Shahzad, A., Abbas, T., Naeem, M., ... & Zahid, M. A. (2023). Thermal inspection for viscous dissipation slip flow of hybrid nanofluid (TiO_2-$Al_2O_3/C_2H_6O_2$) using cylinder, platelet and blade shape features. *Scientific Reports*, 13(1), 8316.

Chapter 12

Magnetohydrodynamic flow of ternary hybrid nanofluids over a wedge

Influence of nanoparticle shape factor in solar energy applications

Pinaki Ranjan Duari and Kalidas Das

12.1 INTRODUCTION

As societies worldwide continue to prioritize sustainability and environmental responsibility, the importance of solar radiation as a clean and renewable energy source becomes increasingly evident. The ongoing transition to solar and other clean energy technologies is crucial for achieving a more sustainable and resilient global energy system. The importance of using solar energy is underscored by its contribution to addressing environmental challenges, promoting economic development and fostering a more sustainable and resilient energy future. In the future, solar energy is expected to play a much greater role in global energy supply as technology advances and costs decline. The integration of nanofluids (Choi, 1995) into solar energy systems is an area of research aimed at enhancing the overall efficiency of energy conversion, particularly in solar thermal applications. The provided excerpt discusses nanofluids, which are colloidal suspensions containing nanoscale particles dispersed in base fluids like water, kerosene, ethylene glycol, toluene, etc. These nanoparticles are often solid metallic, non-metallic, or oxide particles with diameters ranging from 1 to 10^2 nm. The use of nanofluids is explored to address inadequacies in traditional fluids, especially regarding thermal conductivity and transport properties.

The term "nanofluid hybrid" typically refers to a type of nanofluid that combines nanoparticles with a base fluid (Sarkar et al., 2015). A hybrid nanofluid consists of different types of nanoparticles mixed together in a single medium. The goal is to achieve specific and enhanced properties by leveraging the synergistic effects that different nanoparticles may have when combined. The combination of nanoparticles with a fluid creates a colloidal suspension where the particles are dispersed throughout the fluid. These nanoparticles are often engineered to have unique thermal, optical, or other properties that can be beneficial for various applications. On a vertical plate with ramped wall temperatures, Rajesh et al. (2021) investigated

DOI: 10.1201/9781003494454-12

the effects of hybridized nanoliquids on magnetohydrodynamic flow and heat transmission. An experimental study by Khashi'ie et al. (2022) suggests that hybrid nanofluids are deposited between magnetohydrodynamic boundary layers and flow over moving plates. The Akbari-Ganji method (AGM) was used by Zangooee et al. (2019) to investigate magnetohydrodynamic (MHD) nanofluid flows across two discs under radiative stretch rotation. In their study, Acharya and Mabood (2021) found that hybrid nanofluids (Fe_3O_4–graphene-H_2O) increased their Biot number by 74.25%, while conventional nanofluids (Fe_3O_4–H_2O) increased theirs by 69.17%. Recent studies have revealed interesting results using hybrid nanofluids (Das et al., 2022; Chakraborty, et al., 2023).

Over the past few years, interest has grown in developing ternary hybrid nanofluids more extensively. Nanoparticles of three different kinds are combined with a base fluid to create an innovative nanofluid known as a ternary hybrid nanofluid. An increasing number of researchers are interested in the properties of this type of nanofluid. It has been found that ternary hybrid nanofluids exhibit superior thermophysical features, making them of interest to researchers. Heat can be transported more efficiently in ternary hybrid nanofluids than in mono-nanofluids and hybrid nanofluids. Engineers and industrialists have praised nanofluids for their excellent thermal properties, particularly their improved versions (ternary nanofluids). The synergistic effects resulting from the combination of different types of nanoparticles in ternary hybrid nanofluids are multifaceted and contribute significantly to the overall enhancement of thermal and rheological properties. These effects arise from the interaction and cooperation among the various nanoparticles, leading to enhanced performance beyond what can be achieved by individual nanoparticle types alone. The enhancement of thermal and rheological properties in ternary hybrid nanofluids can be attributed to the unique characteristics and interactions of the different types of nanoparticles present. Ternary hybrid nanofluids typically consist of a base fluid, such as water or oil, in which three different types of nanoparticles are dispersed. These nanoparticles can include metallic, metal oxide, carbon-based, or polymer nanoparticles. Each type of nanoparticle contributes to the enhancement of thermal and rheological properties in distinct ways: Metallic nanoparticles, such as Iron oxide (Fe_2O_3), Silver (Ag), and aluminum oxide (Al_2O_3), possess high thermal conductivity and thus play a significant role in enhancing the thermal conductivity of the nanofluid. When dispersed in the base fluid, metallic nanoparticles form conductive pathways that facilitate the transfer of heat, leading to improved thermal conductivity of the nanofluid. By doing so, the fluid's thermal performance is enhanced and heat transfer (HT) rates are increased. A wide range of applications of it are in chemical engineering, applied thermal engineering, mechanical engineering, and biotechnology. A case study using Al_2O_3-CuO-TiO_2/polymer ternary components was carried out by Mahmood and Khan (2023). For the purpose of assessing ternary nanoparticles' thermal properties, Al Oweidi et al. (2022) investigated their effect on HT and liquid movement. A tri-HNF's thermal properties were their objective. In addition to Hall current, thermal radiation, and heat dissipation, flow-described equations were used to investigate several parameters. The relevance of tri-hybrid nanofluid flow in a variety of applications has been explored in several recent studies (Mahmood et al., 2022a, 2022b, 2023).

Non-Newtonian behavior occurs in the handling of liquid plastics, paints, honey, natural liquid motility, food, and slurries. Many different models can be used to predict rheological behaviors, in addition to the power function, viscoelastic, Casson, Carreau, micropolar, Reiner-Philippoff, Giesekus, and Powell-Eyring models (Rehman et al., 2019; Uddin et al., 2021). Literature has developed many non-Newtonian fluid models over the years. In terms of non-Newtonian fluids, Prandtl-Eyring is one of the models to consider. Shear stress and the hyperbolic sine function of strain rate (distortion degree) are linearly related in this fluid model. It is important to examine the materialistic characteristics of visco-inelastic fluids using models like Prandtl-Eyring, which is a valuable approach to understanding and optimizing materials for specific applications, including those related to PVC pipes and bulletproof jackets. The significant increases in HT rates observed in Prandtl-Eyring ternary hybrid nanofluids have profound practical implications across various industries including the electronics industry, automotive, transportation industries aerospace, automotive, and electronics manufacturing. It allows engineers and researchers to predict and tailor the behavior of materials under different conditions, contributing to the development of more efficient and effective products. The Prandtl-Eyring ternary hybrid nanofluids exhibit distinctive features and properties owing to their unique composition and structure. These nanofluids, composed of three different types of nanoparticles, typically show enhanced thermal and rheological characteristics compared to traditional HT fluids. The primary properties of interest include:

i. Prandtl-Eyring ternary hybrid nanofluids often exhibit non-Newtonian rheological behavior. The combination of three nanoparticles can alter the flow characteristics, leading to unique rheological responses under various shear rates. Understanding and controlling these rheological properties are crucial for applications in HT and fluid dynamics.
ii. The thermal and rheological properties of Prandtl-Eyring ternary hybrid nanofluids are often temperature and concentration-dependent. Understanding these dependencies is essential for predicting the nanofluid's performance under varying conditions and tailoring its formulation for specific applications.
iii. Due to their superior thermal properties, Prandtl-Eyring ternary hybrid nanofluids find application in advanced HT systems, such as microelectronics cooling, solar thermal systems, and automotive heat exchangers. Thermal management applications can be met with these nanofluids due to their versatility.

Prandtl-Eyring ternary hybrid nanofluids distinguish themselves from other nanofluids through their intricate composition involving three types of nanoparticles and modern applications, resulting in enhanced thermal properties and more complex rheological behavior. It was discovered that a Prandtl-Eyring fluid could be transported across a sensor surface by Shankar and Naduvinamani (2019) under magnetic force. Prandtl-Eyring fluid warmth transmission in porous media was studied by Khafajy and Kaabi (2021) based on concentration and temperature fluctuations. Prandtl-Eyring nanofluid was studied by Hayat et al. (2021) on stretching cylinder microorganisms cutting-edge a shrinking slip. A new method for solving Prandtl-Eyring fluid flow has been suggested by Munjam et al. (2021), and its effects on Keller box calculations have been evaluated. The most recent progress in this direction can be found in (Abu-Hamdeh et al., 2021; Ullah et al., 2021a; Qureshi, 2021; Qureshi, 2022; Rehman et al., 2018).

Ghadikolaei et al. (2018) have emphasized the influence of shape factors on hybrid nanofluids (Fe_3O_4-Ag-H_2O). After extending the investigation into the effect of nanoparticle shape on 3D MHD flow and porous spaces by considering thermal radiation, the results of the study were discussed in Ghadikolaei and Gholinia (2019). A study by Khashi'ie et al. (2021) suggests that the size and shape of nanoparticles may have an impact on nanofluid thermal conductivity. In a study by Rashid et al. (2021), the shape of nanoparticles (sphere, blade, and lamina) was analyzed for HT in horizontal cylinders that were stretching or shrinking. According to Zahmatkesh et al. (2021), nanoparticle shape contributes to better heat transmission based on the flow regime. Recent studies with effects on nanoparticle shape have revealed interesting results (Imran et al., 2022; Rashid et al., 2020).

Several studies have investigated linear thermal radiation, but the linearity term does not adequately satisfy flow temperature. Thus, authors often refer to the nonlinear term when discussing thermal radiation. In their study of nonlinear thermal radiation, Ghadikolaei et al. (2018a) utilized a swirling stretching channel that contained CNTs-$C_2H_6O_2$-H_2O. The transmission of ultrathin magnetic nanofluid films in a horizontally rotating disk can be affected by nonlinear heat flux, as observed by Shah et al. (2019). The nonlinear radiation emanating from a rotating disk affects the thermal transfer of MHD nanofluids in an experimental study by Prakash and Vinu (2019). A more detailed analysis of the flow behavior for nonlinear heat thermal radiation with the Rosseland approximation can be found in the studies (Acharya et al., 2019; Roy and Pal, 2022; Gorai et al., 2024). Kandasamy et al. (2013) investigated the ability of a copper-water nanofluid to emit solar energy. Nanofluid flow over stretched objects impacted by solar radiation is studied by Das et al. (2014), including hydrothermal slip effects. In a study conducted by Das et al. (2022), hybrid nanofluids under solar energy were examined for slip flow characteristics.

There is an important phenomenon associated with fluid flow due to a moving wedge. It is observed that the fluid velocity and plate velocity are proportional in such a phenomenon. This applies to various engineering processes. It is necessary to process sheet-like substances thermally to prepare paper, wire drawings, linoleum, plastic films, polymeric sheets, fine fiber mats, and metal spinning, among other things. The moment of the sheet is parallel to its plane in all of the processes and applications mentioned above. It is possible that the sheet reduces particle movements in neighboring fluids or that neighboring fluids have parallel forced convection motions. The Casson model can be preserved by using a symmetric wedge as demonstrated by Mukhopadhyay et al. (2013). The authors (Raju and Sandeepa, 2016) demonstrated Falkner-Skan flow in the presence of cross-diffusion for magnetic Carreau fluids. Numerous research works on wedges toward Falkner-Skan flow have been reported (Ullah et al., 2016; Nadeem et al., 2018). Based on nonlinear radiation and an induced magnetic field, Basha et al. (2019) applied solar energy to diamond $C_2H_6O_2$ nanofluid flow. In the presence of a magnetic field, a permeable stretching/shrinking wedge with thermal conductivity and changeable fluid characteristics was studied by Khatun and Islam (2022). The topic of nonlinear radiative heat transport in porous media with magnetic fields was examined in a paper by Jalil et al. (2023).

In light of the literature mentioned above, this study examines the hydrothermal changes when the Prandtl-Eyring ternary hybrid nanofluid Fe_3O_4-Ag-Al_2O_3-H_2O travels past a static and moving wedge under solar radiation heating and magnetic fields. In this study, the shape factor of nanoparticles was taken into account. Researchers can use this type of study to improve solar collectors, solar energy, and solar cells. Simulations are executed using the shooting-based RK-4 procedure. The use of nanofluids and hybrid nanofluids may satisfy engineers' needs, but a fluid with good thermal conductivity remains essential for optimal thermal performance. Consequently, nanocomposite liquids have been developed with greater thermal conductivity than classical and hybrid nanofluids, such as ternary hybrid nanofluids. Hence, the ultimate goal of this text is to present the technological aspects of ternary hybrid nanosuspension. It is known that no studies, such as those discussed in this article, have yet been published, and the literature on Prantl-Eyring ternary hybrid nanofluid issues is quite limited. We are trying to do something new and unique, and that's why we're doing it.

The core questions that are of keen interest and have been explored throughout the investigation are:

- How do the Prandtl-Eyring parameters, magnetic field parameter, wedge parameter, nanoparticle shape effects parameter, Biot number, and Hartree pressure gradient affect the velocity profiles?
- How do the Prandtl-Eyring parameters, thermal radiation, temperature ratio parameter, wedge parameter, Hartree pressure gradient, nanoparticle shape factor, and nanoparticle volume fraction affect the temperature profile?

- How do ternary hybrid nanofluid streamlines behave under an induced magnetic field?
- What are the impacts of interrelated parameters on skin friction and Nusselt numbers?

12.2 MATHEMATICAL FORMATIONS

A steady two-dimensional hydrothermal flow of an incompressible MHD Prandtl-Eyring ternary hybrid nanofluid (Fe_3O_4-Ag-Al_2O_3-H_2O) past a static and moving wedge in the presence of solar radiation is considered. A moving wedge induces fluid flow with velocity, $u_w = U_w x^m$ and a free stream velocity, $u_e = U_e x^m$, with, $0 \le m \le 1$. According to Figure 12.1, $u_w > 0$ compares to a contracting wedge surface velocity, whereas, $u_w < 0$ corresponds to a stretching wedge surface velocity. We assume $\Omega = \beta\pi$ to be the wedge angle, where $\beta = \dfrac{2m}{m+1}$. The magnetic field, $B(x)$ is also applied

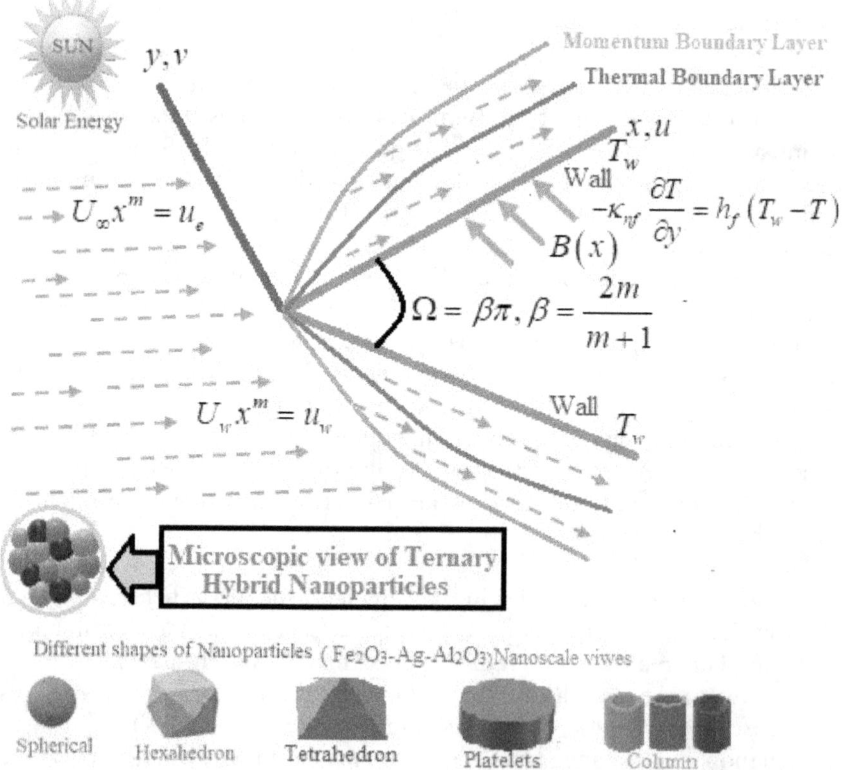

Figure 12.1 Physical model of the flow.

in the y-direction with $B(x) = B_0 x^{(x-1)/2}$, where magnetic field magnitude is B_0. It is assumed that the induced magnetic field is negligible owing to the small magnetic Reynolds number assumption; additionally, Hall current and ohmic dissipation effects are neglected owing to the low magnetic Reynolds number.

The governing equations for the Prandtl-Eyring fluid flow can be formulated using the following stress tensor relation:

$$\tau = \left[\frac{A \text{ arc sinh}\left(\frac{1}{c} \sqrt{\frac{1}{2} \text{trace}\left(R^2\right)} \right)}{\sqrt{\frac{1}{2} \text{trace}\left(R^2\right)}} \right] \tag{12.1}$$

where the symbols A and c represent the material parameters of the fluid. Also, R indicates the first Rivilin-Erickson tensor. Thus, the necessary constituents of Prandtl-Eyring fluid are given by the following relation:

$$\tau_{xy} = \left[\frac{A}{\rho_{Thnf}} \text{arc sinh}\left(\frac{1}{c} \frac{du}{dy} \right) \right] \tag{12.2}$$

However, the term, $\sinh^{-1}\left(\frac{1}{c} \frac{\partial u}{\partial y} \right)$ is expanded through Taylor's series expansion, and after ignoring the higher-order terms except the first two terms, the truncated series is given by the following expression, which is significantly used in the modeling of the present problem.

$$\sinh^{-1}\left(\frac{1}{c} \frac{\partial u}{\partial y} \right) = \frac{1}{c} \frac{\partial u}{\partial y} - \frac{1}{6} \left(\frac{1}{c} \frac{\partial u}{\partial y} \right)^3 \tag{12.3}$$

According to the assumptions above and after applying the usual boundary-layer analysis, despite the presence of a time-dependent magnetic field, the ternary hybrid nanofluid can be described with boundary-layer equations as (Mukhopadhyay et al., 2013; Raju and Sandeepa, 2016; Ullah et al., 2016), which govern mass, momentum, and energy.

$$\frac{\partial u}{\partial x} + \frac{\partial v}{\partial y} = 0 \tag{12.4}$$

$$u \frac{\partial u}{\partial x} + v \frac{\partial u}{\partial y} = u_e \frac{\partial u_e}{\partial x} + \frac{\mu_{Thnf}}{\rho_{Thnf}} \frac{\partial^2 u}{\partial y^2} + \frac{A}{c} \frac{1}{\rho_{Thnf}} \frac{\partial^2 u}{\partial y^2} - \frac{A}{2c^3} \frac{\partial^2 u}{\partial y^2} \left(\frac{\partial u}{\partial y} \right)^2 - \frac{\sigma_{Thnf}}{\rho_{Thnf}} B^2 \left(u - u_e \right) \tag{12.5}$$

$$u\frac{\partial T}{\partial x}+v\frac{\partial T}{\partial y}=\alpha_{Thnf}\frac{\partial^2 T}{\partial y^2}-\frac{1}{\left(\rho C_p\right)_{Thnf}}\left(\frac{A}{c}\left(\frac{\partial u}{\partial y}\right)^2-\frac{A}{6c^3}\left(\frac{\partial u}{\partial y}\right)^4\right)$$
$$-\frac{1}{\left(\rho C_p\right)_{Thnf}}\frac{\partial q_{rad}}{\partial y} \tag{12.6}$$

Rather than radiation, heat is transferred using heat waves instead of direct contact. In Eq. (6), the term q_{rad} represents HT per unit area.

According to the Rosseland approximation for nonlinear radiation, q_{rad} can be written as (Ghadikolaei et al., 2018b; Shah et al., 2019; Prakash and Vinu, 2019; Acharya et al., 2019; Jalil et al., 2023)

$$q_{rad}=-\frac{4\sigma_1}{3k^*}\frac{\partial T^4}{\partial r}=-\frac{16\sigma_1}{3k^*}T^3\frac{\partial T}{\partial r} \tag{12.7}$$

where σ_1 symbolizes the Stefan-Boltzmann constant and k^* describes the mean absorption coefficient.

Now we put T as $\theta=\dfrac{T-T_\infty}{T_w-T_\infty}$ and we get $T=T_\infty\left[1+(\theta_w-1)\theta\right]$ where $\theta_w=\dfrac{T_w}{T_\infty}$

with the boundary conditions:

i. Static wedge

$$\left.\begin{aligned}&u=0, v=0, -\kappa_{nf}\frac{\partial T}{\partial y}=h_f\left(T_w-T\right)\ at\ y=0\\&u\to u_e, T\to T_\infty\ as\ y\to\infty\end{aligned}\right\} \tag{12.8}$$

ii. Moving wedge

$$\left.\begin{aligned}&u=u_w, v=0, -\kappa_{nf}\frac{\partial T}{\partial y}=h_f\left(T_w-T\right)\ at\ y=0\\&u\to u_e, T\to T_\infty\ as\ y\to\infty\end{aligned}\right\} \tag{12.9}$$

Here x and y are, respectively, the separation along the surface of the wedge and normal to it; the velocity components along (x, y) are taken to by (u, v) respectively, T_w = temperature of the wedge wall. α_{Thnf} symbolizes the thermal diffusivity of the ternary hybrid nanofluid. h_f is the convective HT coefficient, and κ_{nf} is the thermal conductivity. $(\ )_\infty$ fluid properties far away from the wedge.

12.2.1 Thermophysical properties

To prepare a ternary hybrid nanofluid, a nanodispersion containing Fe_3O_4, Ag, and Al_2O_3 nanoparticles of 50 nm each is mixed with water. A summary of related thermophysical features can be found in Table 12.1. Table 12.2 illustrates related thermophysical correlations at reference temperatures 20°C–30°C. Prandtl-Eyring ternary hybrid nanofluids are anticipated to exhibit outstanding performance in terms of hydrothermal properties compared to classical mono-nanofluids and hybrid nanofluids. The range of hydrothermal properties where notable improvements are expected includes:

i. The ternary hybrid composition is expected to lead to a notable enhancement in thermal conductivity compared to classical mono-nanofluids and hybrid nanofluids.
ii. Prandtl-Eyring ternary hybrid nanofluids are expected to exhibit improved specific heat capacity compared to classical mono-nanofluids and hybrid nanofluids. This enhancement is crucial for applications requiring effective energy storage and transfer, as higher specific heat capacity contributes to better thermal inertia.
iii. In comparison to mono-nanofluids and hybrid nanofluids, ternary hybrid nanofluids are expected to display superior thermal stability. The presence of three different types of nanoparticles may contribute to enhanced stability under varying temperature conditions, making them suitable for applications involving thermal cycling.
iv. Prandtl-Eyring ternary hybrid nanofluids are likely to exhibit a more tailored and controllable rheological behavior compared to classical mono-nanofluids and hybrid nanofluids.
v Prandtl-Eyring ternary hybrid nanofluids are expected to showcase improved performance under varying temperature conditions. The complex interplay between three different types of nanoparticles may result in a hydrothermal response that remains effective across a broader temperature range compared to classical mono-nanofluid and hybrid nanofluids.

Table 12.1 Thermophysical properties of the base liquid and Fe_3O_4, Ag, and Al_2O_3 nanoparticles (Al Oweidi et al., 2022; Mahmood et al., 2022a, 2022b, 2023)

Physical properties	H_2O	$Fe_3O_4\,(\phi_1)$	$Ag\,(\phi_2)$	$Al_2O_3\,(\phi_3)$
C_p (J/kg K)	4,179	670	235	765
ρ (kg/m³)	997.1	5180	10500	3970
κ (W/mK)	0.613	9.7	429	40
σ (S/m)	0.005	2.5×10^4	6.30×10^7	3.5×10^7

Table 12.2 Thermophysical models of mono-nanofluid, hybrid nanofluid, and ternary hybrid nanofluid (Mahmood et al., 2023; Uddin et al., 2021; Rehman et al., 2019)

Properties	*Mono-nanofluid (Fe_3O_4-H_2O)*
Density	$\rho_{nf} = (1-\phi_1)\rho_f + \phi_1 \rho_{s1}$
Heat capacity	$(\rho C_p)_{nf} = (1-\phi_1)(\rho C_p)_f + \phi_1 (\rho C_p)_{s1}$
Viscosity	$\mu_{nf} = \dfrac{\mu_f}{(1-\phi_1)^{2.5}}$
Thermal conductivity	$\dfrac{\kappa_{nf}}{\kappa_f} = \dfrac{\kappa_{s1} + (p-1)\kappa_f - (p-1)\phi_1(\kappa_f - \kappa_{s1})}{\kappa_{s1} + (p-1)\kappa_f + \phi_1(\kappa_f - \kappa_{s1})}$
Properties	Hybrid nanofluid (Fe_3O_4-Ag-H_2O)
Density	$\rho_{hnf} = (1-\phi_2)\{(1-\phi_1)\rho_f + \phi_1 \rho_{s1}\} + \phi_2 \rho_{s2}$
Heat capacity	$(\rho C_p)_{hnf} = (1-\phi_2)\{(1-\phi_1)(\rho C_p)_f + \phi_1(\rho C_p)_{s1}\} + \phi_2(\rho C_p)_{s2}$
Viscosity	$\mu_{hnf} = \dfrac{\mu_f}{(1-\phi_1)^{2.5}(1-\phi_2)^{2.5}}$
Thermal conductivity	$\kappa_{hnf} = \dfrac{\kappa_{s2} + (p-1)\kappa_{nf} - (p-1)\phi_2(\kappa_{nf} - \kappa_{s2})}{\kappa_{s2} + (p-1)\kappa_{nf} + \phi_2(\kappa_{nf} - \kappa_{s2})} \times \kappa_{nf}$ where $\kappa_{nf} = \dfrac{\kappa_{s1} + (p-1)\kappa_f - (p-1)\phi_1(\kappa_f - \kappa_{s1})}{\kappa_{s1} + (p-1)\kappa_f + \phi_1(\kappa_f - \kappa_{s1})} \times \kappa_f$
Properties	Ternary hybrid nanofluid (Fe_3O_4-Ag-Al_2O_3-H_2O)
Density	$\rho_{Thnf} = (1-\phi_1)\{(1-\phi_2)[(1-\phi_3)\rho_f + \phi_3 \rho_{s3}] + \phi_2 \rho_{s2}\} + \phi_1 \rho_{s1}$
Heat capacity	$(\rho C_p)_{Thnf} = (1-\phi_1)\left\{\begin{array}{l}(1-\phi_2)\left[(1-\phi_3)(\rho C_p)_f + \phi_3(\rho C_p)_{s3}\right] \\ +(\rho C_p)_f + \phi_2(\rho C_p)_{s2}\end{array}\right\} + \phi_1(\rho C_p)_{s1}$
Viscosity	$\mu_{Thnf} = \dfrac{\mu_f}{(1-\phi_1)^{2.5}(1-\phi_2)^{2.5}(1-\phi_3)^{2.5}}$
Thermal conductivity	$\kappa_{Thnf} = \dfrac{\kappa_{s3} + (p-1)\kappa_{hnf} - (p-1)\phi_3(\kappa_{hnf} - \kappa_{s3})}{\kappa_{s3} + (p-1)\kappa_{hnf} + \phi_3(\kappa_{hnf} - \kappa_{s3})} \times \kappa_{hnf}$ where $\kappa_{hnf} = \dfrac{\kappa_{s2} + (p-1)\kappa_{nf} - (p-1)\phi_2(\kappa_{nf} - \kappa_{s2})}{\kappa_{s2} + (p-1)\kappa_{nf} + \phi_2(\kappa_{nf} - \kappa_{s2})} \times \kappa_{nf}$ *and* $\kappa_{hnf} = \dfrac{\kappa_{s1} + (p-1)\kappa_f - (p-1)\phi_1(\kappa_f - \kappa_{s1})}{\kappa_{s1} + (p-1)\kappa_f + \phi_1(\kappa_f - \kappa_{s1})} \times \kappa_f$

Here, the volume fractions of Fe_3O_4, Ag, and Al_2O_3 are symbolized as ϕ_1, ϕ_2, and ϕ_3, respectively. As a result, the suffixes *Thnf*, *hnf*, *nf*, *f*, *s*1, *s*2, and *s*3 are dedicated to understanding ternary hybrid nanofluid, hybrid nanofluid, nanofluid, base fluid, Fe_3O_4, Ag, and Al_2O_3 solid particles, respectively. Table 12.1 shows the experimental values of the properties that were used in this study.

According to the Hamilton-Crosser model (Hamilton and Crosser, 1962; Yang and Xu, 2017), p is the shape factor given by $3/\psi$. The sphericity of nanoparticles is ψ. Sphere-shaped nanoparticles have a sphericity of $\psi=1$ and a shape factor of $p=3$. This can be done by reducing Hamilton-Crosser's model to Maxwell's model based on different nanoparticle shapes, and sphericity values are shown in Table 12.3. High shape factors are associated with higher thermal conductivity in suspensions in this model of conductivity.

Table 12.3 Nanoparticles shape with their shape factor (Ghadikolaei et al., 2018a, 2018b; Ghadikolaei and Gholinia, 2019; Khashi'ie et al., 2021; Rashid et al., 2021; Zahmatkesh et al., 2021; Imran et al., 2022; Rashid et al., 2020)

Nanoparticles type	*Shape structure (nanoscale view)*	*Sphericity* ψ	*Shape factor* $p=3/\psi$
Spherical		1	3
Hexahedron		0.87	3.72
Tetrahedron		0.62	4.0613
Platelets		0.82	5.7
Column		0.4710	6.3698

To solve the system (4)–(6) with (8) and (9), we need to introduce similarity transformations to transform the set of PDEs to ODEs and solve it numerically.

$$\left.\begin{aligned} &\eta = y\sqrt{\tfrac{x^{m-1}u_e}{v_f}}, \psi = f(\eta)\sqrt{v_f x^{m+1} u_e}, \mathrm{u} = u_e x^m f'(\eta) \\ &v = -\sqrt{v_f u_e x^{m-1}}\left(\frac{m+1}{2}\right)\left\{f(\eta) + \frac{m-1}{m+1}\eta f'(\eta)\right\}, \theta = \tfrac{T-T_\infty}{T_w-T_\infty} \end{aligned}\right\} \tag{12.10}$$

So, applying the transformations (10) in (4–8), we obtained the transformed set of ODEs as

$$\left.\begin{aligned} &\frac{A_1}{\vartheta_1\vartheta_2}f''' + ff'' + \frac{A_1A_2}{\vartheta_2}\left(\frac{m+1}{2}\right)f'''(f'')^2 \\ &-\frac{2m}{m+1}\left[(f')^2 - 1\right] - \frac{M}{m+1}(1-f') = 0 \\ &\vartheta_3\theta'' + \vartheta_4 Pr\, f\theta' + PrA_1\ Ec\left[(f'')^2 - \left(\frac{m+1}{6}\right)(f'')^4 A_2\right] \\ &+\frac{4}{3}N_r\frac{d}{d\eta}\left(\left[1+(\theta_w-1)\theta(\zeta)\right]^3\frac{d\theta}{d\eta}\right) = 0 \end{aligned}\right\} \tag{12.11}$$

where $\vartheta_1 = (1-\phi_1)^{2.5}(1-\phi_2)^{2.5}(1-\phi_3)^{2.5}$,

$$\vartheta_2 = (1-\phi_1)\left\{(1-\phi_2)\left[(1-\phi_3) + \phi_3\frac{\rho_{s3}}{\rho_f}\right] + \phi_2\frac{\rho_{s2}}{\rho_f}\right\} + \phi_1\frac{\rho_{s1}}{\rho_f}$$

$$\begin{aligned} \vartheta_3 = &\frac{\kappa_{s3} + (p-1)\kappa_{hnf} - (p-1)\phi_3(\kappa_{hnf} - \kappa_{s3})}{\kappa_{s3} + (p-1)\kappa_{hnf} + \phi_3(\kappa_{hnf} - \kappa_{s3})} \\ &\times\frac{\kappa_{s2} + (p-1)\kappa_{nf} - (p-1)\phi_2(\kappa_{nf} - \kappa_{s2})}{\kappa_{s2} + (p-1)\kappa_{nf} + \phi_2(\kappa_{nf} - \kappa_{s2})} \\ &\times\frac{\kappa_{s1} + (p-1)\kappa_f - (p-1)\phi_1(\kappa_f - \kappa_{s1})}{\kappa_{s1} + (p-1)\kappa_f + \phi_1(\kappa_f - \kappa_{s1})} \end{aligned}$$

$$\vartheta_4 = (1-\phi_1)\left\{(1-\phi_2)\left[(1-\phi_3) + \phi_3\frac{(\rho C_p)_{s3}}{(\rho C_p)_f}\right] + \phi_2\frac{(\rho C_p)_{s2}}{(\rho C_p)_f}\right\} + \phi_1\frac{(\rho C_p)_{s1}}{(\rho C_p)_f}$$

And

(i) Static wedge

$$\left.\begin{array}{l} f(0)=0, f'(0)=0, \theta'(0)=-B_i\left(1-\theta(0)\right) \\ f\rightarrow 1, \theta\rightarrow 0 \ as \ \eta\rightarrow\infty \end{array}\right\} \tag{12.12}$$

(ii) Moving Wedge

$$\left.\begin{array}{l} f(0)=0, f'(0)=S, \theta'(0)=-B_i\left(1-\theta(0)\right) \\ f\rightarrow 1, \theta\rightarrow 0 \ as \ \eta\rightarrow\infty \end{array}\right\} \tag{12.13}$$

Here, $M=\frac{\sigma_f B_0^2}{U_e\rho_f}$ magnetic field parameter, $A_1=\frac{A}{c\nu_f\rho_f}$ and $A_2=\frac{a^3}{2c^2\nu_f}$ are Prandtl-Eyring parameters, and $Pr=\frac{\nu_f}{\alpha_{Thnf}}$ Prandtl number. Thermal radiation parameter $N_r=\frac{4\sigma_1 T_\infty^3(T_w-T_\infty)}{3k^*\kappa_{hbf}}$, and Temperature ratio $\theta_w=\frac{T_w}{T_\infty}$. A moving wedge is defined by S as a moving wedge in the same and inverse direction to the free stream accordingly $S>0$ and $S<0$, while a static wedge is defined by $S=0$ and $B_i=\frac{h_f}{\kappa_{nf}}\sqrt{\frac{\nu_f}{x^{m-1}u_e}}$ is the Biot number.

For engineering benefits, we need to calculate the shear stress and heat transmission at the surface of the sphere, which can be represented as

Skin friction: $C_f=\frac{2\tau_s}{\rho_f u_e^2}, \tau_s=\left(\frac{A}{c}\left(\frac{\partial u}{\partial y}\right)-\frac{A}{6c^3}\left(\frac{\partial u}{\partial y}\right)^3\right)_{y=0}$ which transforms into reduced skin friction $C_{fr}=\frac{1}{2}Cf\,\mathrm{Re}^{\frac{1}{2}}=\sqrt{\frac{m+1}{2}}\left[A_1 f''(0)-A_1A_2\left(f''(0)\right)^3\right]$

Nusselt number: $Nu=\frac{xq_s}{\kappa(T_w-T_\infty)}, q_s=-\kappa\left(\frac{\partial T}{\partial y}\right)_{y=0}$ which becomes

$$Nur=NuRe^{-\frac{1}{2}}=-\sqrt{\frac{m+1}{2}}\left[1+\frac{4}{3}N_r\left[1+(\theta_w-1)\theta(0)\right]^3\right]\theta'(0)$$

12.2 NUMERICAL SOLUTION METHODOLOGY

It has been mathematically addressed using two efficient mathematical strategies, the shooting technique with the Fehlberg equation and Newton's strategies, to solve the nonlinear ordinary differential equation (12.11) with appropriate boundary conditions (12.12) and (12.13). The choice of the

Table 12.4 Values of skin friction for different values of '*m*'

m	*Mukhopadhyay et al. (2013)*	*Raju and Sandeepa (2018)*	*Ullah et al. (2018)*	*Present result*
−0.05	0.213802	0.2134791	0.21321	0.2134561
0.0	0.332206	0.3332321	0.33206	0.3334591
0.333	0.757586	0.7574542	0.75743	0.7573243
1	1.232710	1.23258887	1.23259	1.2325442

Runge-Kutta-Fehlberg (RKF) numerical method for solving the governing equations in the context of the study offers several advantages that are particularly relevant to the specific requirements and objectives of the research. One of the key advantages of the RKF method is its ability to adaptively adjust the step size during the numerical integration process. This method offers high accuracy and stability across a wide range of problem scenarios. In comparison to other numerical methods commonly used in similar research, such as finite difference, finite element, or finite volume methods, the RKF method often exhibits superior performance in terms of accuracy, efficiency, and computational cost. While each numerical method has its own strengths and limitations, the adaptive nature and high accuracy of the RKF method make it particularly. The simulation is run using MAPLE-17 software. A thorough mathematical calculation is performed for various upsides of the relevant boundaries, in particular the magnetic field parameter M, Prandtl-Eyring parameters A_1 and A_2, and thermal radiation Nr. In select cases, a correlation with the current writing is also performed to demonstrate the credibility of the mathematical outcomes. The measurements of Cfr for various values of 'm' are calculated numerically from the first equation of (12.11) considering $M = A_{1=}A_2\ Nr = Ec = 0$ and compared with the results deduced by Mukhopadhyay et al. (2013), Raju and Sandeepa (2018) (Table 12.4). Here we observe (Table 12.4) a high level of agreement with what was previously mentioned. A flow chart is illustrated in Figure 12.2.

12.3 RESULT AND DISCUSSION

The nonlinear system of differential equations (12.10) along with the boundary conditions (12.12)–(12.13) is solved numerically. The outcomes of the arduous calculations are staged here through proper graphs and charts, considering $m = 0.25$ and $Pr = 6.2$ otherwise. Our main purpose is to observe the influence of Prandtl-Eyring parameters (A_1 and A_2), magnetic field parameter (M), thermal radiation parameter (Nr), temperature ratio (θ_w), wedge parameter (S), nanoparticle shape effects parameter (p), Biot number (B_i), Hartree pressure gradient m, and nanoparticle volume

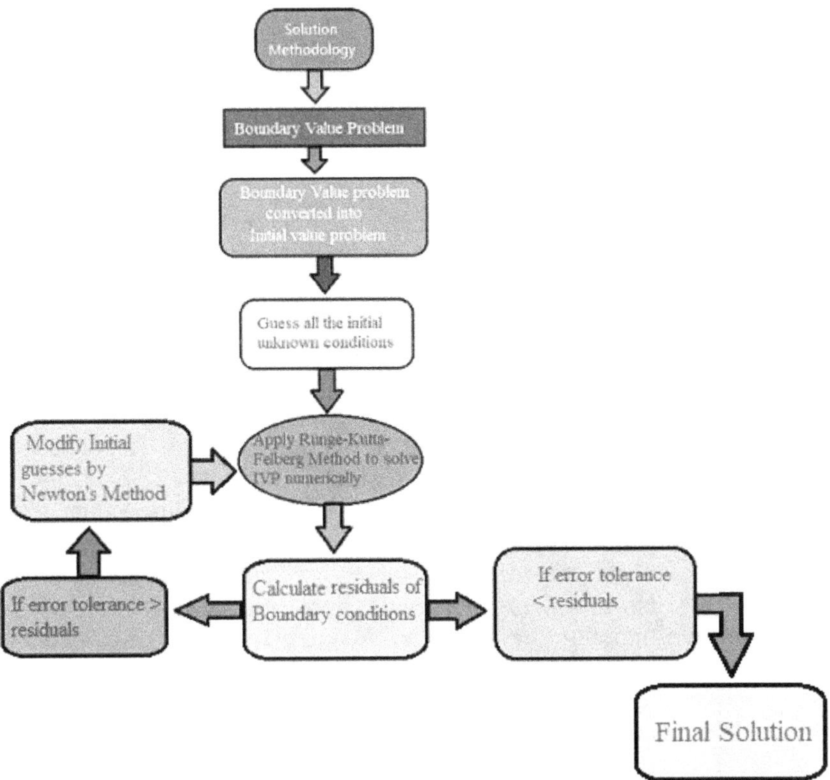

Figure 12.2 Flow chart of the RK-4 base shooting method.

fraction $(\phi_1 = \phi_2 = \phi_3)$ on the fluid velocity $f'(\eta)$, temperature $(\theta(\eta))$ and sum up the overall performance of the flow system to optimize the decay in the process. Velocity and temperature profiles for mono, hybrid, and ternary hybrid nanofluids are plotted with solid, dotted, and dashed curves, respectively, and these curves are shown in Figures 12.3–12.5.

12.3.1 Influence of parameters on the velocity profile

Figures 12.3a–12.3f highlight the presentation of velocity profile curves $f'(\eta)$ of mono-nanofluid (Fe_3O_4-H_2O), hybrid nanofluid (Fe_3O_4-Ag-H_2O), and ternary hybrid nanofluid (Fe_3O_4-Ag-Al_2O_3-H_2O) against the variation of the Prandtl-Eyring parameters (A_1 and A_2), magnetic field parameter (M), Hartree pressure gradient m, nanoparticle volume fraction $(\phi_1 = \phi_2 = \phi_3)$, and wedge parameter ($S$), respectively. It can be seen in Figure 12.3a that as

Prandtl-Eyring parameter A_1 increases, the angle of slope of fluid velocity is trimmed down, causing the hydrodynamic boundary layer width to decrease rapidly. In the context of the Prandtl-Eyring ternary hybrid nanofluid model, the velocity field is determined by the balance between the applied shear stress and the rheological response of the fluid. The Prandtl-Eyring fluid model relates shear rate (γ), shear stress (τ), and dynamic viscosity (ν) through a constitutive equation, $\tau = \nu\left(\frac{\gamma}{\lambda}\right)\left(1 + c\exp\left(\frac{\sigma}{\nu}\right)\right)$. The Prandtl-Eyring fluid parameters collectively influence the balance between viscous and elastic behavior in a fluid. These parameters impact how the fluid responds to applied shear stress and deformations, which, in turn, affects the velocity field. It is vividly clear from the figure that the reduction rate is much quicker for hybrid nanofluid (Fe_3O_4-Ag-H_2O) in comparison to mono-nanofluid (Fe_3O_4-H_2O) and ternary hybrid nanofluid (Fe_3O_4-Ag-Al_2O_3-H_2O). It has been observed that the same effect is occurring for Prandtl-Eyring fluid parameter $A2$ as well (Figure 12.3b). Thus, the velocity profile shows a diminishing trend for the increasing values of A_1 and A_2, and consequently, the momentum boundary layer decreases.

The magnetic field parameter can have significant effects on nanofluid flow. It is possible to induce a magnetic dipole moment in nanoparticles suspended in a nanofluid by applying a magnetic field, which can change its flow behavior. Due to the formation of chains or clusters of nanoparticles, a magnetic field can increase nanofluid viscosity. These structures can obstruct the fluid flow and increase the fluid's resistance to deformation, leading to an increase in viscosity. As a result, the flow velocity gradually lessens with more powerful for M, as shown in Figure 12.3c. When nanofluids are subjected to a magnetic field, they exhibit MHD behavior. This means that the fluid motion is influenced by electromagnetic forces. Nanofluids' velocity is altered by the Lorentz force, in particular. There are a number of factors that can affect nanofluid velocity, including the magnitude, orientation, and properties of the magnetic field, nanoparticle properties, and the design of the system. Researchers study these effects to understand and control nanofluid behavior in applications ranging from HT enhancement to drug delivery systems and microfluidics. A close examination of the figure demonstrates that the reduction rate is much faster for hybrid nanofluids (Fe_3O_4-Ag-H_2O) compared to mono-nanofluids (Fe_3O_4-H_2O) and ternary hybrid nanofluids (Fe_3O_4-Ag-Al_2O_3-H_2O). Figure 12.3d illustrates how the Hartree pressure gradient (m) is influenced by velocity. The fluid velocity seems to escalate for Hartree pressure gradient m.

A plot of the velocity variations for nanoparticle concentration is shown in Figure 12.3e. Velocity seems to incline toward higher concentrations. A highly concentrated hybrid nanofluid (Fe_3O_4-Ag-H_2O) becomes denser and has a greater viscosity in comparison to mono-nanofluid (Fe_3O_4-H_2O) and ternary hybrid nanofluid (Fe_3O_4-Ag-Al_2O_3-H_2O), which impedes smooth transport over the intricacies of geometries. Consequently, hybrid

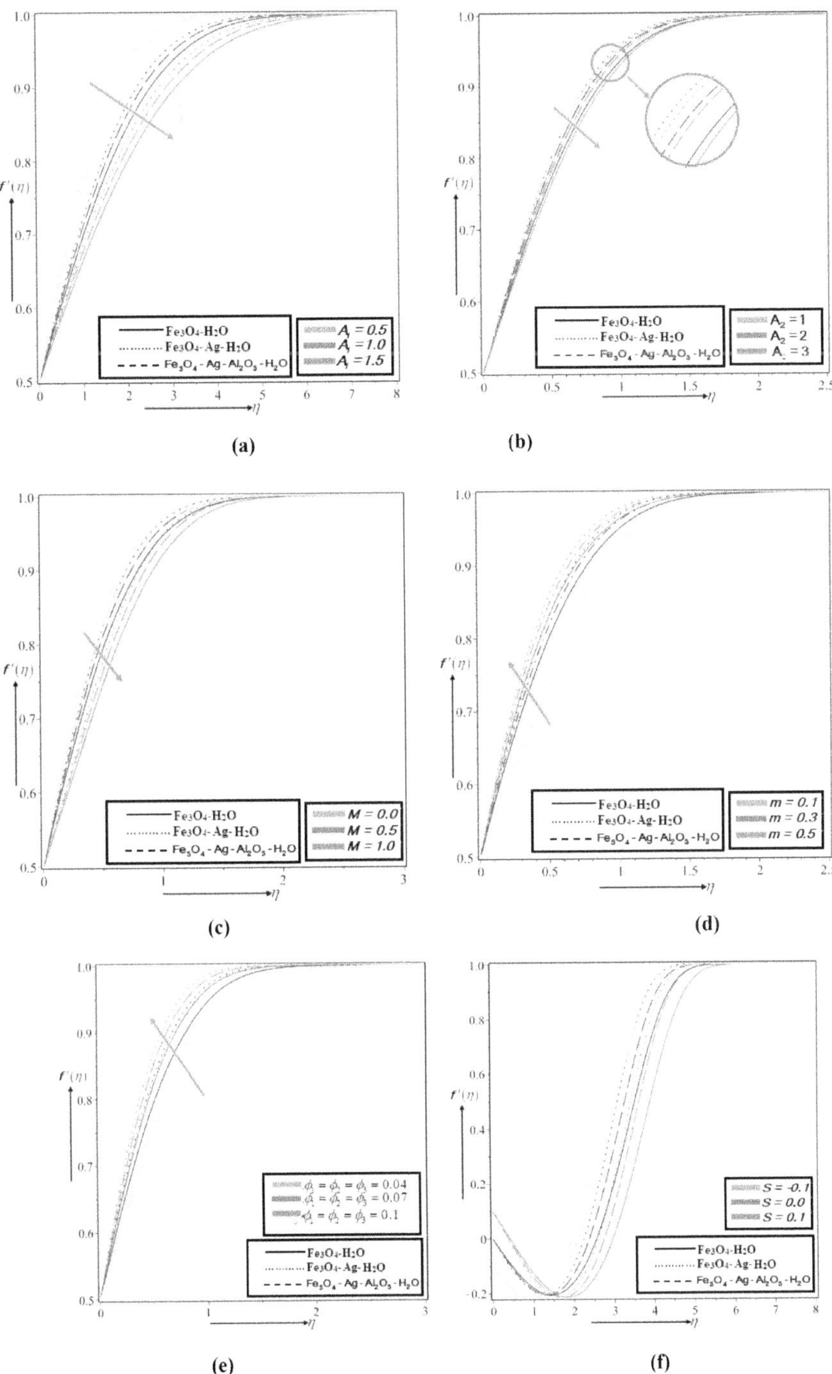

Figure 12.3 velocity profiles for (a) A_1, (b) A_2, (c) M, (d) m, (e) $\phi_1 = \phi_2 = \phi_3$, (f) S.

nanofluids exhibit high velocity compared to mono-nanofluids and ternary hybrid nanofluids because existing double-tiny ingredients (Fe_3O_4 and Ag) make hybrid nanofluids comparatively much denser. The effects are clear and distinct. Typically, the velocity variation with nanoparticle concentration can vary depending on factors such as nanoparticle type, concentration range, base fluid properties, and flow conditions.

The influence of the wedge parameter (S) on velocity profiles is portrayed in Figure 12.3f. The fluid velocity seems to incline for the wedge parameter $(0 \leq \eta \leq 1)$ and decline in the region $\eta > 1$. The wedge parameter, often denoted as S, is a geometric parameter that characterizes the shape of an obstacle or body in fluid flow, particularly in the context of fluid mechanics and aerodynamics. It is commonly used when studying flow around solid objects, such as wedges, air foils, or other streamlined or non-streamlined shapes. The effect of the wedge parameter on nanofluid velocity would depend on several factors, including the specific flow conditions, fluid properties, and the shape and orientation of the wedge. Wedges are often described by wedge angles (Ω), which are angles between their leading edges and flow directions. There can be a significant variation in flow patterns around the wedge as the wedge angle changes. For small wedge angles (Ω close to 0), the flow may resemble more of a flat plate flow. For larger wedge angles, the flow separation and recirculation regions may become more prominent. It is vividly clear from the figure that the increment rate is much quicker for hybrid nanofluid (Fe_3O_4-Ag-H_2O) in comparison to mono-nanofluid (Fe_3O_4-H_2O) and ternary hybrid nanofluid (Fe_3O_4-Ag-Al_2O_3-H_2O).

12.3.2 Influence of parameters on the temperature profile

Figure 12.4 delineates the presentation of temperature profile curves $\theta(\eta)$ of mono-nanofluid (Fe_3O_4-H_2O), hybrid nanofluid (Fe_3O_4-Ag-H_2O), and ternary hybrid nanofluid (Fe_3O_4-Ag-Al_2O_3-H_2O) against the variation of the Prandtl-Eyring fluid parameters, Biot number, temperature ratio parameter, Hartree pressure gradient, thermal radiation, nanoparticle shape factor, nanoparticle volume fraction, and wedge parameter, respectively, in Figure 12.4a–i.

The effect of the Prandtl-Eyring fluid parameters on a nanofluid can enhance its thermal conductivity due to the alignment of the nanoparticles in the direction of the magnetic field. This can lead to improved HT performance in various applications. As illustrated in Figure 12.4a and b, higher $A1$ and $A2$ values raise the flow temperature. It's important to note that the specific temperature dependence of these parameters can vary widely depending on the nature of the fluid being modeled. Empirical data, experimentation, or additional temperature-dependent models may be required to accurately capture and predict the behavior of Prandtl-Eyring fluids under different temperature conditions. In practical applications, understanding the temperature sensitivity of fluid parameters is crucial, especially when dealing with non-Newtonian fluids that may exhibit complex behavior under varying thermal conditions. It is vividly clear from the figure that the increment rate is much quicker for ternary hybrid nanofluid (Fe_3O_4-Ag-Al_2O_3-H_2O) in comparison to mono-nanofluid (Fe_3O_4-H_2O) and hybrid nanofluid (Fe_3O_4-Ag-H_2O).

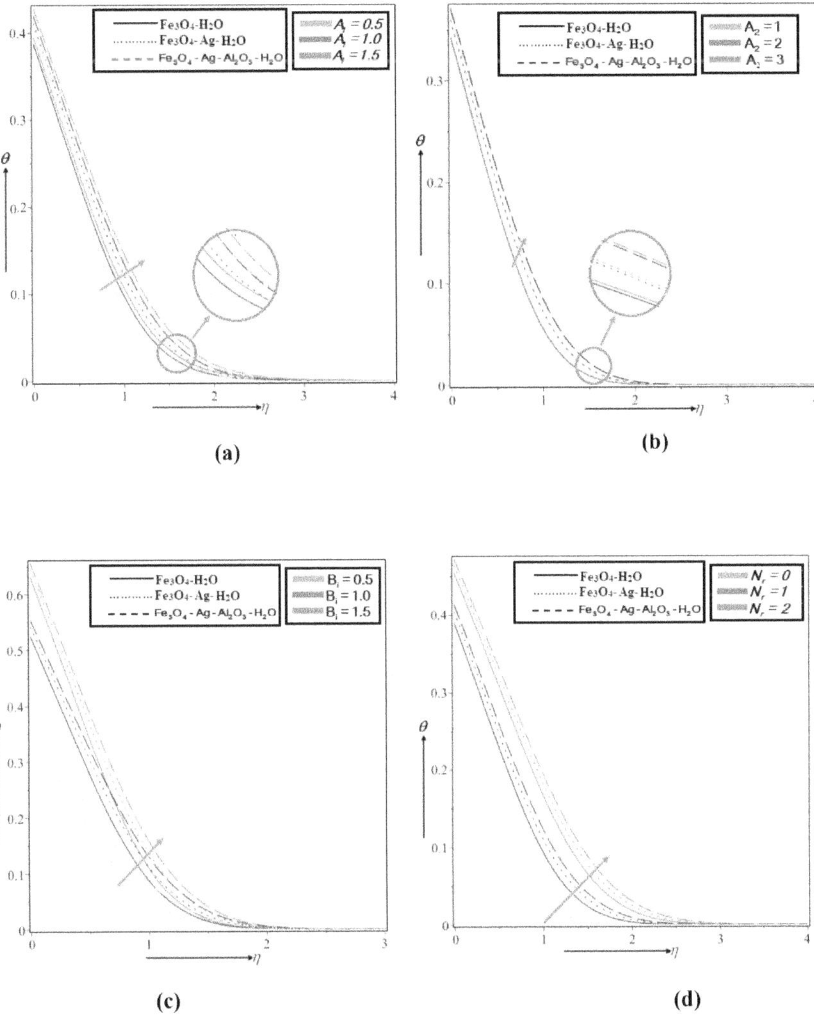

Figure 12.4 Temperature profiles for (a) A_1, (b) A_2, (c) B_i, (d) Nr, (e) θ_w, (f) m, (g) p, (h) $\phi_1 = \phi_2 = \phi$ and (i) S.

(Continued)

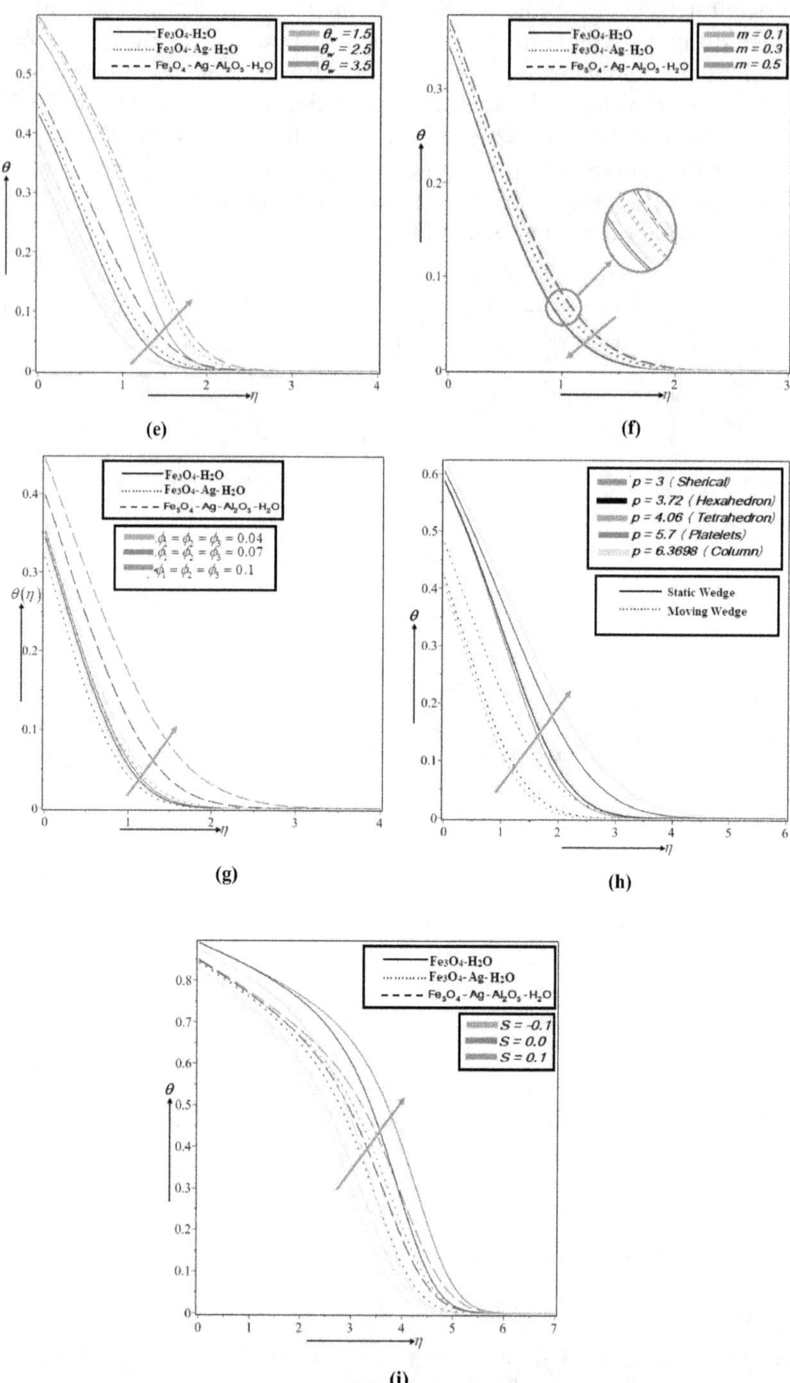

Figure 12.4 (Continued) Temperature profiles for (a) A_1, (b) A_2, (c) B_1, (d) Nr, (e) θ_w, (f) m, (g) p, (h) $\phi_1 = \phi_2 = \phi_3$ and (i) S.

Figure 12.4c illustrates how Biot number (B_i) affects the temperature field $\theta(\eta)$. As expected, the temperature field is increasing for rising values of B_i. The Biot number is used to assess the dominance of conduction or convection in a HT process. In practical terms, understanding the Biot number is crucial when analyzing HT between solids and fluids, such as in situations involving heat exchangers, cooling of electronic components, or any system where a solid is in contact with a fluid. The Biot number helps engineers and scientists determine which mode of HT predominates and aids in designing efficient heat exchange systems. There is a significant difference in the Biot number between mono-nanofluids (Fe_3O_4-H_2O), hybrid nanofluids(Fe_3O_4-Ag-H_2O), and ternary hybrid nanofluids (Fe_3O_4-Ag-Al_2O_3-H_2O).

Figure 12.4d describes thermal radiation's impression on the thermal profile. Temperature seems to escalate for radiation. Ternary hybrid nanofluids (Fe_3O_4-Ag-Al_2O_3-H_2O) explore a higher thermal profile compared to mono-nanofluids (Fe_3O_4-H_2O) and hybrid nanofluids (Al_2O_3-Ag-H_2O). A heightening in thermal radiation speeds the frequent migration of nanoparticles, and thus random frictional collisions among the metallic ingredients are increased. The kinetic energy generated by molecular friction is converted into thermal energy by continuing friction. As a result, the temperature rises. Similarly, Figure 12.4e shows a swell in the temperature ratio factor.

The Hartree pressure gradient temperature variations are shown in Figure 12.4f. The fluid temperature seems to be a diminishing trend for the Hartree pressure gradient m. Figure 12.4g depicts the temperature variations for nanoparticle concentration. The thermal trajectory swells for nanoparticle concentration. Ternary hybrid nanofluids (Fe_3O_4-Ag-Al_2O_3-H_2O) and mono-nanofluids (Fe_3O_4-H_2O) exhibit an elevated temperature, while hybrid nanofluids (Fe_3O_4-Ag-H_2O) exhibit a declining temperature. Triple nanoparticles' existence maximizes the thermal effects, and hence ternary hybrid nanofluidic mediums receive a high thermal magnitude compared to mono-nanofluids and hybrid nanofluids. Figure 12.4h depicts the behavior of different shape factors (p) such as spherical, hexahedrons, tetrahedrons, platelets, and columns, as a function of temperature. Several shape factors are found to have a positive impact on the energy profile of ternary hybrid nanofluids in the presence of static and moving wedges. The Impact of the wedge parameter (S) on the temperature field $\theta(\eta)$ is highlighted in Figure 12.4i. As expected, the temperature field is increasing for rising values of S. Mono-nanofluids (Fe_3O_4-H_2O) acquire a high magnitude compared to ternary hybrid nanofluids (Fe_3O_4-Ag-Al_2O_3-H_2O) and hybrid nanofluids (Fe_3O_4-Ag-H_2O).

12.3.3 Analysis of Skin friction coefficients(C_{fr})

In Figures 12.5a–12.5c, the skin friction C_{fr} is displayed in 3D color plots for numerous parameters M and A_1 for mono-nanofluids (Fe_3O_4-H_2O), hybrid nanofluids (Fe_3O_4-Ag-H_2O), and ternary hybrid nanofluids

(Fe_3O_4-Ag-Al_2O_3-H_2O), respectively. Enlargement of C_{fr} can be observed with declining values of $M(0 \le M \le 1)$ and $A_1(0.5 \le A_1 \le 1.5)$ for Fe_3O_4-H_2O, Fe_3O_4-Ag-H_2O and Fe_3O_4-Ag-Al_2O_3-H_2O in Figure 12.5a–c. It can be seen from the figure that the skin friction rate is maximum between M (= 0 approx.) and A_1 (= 0.5 approx.) and the maximum value is 0.4833 for ternary hybrid nanofluids (Fe_3O_4-Ag-Al_2O_3-H_2O). It is noted that the skin friction rate has a greater effect on ternary hybrid nanofluid (Fe_3O_4-Ag-Al_2O_3-H_2O) flow compared to the mono-nanofluids (Fe_3O_4-H_2O) and hybrid nanofluids (Fe_3O_4-Ag-H_2O) owing to the presence of triple tiny particles. In the framework of magnetohydrodynamics (MHD), the study of the magnetic properties

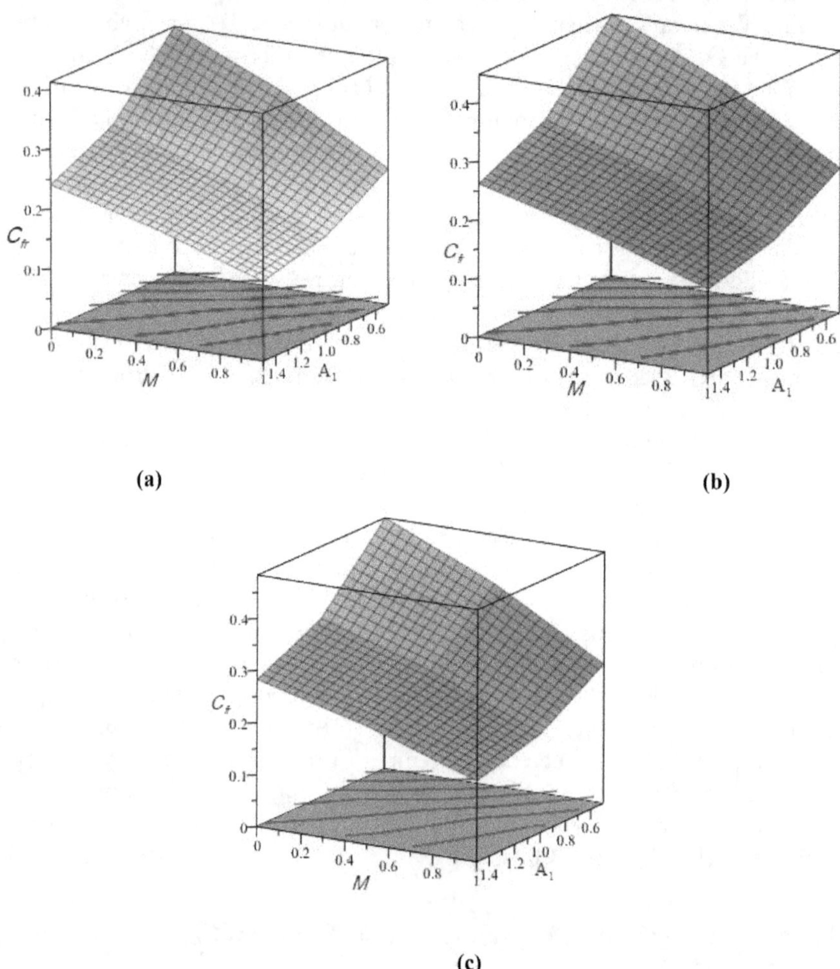

Figure 12.5 3D view of skin friction for (a) mono-nanofluids, (b) hybrid nanofluids, and (c) ternary hybrid nanofluids M vs. A_1.

and behavior of electrically conducting fluids, including as plasmas, liquid metals, and certain electrolytes, the relationship between skin friction and a magnetic field can be investigated. The interaction between a magnetic field and a flowing electrically conductive fluid can influence the skin friction or drag force experienced by the fluid. This relationship is particularly significant in MHD applications, where the interaction between magnetic fields and conducting fluids can lead to unique flow phenomena and influence skin friction patterns.

Figure 12.6a–12.6c delivers the 3D view of the drag friction C_{fr} for S and A_2. It is noticed that drag friction C_{fr} enlarges for decreasing values of A_2

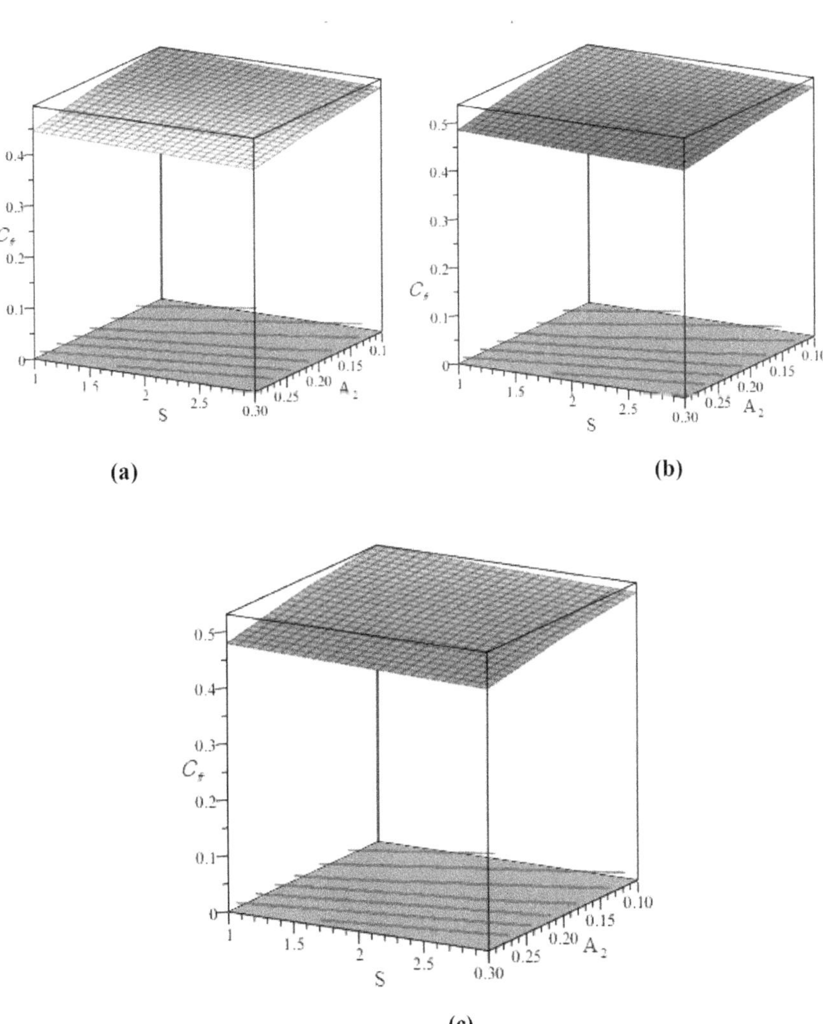

Figure 12.6 3D view of skin friction for (a) mono-nanofluids, (b) hybrid nanofluids, and (c) ternary hybrid nanofluids S vs. A_2.

and S. Thus, by swelling A_2, comparatively less drag is qualified by fluid. It can be seen from the figure that the skin friction rate is maximum between S (= 1 approx.) and A_2 (= 0.1 approx.) and the maximum value is 0.5368 for hybrid nanofluids (Fe_3O_4-Ag-H_2O). It's important to note that the specific relationship between skin friction and a wedge is highly dependent on the specific flow conditions, including the Reynolds number, Mach number, and the characteristics of the boundary layer. Analytical methods, empirical correlations, and numerical simulations are often employed to study and predict skin friction on wedge-shaped objects in different flow scenarios. The study of skin friction is essential in aerodynamics, fluid dynamics, and the design of vehicles and aerodynamic surfaces.

12.3.4 Analysis of local Nusselt number (*Nur*)

Figures 12.7a–c decorate the graphical interaction between mono, hybrid, and ternary nanofluid parameters Nr and θ_w on Nur. Reduction of Nur can be observed with growing values of $Nr\,(0 \le S \le 2)$ and declining values of $\theta_w\,(1.5 \le \beta \le 3.5)$ for Fe_3O_4-H_2O, Fe_3O_4-Ag-H_2O, and Fe_3O_4-Ag-Al_2O_3-H_2O in Figure 12.7a–c. It can be seen from the figure that the HT rate is maximum between Nr (= 0 approx.) and θ_w (=1.5 approx.) and the maximum rate is 55.90% for ternary hybrid nanofluids (Fe_3O_4-Ag-Al_2O_3-H_2O), whereas for HT, rates of Fe_3O_4-H_2O and Fe_3O_4-Ag-H_2O are 54.75% and 54.56%, respectively. The relationship between the Nusselt number and thermal radiation is typically associated with combined modes of HT, where both convection and radiation contribute to the overall HT. This often occurs in scenarios where surfaces exchange heat with their surroundings at different temperatures. It's important to note that the contribution of radiation to HT becomes more significant at higher temperatures or in scenarios with transparent media. In practical applications, the combined Nusselt number is useful for assessing the overall HT performance in situations where both convection and radiation are important.

Three-dimensional color plots of Nur are shown in Figure 12.8a–c for active parameters B_i and A_1 for Fe_3O_4-H_2O, Fe_3O_4-Ag-H_2O, and Fe_3O_4-Ag-Al_2O_3-H_2O, respectively. Enlargement of Nur can be observed with growing values of $B_i\,(0.5 \le Nr \le 1.5)$ and decline values of $A_1\,(0.5 \le \beta \le 1.5)$ for Fe_3O_4-H_2O, Fe_3O_4-Ag-H_2O, and Fe_3O_4-Ag-Al_2O_3-H_2O. The rate of HT is maximum for mono-nanofluids, which is 53.15%, whereas for hybrid nanofluids, it is 51.34%, and for ternary hybrid nanofluids, it is 50.30%. Therefore, the HT rate is higher for mono-nanofluids compared to hybrid and ternary hybrid nanofluids. It's important to note that the specific relationship between the Biot number and the HT rate depends on the configuration of the system, the nature of the materials involved, and the

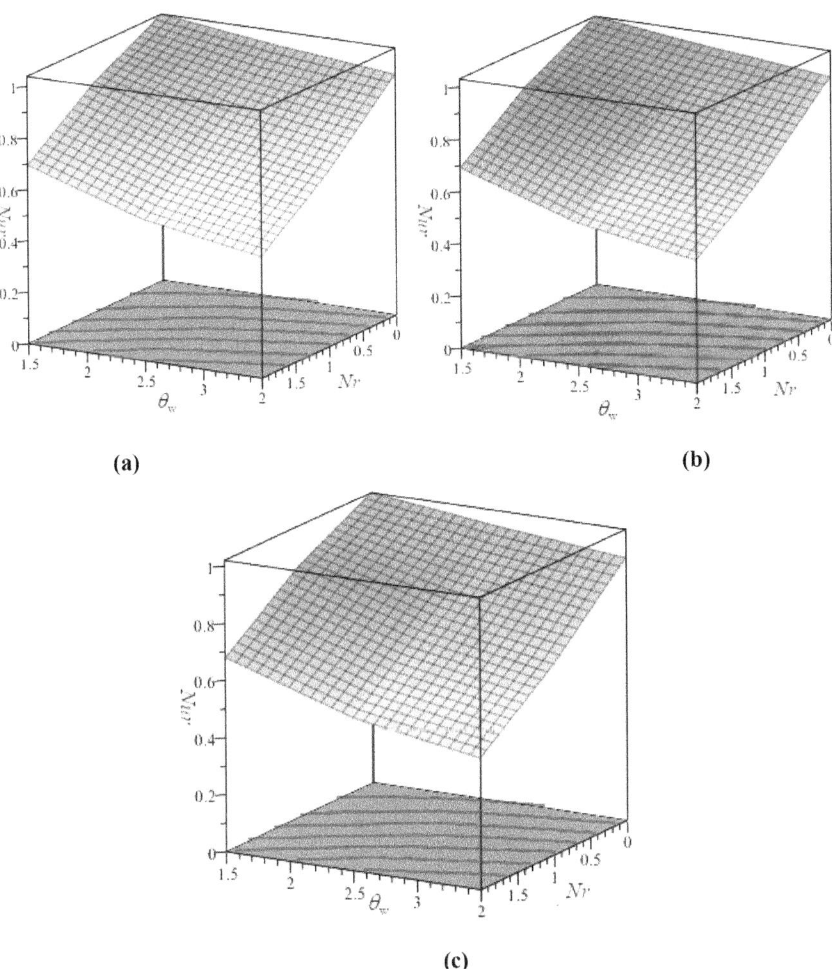

Figure 12.7 3D view of Nusselt numbers for (a) mono-nanofluids, (b) hybrid nanofluids, and (c) ternary hybrid nanofluids Nr vs. θ_w.

boundary conditions. Analyzing the Biot number is essential in designing and understanding HT processes, especially when considering the interactions between conduction and convection in a system.

Figure 12.8a–c illustrates two-dimensional streamline variation for ternary hybrid magnetized nanofluidic transport, showing that stream function values are reduced for high Prandtl-Eyring fluid parameters ($A1$) (Figure 12.9).

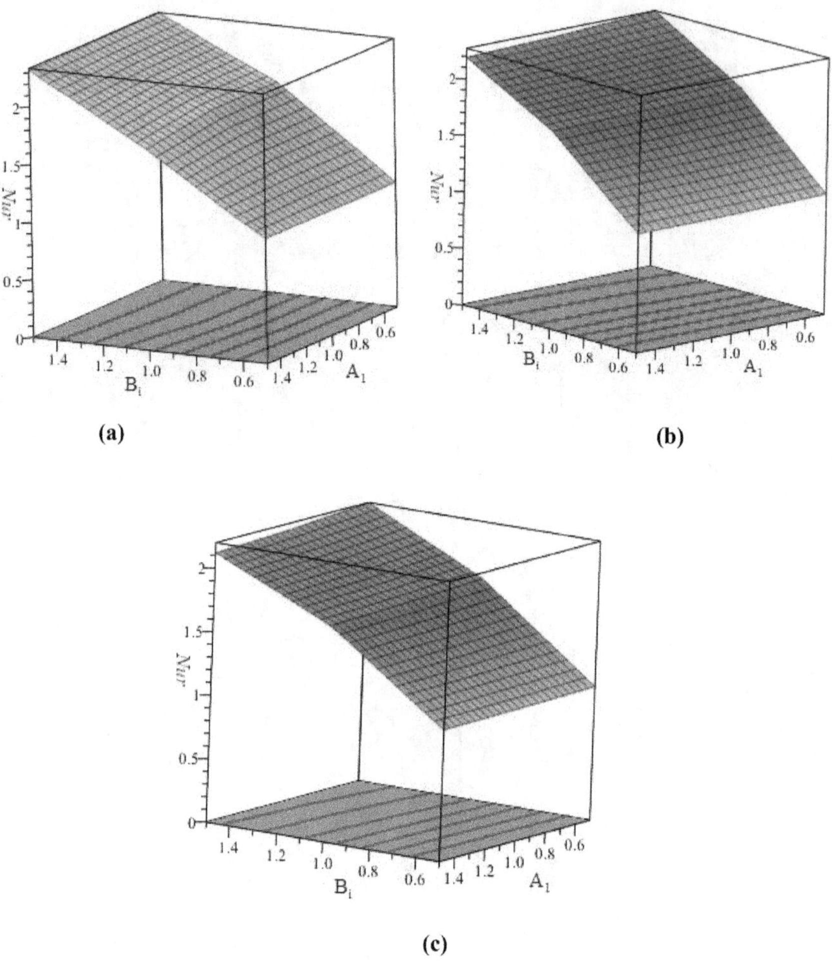

Figure 12.8 3D view of Nusselt numbers for (a) mono-nanofluids, (b) hybrid nanofluids, and (c) ternary hybrid nanofluids B_i vs. A_l.

12.4 CONCLUSIONS

The Fe_3O_4-Ag-Al_2O_3-H_2O-based incompressible MHD Prandtl-Eyring ternary hybrid nanofluidic transport over a static and moving wedge, assuming the existence of magnetic fields, is examined numerically throughout the study. Thermal radiation and nanoparticle shape effects are the external factors whose influence is considered during the modeling and simulation of the work. After reducing the foremost equations into dimensionless form, the shooting-based RKF technique is employed to run the simulation. Several streamlines, 3D plots, and graphs are illustrated to enlighten the

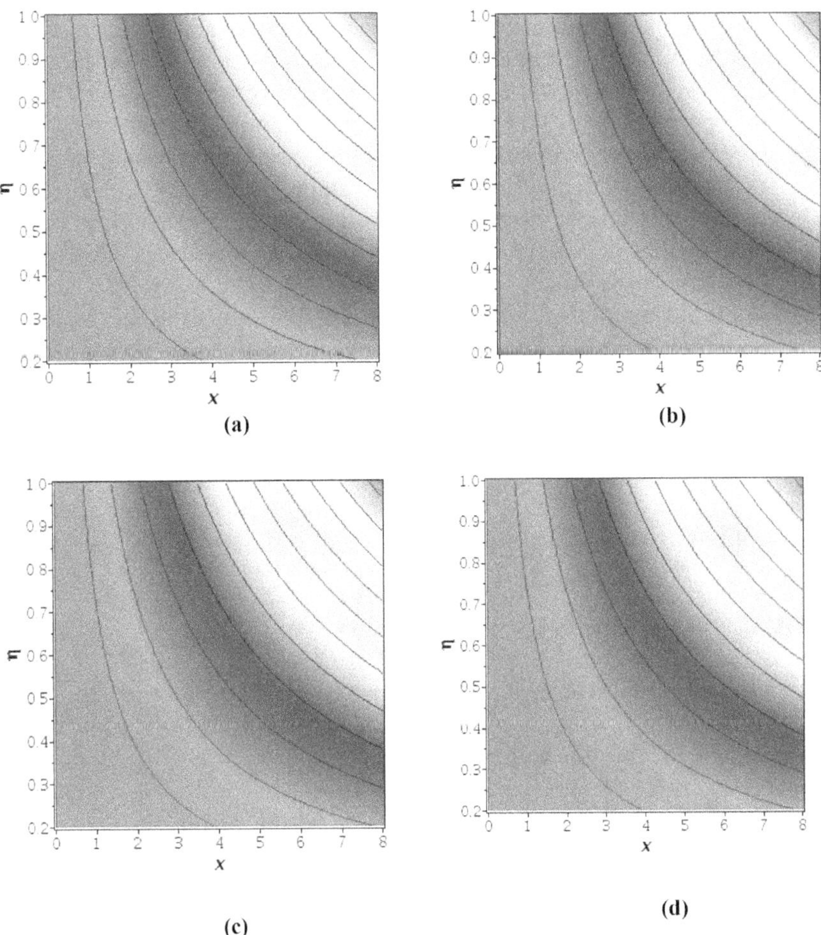

Figure 12.9 Streamline designs for (*a*) $A_1 = 0.5$, (*b*) $A_1 = 1.5$, (*c*) $A_1 = 2.5$ and (*d*) $A_1 = 3.5$.

noteworthy fallouts of the investigation. The core conclusions that can be drawn from the study are as follows:

- The velocity is a decreasing function of the Prandtl-Eyring fluid parameters and the magnetic field parameter. The reduction rate is much quicker for hybrid nanofluid (Fe_3O_4-Ag-H_2O) in comparison to mono-nanofluid (Fe_3O_4-H_2O) and ternary hybrid nanofluid (Fe_3O_4-Ag-Al_2O_3-H_2O). However, velocity explores the reverse scenario for the Hartree pressure gradient, wedge parameter, and nanoparticle volume fraction.

- Thermal profiles are inclined for the Prandtl-Eyring fluid parameters, Biot number, thermal radiation parameter, temperature ratio parameter, nanoparticle shape effect parameter, nanoparticle concentration, and wedge parameter, while the Hartree pressure gradient exhibits the opposite trend. Ternary hybrid nanofluids depict high magnitudes.
- Heat transference upsurges for radiation and Biot number, while the reverse trend is marked for temperature ratio and Prandtl-Eyring fluid parameter (A_1). Ternary hybrid nanofluids explore the highest magnitude in heat transport and thus prove to be the better candidate for heat transmission.
- The fluid qualifies the swelling Prandtl-Eyring parameters as having low drag. The skin friction rate has a greater effect on ternary hybrid nanofluid flow compared to mono-nanofluids and hybrid nanofluids, owing to the presence of triple tiny particles.

The proposed study should be extended to include fuzzy volume fraction, ternary hybrid nanofluid flow toward stretching wedges, and entropy generation analysis in the future.

ACKNOWLEDGMENT

The authors wish to extend their heartfelt gratitude to the Honorable Reviewers for their valuable comments and advice, which have significantly improved the quality of the present paper.

REFERENCES

Abu-Hamdeh, N. H., Alsulami, R. A., Rawa, M. J., Alazwari, M. A., Goodarzi M., and Safaei, M. R. A. (2021). Significant solar energy note on Powell-Eyring nanofluid with thermal jump conditions: Implementing Cattaneo-Christov heat flux model. *Mathematics*, 9(21), 2669.

Acharya, N., and Mabood, F. (2021). On the hydrothermal features of radiative Fe_3O_4-graphene hybrid nanofluid flow over a slippery bended surface with heat source/sink. *Journal of Thermal Analysis and Calorimetry*, 143, 1273–1289.

Acharya, N., Maity, S., and Kundu, P. K. (2019). Framing the hydrothermal features of magnetized TiO_2-$CoFe_2O_4$ water-based steady hybrid nanofluid flow over a radiative revolving disk. *Multidiscipline Modeling in Materials and Structures*, 16(4), 765–790.

Al Oweidi, K. F., Shahzad, F., Jamshed, W., Usman, W. Ibrahim, R., El Sayed M., El Din T., and M. AlDerea, A.(2022). Partial differential equations of entropy analysis on ternary hybridity nanofluid flow model via rotating disk with hall current and electromagnetic radiative influences. *Scientific Reports*, 12(1), 20692.

Basha, T., Sivaraj, R., Reddy, A. S., and Chamkha, A. J. (2019). SWCNH/diamond-ethylene glycol nanofluid flow over a wedge, plate and stagnation point with induced magnetic field and nonlinear radiation-solar energy application. *The European Physical Journal Special Topics*, 228, 2531–2551.

Chakraborty, T., Duari, P. R., and Acharya, N. (2023). Unfolding flow features of MHD hybrid nanofluid (Ag-Al_2O_3-H_2O) and mono-nanofluid (Al_2O_3-H_2O) flow over exponentially expanded sheet soaked in a Darcy-Forchheimer *absorbent medium coexisting non-uniform heat generation/*absorption. *Waves in Random and Complex Media.* https:/doi.org/10.1080/17455030.2023.2171153.

Choi, S.U.S. (1995). Enhancing thermal conductivity of fluids with nanoparticles in developments and applications of non-Newtonian flows. *FED* 231/*MD*, 66, 99–105.

Das, K., Acharya, N., Md SK, T., Duari, P. R., and Chakraborty T. (2022). Slip flow of hybrid nanofluid in presence of solar radiation. International *Journal of Modern Physics C*, 33(2), 2250017.

Das, K.. Duari, P. R., and Kundu, P. K. (2014). Solar radiation effect on Cu-water nanofluid flow over a stretching sheet with surface slip and temperature jump. Arabian *Journal for Science and Engineering*, 39, 9015–23.

Ghadikolaei, S. S., and Gholinia, M. (2019). Terrific effect of H_2 on $_3$D free convection MHD flow of $C_2H_6O_2$-H_2O hybrid base fluid to dissolve Cu nanoparticles in a porous space considering the thermal radiation and nanoparticle shapes effects. *International Journal of Hydrogen Energy*, 44, 17072–17083.

Ghadikolaei, S. S., Hosseinzadeh, Kh., Hatami, M., Ganji, D. D., Armin, M. (2018a). Investigation for squeezing flow of ethylene glycol ($C_2H_6O_2$) carbon nanotubes (CNTs) in rotating stretching channel with nonlinear thermal radiation. *Journal of Molecular Liquids*, 263, 10–21.

Ghadikolaei, S. S., Hosseinzadeh, Kh., and Ganji, D. D. (2018b). Investigation on three-dimensional squeezing flow of mixture base fluid (ethylene glycol-water) suspended by hybrid nanoparticle (Fe_3O_4-Ag) dependent on shape factor. *Journal of Molecular Liquids*, 262(15), 376–388.

Gorai, D., Duari, P. R., and Das, K. (2024). Influence of variable characteristic of the porous medium on unsteady nanofluid flow with melting heat transference. *Numerical Heat Transfer, Part A: Applications*, 85(5), 702–718.

Hamilton, R., Crosser, O. (1962). Thermal conductivity of heterogeneous two-component systems. *Industrial and Engineering Chemistry Fundamentals*, 1, 187–191.

Hayat, T., Ullah, I., Muhammad K., and Alsaedi, A. (2021). Gyrotactic microorganism, and bio-convection during flow of Prandtl-Eyring nanomaterial. *Nonlinear Engineering*, 10(1), 201–212.

Imran, M., Yasmin, S., Waqas, H., Ali Khan, S., Muhammad, T., Alshammari, N., Hamadneh, N. N., and Khan, I. (2022). Computational analysis of nanoparticle shapes on hybrid nanofluid flow due to flat horizontal plate via solar collector. *Nanomaterials*, 12, 663.

Jalili, P., Azar, A.A., Jalili, B., and Ganji, D. D. (2023). Study of nonlinear radiative heat transfer with magnetic field for non-Newtonian Casson fluid flow in a porous medium. *Results in Physics*, 48, 106371.

Kandasamy, R., Muhaimin, I., Khamis, A. B., and Roslan, R. B. (2013). Unsteady Heimenz flow of Cu nanofluid over a porous wedge in the presence of thermal stratification due to solar energy radiation: Lie group transformation. *International Journal of Thermal Sciences*, 65, 196–205.

Khafajy, D. G. S. A., and Kaabi, W. A. (2021). Radiation and mass transfer effects on inclined MHD oscillatory flow for Prandtl-Eyring fluid through a porous channel. *Al-Qadisiyah Journal of Pure Science*, 26(4), 347- 363.

Khashi'ie, N. S., Md Arifin, N., Sheremet, M., Pop, I. (2021). Shape factor effect of radiative Cu-Al_2O_3/H_2O hybrid nanofluid flow toward an EMHD plate. *Case Studies in Thermal Engineering*, 26, 101199.

Khashi'ie, N.S., Md Arifin, N., and Pop I. (2022). Magnetohydrody.namics (MHD) boundary layer flow of hybrid nanofluid over a moving plate with Joule heating. *Alexandria Engineering Journal*, 61(3), 1938–1945.

Khatun, A., and Islam, T. (2022). Influence of magnetic field and heat generation/absorption on unsteady MHD convective flow along a permeable stretching/shrinking wedge with thermophoresis and variable fluid properties. *International Journal of Thermofluids*, 16, 100204.

Mahmood Z., and Khan, U. (2023). Unsteady three-dimensional nodal stagnation point flow of polymer-based ternary-hybrid nanofluid past a stretching surface with suction and heat source. *Science Progress*, 106(1), 0036850 4231152741.

Mahmood, Z., Ahammad, N., Alhazmi, S. E., Khan, U., and Bani-Fwaz, M. Z. (2022a). Ternary hybrid nanofluid near a stretching/ shrinking sheet with heat generation/absorption and velocity slip on unsteady stagnation point flow. *International Journal of Modern Physics B*, 36(29), 2250209.

Mahmood, Z., Alhazmi, S. E., Khan, U., Bani-Fwaz, M. Z., Galal, A.M. (2022b). Unsteady MHD stagnation point flow of ternary hybrid nanofluid over a spinning sphere with Joule heating. *International Journal of Modern Physics B*, 36(32), 2250230.

Mahmood, Z., Khan, U., Saleem, S., Rafique, K., and Eldin, S. M.(2023). Numerical analysis of ternary hybrid nanofluid flow over a stagnation region of stretching/shrinking curved surface with suction and Lorentz force. *Journal of Magnetism and Magnetic Materials*, 573, 170654.

Mukhopadhyay, S., Mondal, I. C., Chamkha, A. J. (2013). Casson fluid flow and heat transfer past a symmetric wedge. *Heat Transfer Research*, 42(8), 665–675.

Munjam, S. R., Gangadhar, K., Seshadri, R., and Rajeswar, M. (2021). Novel technique MDDIM solutions of MHD flow and radiative Prandtl-Eyring fluid over a stretching sheet with convective heating. *International Journal of Ambient Energy*, 43(1), 4850–4859.

Nadeem, S., Ahmad, S., and Muhammad, N. (2018). Computational study of Falkner-Skan problem for astatic and moving wedge. *Sensors and Actuators B: Chemical*, 263(15), 69–76.

Prakash, D., and Vinu, A. (2019). Effect of nonlinear radiation on MHD nanofluid flow and heat transfer due to rotating disk. *AIP Conference Proceedings*, 2112, 020094.

Qureshi, M. A. (2021). A case study of MHD driven Prandtl-Eyring hybrid nanofluid flow over a stretching sheet with thermal jump conditions. *Case Studies in Thermal Engineering*, 28, 101581.

Qureshi, M. A. (2022). Thermal capability and entropy optimization for Prandtl-Eyring hybrid nanofluid flow in solar aircraft implementation. *Alexandria Engineering Journal*, 61, 5295–5307.

Rajesh, V., Sheremet, M. A., and Öztop, H. F. (2021). Impact of hybrid nanofluids on MHD flow and heat transfer near a vertical plate with ramped wall temperature. *Case Studies in Thermal Engineering*, 28, 101557.

Raju C. S. K., and Sandeepa, N. (2016). Falkner-Skan flow of a magnetic-Carreau fluid past a wedge in the presence of cross-diffusion effects. *The European Physical Journal Plus*, 131, 267.

Rashid, U., Abdeljawad, T., Liang, H., Iqbal, A., Abbas, M., Mohd . M., and Siddiqui, J. (2020). The shape effect of gold nanoparticles on squeezing nanofluid flow and heat transfer between Parallel Plates. *Mathematical Problems in Engineering*, 2020, 9584854.

Rashid, U., Liang, H., Ahmad, H., Abbas, M., Iqbal, A., and Hamed, Y. S. (2021). Study of (Ag and TiO_2)/water nanoparticles shape effect on heat transfer and hybrid nanofluid flow toward stretching shrinking horizontal cylinder. *Results in Physics*, 21, 103812.

Rehman, K.U., Malik, A. A., Malik, M. Y., Tahir M., and Zehra, I. (2018). On new scaling group of transformation for Prandtl-Eyring fluid model with both heat and mass transfer. *Results in Physics*, 8, 552–558.

Rehman, Ur . K., Awais, M., Hussain, A., Kousar, N., and Malik, M. Y. (2019). Mathematical analysis on MHD Prandtl-Eyring nanofluid new mass flux conditions. *Mathematical Methods in the Applied Sciences*, 42(1), 24–38.

Roy, N., D. and Pal, D. (2022). Influence of activation energy and nonlinear thermal radiation with ohmic dissipation on heat and mass transfer of a casson nanofluid over stretching sheet. *Journal of Nanofluids*, 11, 819–832.

Sarkar, J., Ghosh, P., and Adil, A. (2015). A review on hybrid nanofluids: Recent research, development and applications. *Renewable and Sustainable Energy Reviews*, 43, 164–177.

Shah, Z., Dawar, A., Kumam, P. (2019). Impact of nonlinear thermal radiation on MHD nanofluid thin film flow over a horizontally rotating disk. *Applied Sciences*, 9(8), 1533.

Shankar, U., and Naduvinamani, N. (2019). Magnetized squeezed flow of time-dependent Prandtl-Eyring fluid past a sensor surface. *Heat Transfer*, 8(6), 2237–2261.

Uddin, I., Ullah, I., Ali, R., Khan, I., and Nisar, K. S. (2021). Numerical analysis of nonlinear mixed convective MHD chemically reacting flow of Prandtl-Eyring nanofluids in the presence of activation energy and Joule heating. *Journal of Thermal Analysis and Calorimetry*, 145(2), 495–505.

Ullah, I., Khan, I., and Shafie, S. (2016). Hydromagnetic Falkner-Skan flow of Casson fluid past a moving wedge with heat transfer. *Alexandria Engineering Journal*, 55, 2139–2148.

Ullah, Z., Ullah, I., Zaman, G., Khan H., and Muhammad, T. (2021). Mathematical modeling and thermodynamics of Prandtl-Eyring fluid with radiation effect: A numerical approach. *Scientific Reports*, 11(1), 22201, 2021.

Yang, L., X. Xu, X. (2017). A renovated Hamilton-Crosser model for the effective thermal conductivity of CNTs nanofluids. *International Communications in Heat and Mass Transfer*, 81, 42–50.

Zahmatkesh, I., Sheremet, M., Yang, L., Heris, S. Z., Sharifpur, M., Meyer, J. P., Ghalambaz, M., Wongwises, S., Jing, D., and Mahian, O. (2021). Effect of nanoparticle shape on the performance of thermal systems utilizing nanofluids: A critical review. *Journal of Molecular Liquids*, 321, 114430.

Zangooee, M. R., Hosseinzadeh, Kh., and D. D. Ganji, D.D. (2019). Hydrothermal analysis of MHD nanofluid (TiO_2-GO) flow between two radiative stretchable rotating disks using AGM. *Case Studies in Thermal Engineering*, 14, 100460.

Chapter 13

Flow and thermal characteristics of diathermic oil-based tri-hybrid nanofluid

Bhagyashri Patgiri and Ashish Paul

13.1 INTRODUCTION

The performance of conventional fluids as a heat-transferrable medium in various applications is restricted by their less effective thermal conductivity. Therefore, various types of nanoparticles (such as oxide nanoparticles, ceramic nanoparticles, polymeric nanoparticles, carbon-based nanoparticles, etc.) are mixed in the conventional working fluids to amplify their thermal feature. Tri-hybrid nanofluids are the amalgamation of four distinct nanoparticles in the same conventional working fluid. Tri-hybrid nanofluids have astounding applications such as space vehicles, solar water heating, nuclear reactors, etc. because of their marvelous thermal conductivity. (Palanisamy et al., 2021) discussed the thermophysical attributes, characterization, and synthesis of $Al_2O_3 + SiO_2 + TiO_2 / H_2O$ tri-hybrid nanofluid. Sajid et al. (2022) elaborated on the trace of activation energy on tri-hybrid nanofluid's flow past a wedge. Adnan and Ashraf (2022) inspected the thermal efficiency of $Al_2O_3 + CuO + Cu / H_2O$ tri-hybrid nanofluid. Thakur and Sood (2023) studied the tri-hybrid nanofluid's flow through a stretchable heated Riga plate. Several researchers (Ahmed et al., 2021; Sundar et al., 2021; Cao et al., 2022; Sarada et al., 2022; Alharbi et al., 2022; Nasir et al., 2022) have inspected the flow and thermal features of various tri-hybrid nanofluids.

Previously, many flow geometries involving the Casson model were explored by several researchers because of advantageous implications in biological treatments, lubrication, drilling processes, etc. (Hamid et al., 2019) explored the stability inspection and dual solutions with heat deliverance of the flow of Casson fluid past a stretching sheet. Sulochana and Poornima (2019) discussed the unsteady MHD flow of a Casson fluid across a vertical plate accompanied by the hall current. Loganathan and Deepa (2019) considered radiative and electromagnetic Casson fluid flow induced by one vertical permeable Riga-plate (Nandeppanavar et al. 2020) theoretical inspected the thermal attributes of the flow of Casson nanofluid brought by an exponentially stretchable sheet in one Darcy porous media (Naqvi et al. 2020).

DOI: 10.1201/9781003494454-13

The hydromagnetic flow of Casson nanofluid over a porous stretching cylinder with Newtonian heat and mass conditions (Naqvi et al. 2020) elucidated the hydromagnetic Casson nanofluid flow through a stretching cylinder. The Newtonian mass and heat stipulations are also incorporated by Naqvi et al. (2020). Krishna et al. (2021) addressed the Radiative Casson hybrid nanofluid flow through one exponentially accelerated infinite vertical porous surface. Das et al. (2021) explained the dispersion of a solute in the transient flow of Casson fluid pasta stenotic tube accompanied by in-between exchange of phases. Obalalu et al. (2022) analyzed the minimization of entropy production on radiative electromagnetohydrodynamic Casson nanofluid flow through one melting Riga plate. Basha and Sivaraj (2022) conducted the stability inspection of the flow of Casson nanofluid through a contracting or extending wedge. Casson fluid's MHD flow across one porous vertical plate encountered by chemical reaction and thermal diffusion was addressed by Kodi et al. (2021). Priam and Nasrin (2022) numerically appraised the Casson fluid's unsteady peristaltic duct flow. Prameela et al. (2022) expounded convective (free) Casson (non-Newtonian) fluid flow brought by a vertical oscillating plate. Umavathi et al. (2023) discussed the flow of squeezing Casson nanofluid amidst convectively parallel heated disks. Paul et al. (2023) elaborated on mass relocation and heat relocation of Maxwell-Casson hybrid nanofluid induced by a one-time reliant horizontal cylinder with temperature-reliant thermal conductivity. Majeed et al. (2023) explained heat deliverance attributes and mass deliverance attributes in the MHD flow of Casson fluid across a cylinder in one wavy channel. They implemented FEM computations of higher order. Paul et al. (2024) numerically address the MHD mass-thermo Casson tri-hybrid nanoliquids flow past one exponentially stretching cylinder.

Recently, many researchers (Malik et al., 2016; Arifuzzaman et al., 2018; Lund et al., 2019; Manjunatha et al., 2020; Megahed et al., 2021; Kotha et al., 2020; Jamshed, 2021; Asjad et al., 2021; Gouran et al., 2022; Ali et al., 2022a; Jalili et al., 2022; Bejawada et al., 2023; Reddy et al., 2023; Patel et al., 2023) have inspected the MHD flow across various configurations owing to their usefulness in cooling industries, nuclear reactors, medical industries (identifying diseases), etc.

Researchers (Hayat et al., 2015; Rehman et al., 2018; Ijaz et al., 2019; Prasad et al., 2019; Devi and Mabood, 2020; Tassaddiq et al., 2020; Zhang et al., 2021; Mabood et al., 2021; Jyothi et al., 2021; Khan et al., 2021; Kumar et al., 2021; Jawad et al., 2021; Acharya et al., 2022; Ramasekhar and Reddy, 2022; Al Nuwairan et al., 2022; Saraswathy et al., 2022; Ali et al., 2023) have scrutinized the fluid flow transfer over spinning or rotating disks because of their industrial significance such as air cleaning machinery, rotating machinery, medical equipment, etc.

Diathermic oils are heat transmissive fluids majorly employed in several temperature regulating systems owing to their low pour points, low vapor pressure, and high boiling points, and therefore they have been used as heat transporters in several heat transmissive systems. These thermal attributes provide diathermic oil a superb thermal stability, making it an ideal replacement for vapor and water in thermoregulation and heat transmissive systems, for various utilities incorporating temperature in a large range. Diathermic oils are widely applicable as a means of transmitting thermal energy generated in a source in a large range of industrial sectors like textiles, metal processing, food processing, wood processing, marine applications, production of plastics, chemical plants, production of rubbers, etc. Silicon carbide nanoparticles possess features like good wear resistance, excellent thermal conductivity, small thermal expansion coefficient high purity, and high stability, etc. Silicon carbide nanoparticles are highly employed in many applied fields such as grinding material manufacturing, rubber tires' manufacturing, spray nozzles (high temperature) various ceramic parts, special substances for polishing abrasive, high-frequency ceramics, textile ceramics, substrates for ICs, manufacturing of resistance heating element, sealing valves' manufacturing, mirror coatings for high ultraviolet environments, ceramic bearing, etc. Silicon carbide nanoparticles are very essential material in the sectors of industrial goods manufacture. Research attempts in the past few years have resulted in numerous valuable applications of Al_2O_3 nanoparticles owing to their superior rheological efficacies. Al_2O_3 nanoparticles are used in several practical applications including EP-ROM windows, YAG laser crystals, sodium lamps (high pressure), integrated circuit baseboards (to enhance performance), transparent ceramics, optical systems (as a vital component), data storage technologies, laser systems (as a vital component), plastic additives, advanced waterproof materials, catalyst carriers, analytical reagents, fluorescent materials, vapor deposition materials, etc. TiO_2 nanoparticles possess diverse utilizations when compared to several other nanoparticles. Researchers and scholars have found worthy applications of TiO_2 nanoparticles in several industries like the chemical industry, paint industry, coating industry, energy industry, biomedical or pharmaceutical industry, textile industry, etc. They are employed as photocatalysts, air deodorizers, antibacterial agents, pigmenting agents, antifungal agents, antiviral, drug delivery, anticancer agents, antioxidants, etc.

Buoyed by the applications of diathermic oils, TiO_2, Al_2O_3 and SiC nanoparticles and fluid's flow through spinning disks and also by reviewing the related published articles as mentioned above, this inspection includes the predicament of 3D, MHD flow of $SiC + Al_2O_3 + TiO_2$ / Diathermic oil tri-hybrid nanofluids through a spinning disk under slip condition. The famous non-Newtonian Casson model is also implemented in the chosen flow model. As per the author's knowledge, no researcher or academician has scrutinized the flow of $SiC + Al_2O_3 + TiO_2$ / Diathermic oil tri-hybrid nanofluids through a spinning disk.

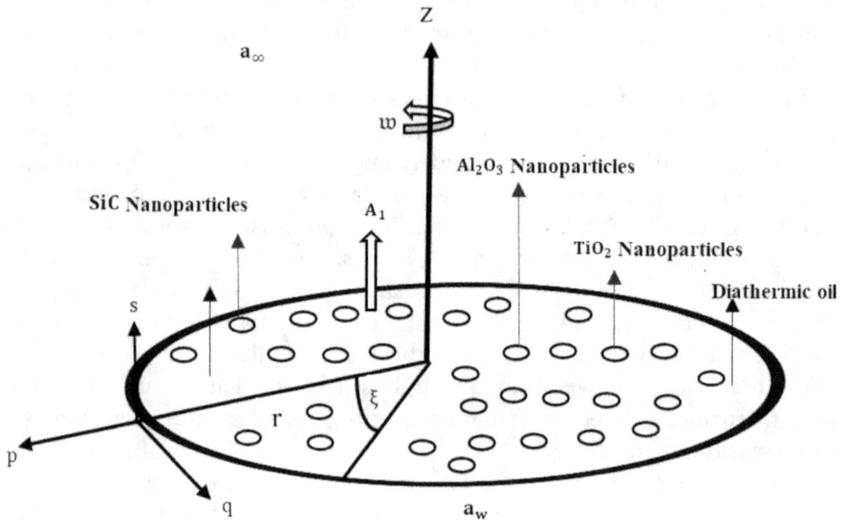

Figure 13.1 Flow diagram.

13.2 PROBLEM FORMULATION

An incompressible 3D, MHD flow of $SiC + Al_2O_3 + TiO_2$ / Diathermic oil tri-hybrid nanofluid across a spinning disk is inspected in this research. The attributes of thermal radiation, slip conditions, and changeable thermal conductivity are being studied. The famous Casson shear thinning fluid model is also employed to delineate the non-Newtonian fluid characteristic. $z = 0$ is the axis of rotation. The stretching disk is supposed to rotate with the fixed angular velocity $\mathfrak{w}$. a_w is the surface temperature while a_∞ is the ambient temperature of the tri-hybrid nanofluid, respectively. Also, p is picked as the dimensional radial velocity constituent acting along r direction, q is picked as the dimensional tangential velocity constituent acting along ξ direction, and s is picked as the dimensional axial velocity constituent acting along z direction. A_1 is presumed to be the magnetic strength. Figure 13.1 is sketched to comprehend the flow of $SiC + Al_2O_3 + TiO_2$ / Diathermic oil tri-hybrid nanofluid across a spinning disk. Now, the governing equations of the chosen flow system are typified as follows (Mabood et al., 2021; Waqas et al., 2021):

$$\frac{\partial p}{\partial r} + \frac{p}{r} + \frac{\partial s}{\partial z} = 0 \tag{13.1}$$

$$p\frac{\partial p}{\partial r} + s\frac{\partial p}{\partial z} - \frac{q^2}{r} = \frac{\mathcal{M}_{thnf}}{\wp_{thnf}}\left(1 + \frac{1}{P_1}\right)\left(\frac{\partial^2 p}{\partial r^2} + \frac{1}{r}\frac{\partial p}{\partial r} - \frac{u}{r^2} + \frac{\partial^2 p}{\partial z^2}\right) - \frac{\sigma_{thnf}}{\rho_{thnf}} pA_1^2 \tag{13.2}$$

$$p\frac{\partial q}{\partial r}+w\frac{\partial q}{\partial z}-\frac{pq}{r}=\frac{\mathcal{M}_{thnf}}{\wp_{thnf}}\left(1+\frac{1}{P_1}\right)\left(\frac{\partial^2 q}{\partial r^2}+\frac{1}{r}\frac{\partial q}{\partial r}-\frac{q}{r^2}+\frac{\partial^2 q}{\partial z^2}\right)-\frac{\sigma_{thnf}}{\rho_{thnf}}qA_1^2 \tag{13.3}$$

$$p\frac{\partial s}{\partial r}+s\frac{\partial s}{\partial z}=\frac{\mathcal{M}_{thnf}}{\wp_{thnf}}\left(1+\frac{1}{P_1}\right)\left(\frac{\partial^2 s}{\partial r^2}+\frac{1}{r}\frac{\partial s}{\partial r}+\frac{\partial^2 s}{\partial z^2}\right) \tag{13.4}$$

$$p\frac{\partial a}{\partial r}+s\frac{\partial a}{\partial z}=\left[\frac{\varkappa_{thnf}(a)}{(\wp c_p)_{thnf}}+\frac{1}{(\wp c_p)_{thnf}}\left(\frac{16\,\chi^{*}\,a_{\infty}{}^{3}}{3\tau^{*}}\right)\right]\left(\frac{\partial^2 a}{\partial r^2}+\frac{1}{r}\frac{\partial a}{\partial r}+\frac{\partial^2 a}{\partial z^2}\right) \tag{13.5}$$

where the changeable thermal conductivity is perceived as

$$\varkappa_{thnf}(a)=\varkappa_{thnf}\left[1+P_3\left(\frac{a-a_{\infty}}{a_{W}-a_{\infty}}\right)\right]$$

Boundary stipulations are (Waqas et al., 2021) as follows:

$$\left.\begin{array}{l} p=m\left(\frac{\partial p}{\partial z}\right),\ q=r\mathfrak{w}+m\left(\frac{\partial q}{\partial z}\right),\ s=0\ ,\ a=a_W, b=b_W\ \ at\ \ z=0 \\ p\to 0,\ q\to 0,\ s\to 0,\ a\to a_{\infty},\ b\to b_{\infty}\ at\ z\to\infty \end{array}\right\} \tag{13.6}$$

The subscripts thnf mean the tri-hybrid nanofluid. Besides, the symbols $\wp$, P_1, P_3, $\varkappa$, χ^{*}, A_1, $\mathcal{M}$, τ^{*}, σ, m, $\mathfrak{w}$, c_p are depicted in the nomenclature part. The physical traits of diathermic oil, SiC, Al_2O_3 and TiO_2 nanoparticles are elucidated in Table 13.1.

Table 13.1 Thermophysical properties (Gul et al., 2022; Ali et al., 2022b)

Physical properties	*SiC*	*Al_2O_3*	*TiO_2*	*Diathermic oil*
Thermal conductivity $(\varkappa)$ (W/mK)	150	40	8.4	0.133
Heat capacity (c_p) (J/kg K)	1,340	765	692	2,030
Density $(\wp)$ (Kk/m^3)	3,370	3,970	4,230	855

Further, the mathematical bond relating the diathermic oil, TiO_2/diathermic oil, $Al_2O_3 + TiO_2$/diathermic oil, and $SiC + Al_2O_3 + TiO_2$/diathermic oil are

$$7_{nf} = 7_f \rho \left\{ \frac{7_\alpha + 27_f - 2\psi_\alpha \left(7_f - 7_\alpha\right)}{7_\alpha + 27_f + \psi_\alpha \left(7_f - 7_\alpha\right)} \right\}, \; 7_{hnf} = 7_f \left\{ \frac{7_\beta + 27_{nf} - 2\psi_\beta \left(7_{nf} - 7_\beta\right)}{7_\beta + 27_{nf} + \psi_2 \left(7_{nf} - 7_\beta\right)} \right\}$$

$$7_{thnf} = 7_f \left\{ \frac{7_\gamma + 27_{hnf} - 2\psi_\gamma \left(7_{hnf} - 7_\gamma\right)}{k_\gamma + 27_{hnf} + \psi_\gamma \left(7_{hnf} - 7_\gamma\right)} \right\}$$

$$\mathcal{M}_{nf} = \frac{\mathcal{M}_f}{\left(1 - \psi_\alpha\right)^{2.5}}, \; \mathcal{M}_{nf} = \frac{\mathcal{M}_f}{\left(1 - \psi_\alpha\right)^{2.5} \left(1 - \psi_\beta\right)^{2.5}},$$

$$\mathcal{M}_{thnf} = \frac{\mathcal{M}_f}{\left(1 - \psi_\alpha\right)^{2.5} \left(1 - \psi_\beta\right)^{2.5} \left(1 - \psi_\gamma\right)^{2.5}}$$

$$\wp_{nf} = \left(1 - \psi_\alpha\right) \wp_f + \psi_\alpha \wp_\alpha, \quad \wp_{hnf} = \left(1 - \psi_\beta\right) \left\{ \left(1 - \psi_\alpha\right) \wp_f + \psi_\alpha \wp_\alpha \right\} + \psi_\beta \wp_\beta$$

$$\wp_{thnf} = \left(1 - \psi_\gamma\right) \left[\left(1 - \psi_\beta\right) \left\{ \left(1 - \psi_\alpha\right) \wp_f + \psi_\alpha \wp_\alpha \right\} + \psi_\beta \wp_\beta \right] + \psi_\gamma \wp_\gamma$$

$$\left(\wp_{c_p}\right)_{nf} = \left(1 - \psi_\alpha\right) \left(\wp_{c_p}\right)_f + \psi_\alpha \left(\wp_{c_p}\right)_{\alpha,}$$

$$\left(\wp_{c_p}\right)_{hnf} = \left(1 - \psi_\beta\right) \left\{ \left(1 - \psi_\alpha\right) \left(\wp_{c_p}\right)_f + \psi_\alpha \left(\wp_{c_p}\right)_\alpha \right\} + \psi_\beta \left(\wp_{c_p}\right)_\beta$$

$$\left(\wp_{c_p}\right)_{thnf} = \left(1 - \psi_\gamma\right) \left[\left(1 - \psi_\beta\right) \left\{ \left(1 - \psi_\alpha\right) \left(\wp_{c_p}\right)_f + \psi_\alpha \left(\wp_{c_p}\right)_\alpha \right\} + \psi_\beta \left(\wp_{c_p}\right)_\beta \right] + \psi_\gamma \left(\wp_{c_p}\right)_\gamma$$

Here ψ_α, ψ_β, and ψ_γ refer to the volume fractions of TiO_2, Al_2O_3, and SiC nanoparticles and suffixes α, β, and γ specify TiO_2, Al_2O_3, and SiC nanoparticles.

To change dimensional Eqs. (13.1)–(13.6) into one dimensionless form, we consider the below-mentioned similarity reformation (Mustafa, 2017):

$$\left.\begin{aligned}\eta &= \left(\sqrt{\frac{2\mathfrak{w}}{\vartheta}}\right)z, \quad p = r\mathfrak{w}u'(\eta),\ q = r\mathfrak{w}v(\eta),\ s = -\sqrt{2\mathfrak{w}\vartheta}\ \ u(\eta),\ w(\eta)\\ &= \frac{(a-a_\infty)}{(a_w-a_\infty)}\end{aligned}\right\} \tag{13.7}$$

By using Eq. (13.7) in Eqs. (13.1)–(13.6), we get:

$$\frac{2\left(1+\frac{1}{P_1}\right)}{\mathbb{N}_1\mathbb{N}_2}u''' - \left[(u')^2 - 2uu'' - v^2\right] - \frac{2\mathbb{N}_5}{\mathbb{N}_2}\ P_2\ u' = 0 \tag{13.8}$$

$$\frac{2\left(1+\frac{1}{P_1}\right)}{\mathbb{N}_1\mathbb{N}_2}\ v'' - \left[\ 2u'v - 2uv'\right] - \frac{2\mathbb{N}_5}{\mathbb{N}_2}\ P_2\ v = 0 \tag{13.9}$$

$$\left[\mathbb{N}_4(1+P_3w)+4P_4\right]w'' + \mathbb{N}_3P_nuw' = 0 \tag{13.10}$$

And

$$\left.\begin{aligned}u'(0) = P_5\ u''\ (0),\ v(0) = 1 + P_5\ v'(0)\ ,\ u(0) = 0, w(0) = 1\ \text{ at }\ \eta = 0\\ u(\infty)\to 0,\ u'(\infty)\to 0, v(\infty)\to 0,\ w(\infty)\to 0\ \text{ at }\ \eta \to \infty\end{aligned}\right\} \tag{13.11}$$

Here,

$$\mathbb{N}_1 = \frac{\mathcal{M}_{thnf}}{\mathcal{M}_f},\quad \mathbb{N}_2 = \frac{\wp_{thnf}}{\wp_f},\quad \mathbb{N}_3 = \frac{(\wp c_p)_{thnf}}{(\wp c_p)_f},\quad \mathbb{N}_4 = \frac{\daleth_{thnf}}{\daleth_f}$$

Also,

$$\vartheta = \frac{\mathcal{M}_f}{\wp_f}(\text{Kinematic viscosity}),\ \ P_5 = m\left(\sqrt{\frac{2\mathfrak{w}}{\vartheta}}\right)(\text{Velocity slip parameter})$$

$$P_2 = \frac{\sigma_f A_1^2}{2\mathfrak{w}\,\wp_f}(\ \text{Magnetic parameter}),\ P_n = \frac{\vartheta(\wp c_p)_f}{\daleth_f}(\text{Prandtl number})$$

$$\text{Re} = \frac{\mathfrak{w}r^2}{\vartheta}(\text{Reynold's number}),\ P_4 = \frac{4\ \chi^*\ a_\infty^{\ 3}}{3\tau^*\daleth_f}(\text{Radiation parameter})$$

The dimensionless pertinent engineering quantities are as follows:

- Skin friction coefficient:
- $Ck_1 \mathrm{Re}^{0.5} = \left(1 + \frac{1}{P_1}\right) \frac{\sqrt{\left(u''(0)\right)^2 + \left(v(0)\right)^2}}{\left(1-\psi_\alpha\right)^{2.5}\left(1-\psi_\beta\right)^{2.5}\left(1-\psi_\gamma\right)^{2.5}}$
- Local Nusselt number (heat transmissive rate):

$$Nk_1 \mathrm{Re}^{-0.5} = -\left(\frac{\daleth_{thnf}}{\daleth_f}\right) w'(0) = -\mathrm{N}_4 w'(0)$$

13.3 NUMERICAL METHODOLOGY AND CODE AFFIRMATION

Eqs. (8)–(11) are numerically dealt with using the fourth-order Runge-Kutta shooting methodology. Besides, Table 13.2 presents the comparison outcomes that we attained with the outcomes procured by Mustafa (2017) for $u'(0)$ by considering $P_2 = 0.5$ and $P_5 = 0.2$, 0.6, and it is noticeable that all outputs are within the justified range.

13.4 RESULTS AND DISCUSSION

This segment provides a discussion of the impacts of elegant flow factors on temperature $w(\eta)$, tangential velocity $v(\eta)$, radial velocity $u'(\eta)$, skin friction $Ck_1 \mathrm{Re}^{0.5}$, and Nusselt number $Nk_1 \mathrm{Re}^{-0.5}$.

Figures 13.2–13.6 sketched the repercussions of parameters of magnetic P_2, Casson P_1, and slip P_5 upon the velocity constituents $u'(\eta)$ and $v(\eta)$.

Figures 13.2 and 13.3 elucidate that by upgrading the magnetic parameter P_2, both the velocities $u'(\eta)$ and $v(\eta)$ decelerate. As P_2 escalates, the associated Lorentz force enhances which repels the tri-hybrid nanofluid motion, and, correspondingly, velocities $u'(\eta)$ and $v(\eta)$ decreases.

Figures 13.4 and 13.5 disclose that bigger amplitudes of the Casson parameter P_1 deplete both the velocities $u'(\eta)$ and $v(\eta)$. Larger amplitudes of P_1 oppose the tri-hybrid nanofluid flow by lessening the yield stress and thus both $u'(\eta)$ and $v(\eta)$ decrease as P_1 increases.

Table 13.2 Comparison of $u'(0)$ for different P_5' s values with (Mustafa, 2017)

P_2	P_5	$u'(0)$ *(Mustafa, 2017)*	$u'(0)$ *(Our output)*
0.5	0.2	0.028608	0.028608
0.5	0.6	0.043151	0.043151

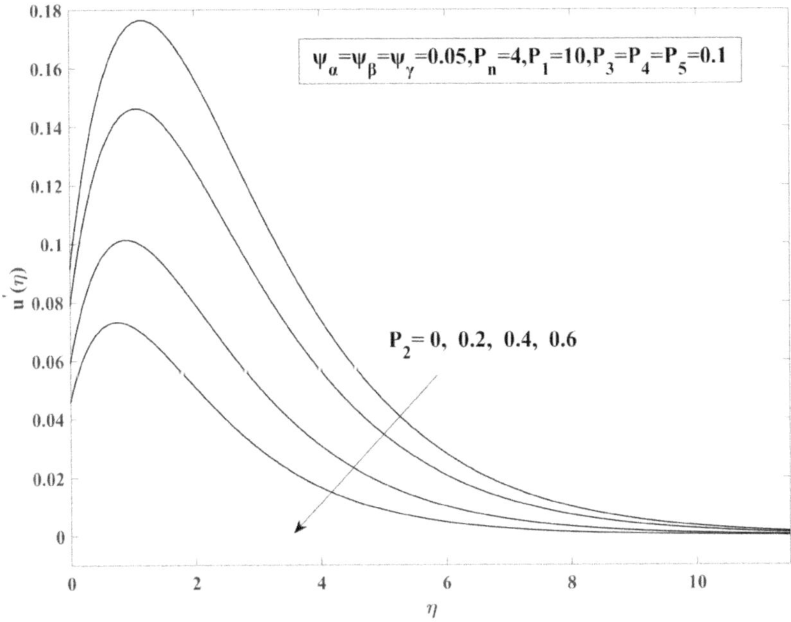

Figure 13.2 Radial velocity $u'(\eta)$ for $SiC + Al_2O_3 + TiO_2$ / diathermic oil tri-hybrid nanofluid for various amplitudes of the magnetic parameter P_2.

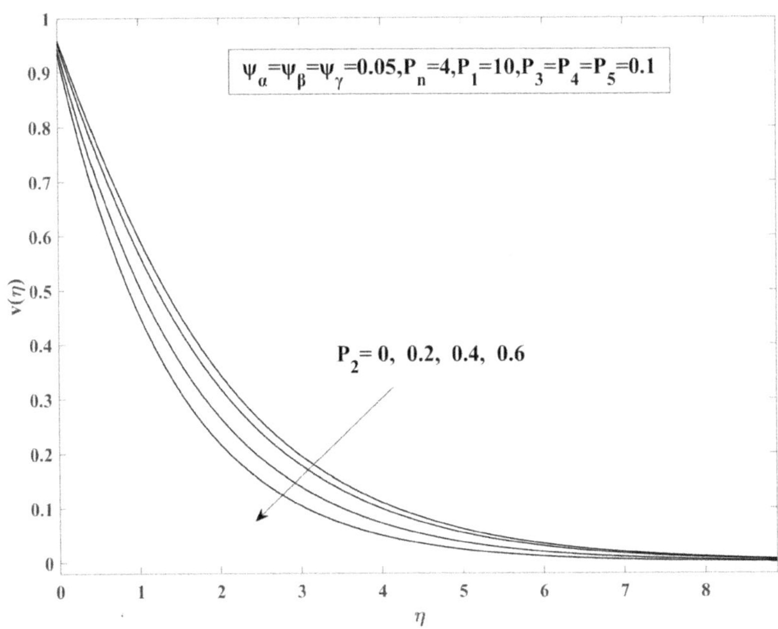

Figure 13.3 Tangential velocity $v(\eta)$ for $SiC + Al_2O_3 + TiO_2$ / diathermic oil tri-hybrid nanofluid for various amplitudes of the magnetic parameter P_2.

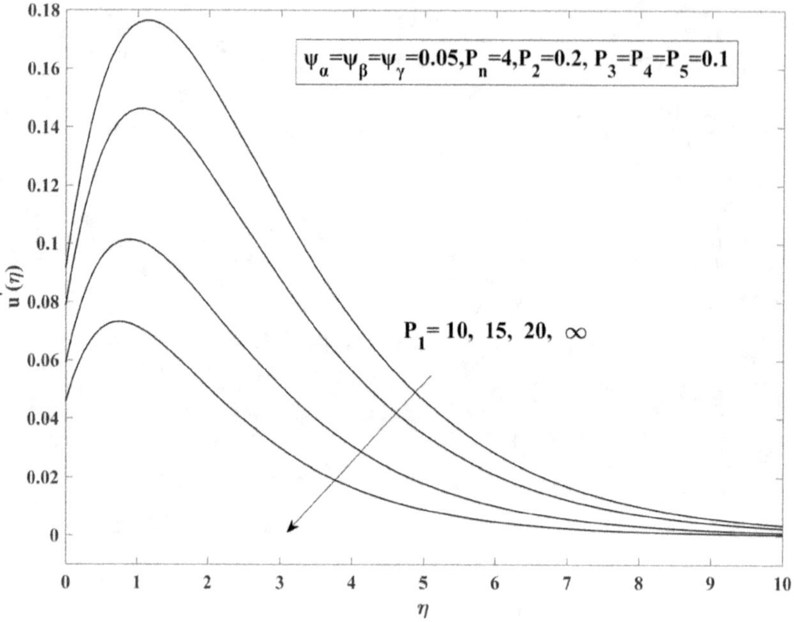

Figure 13.4 Radial velocity $u'(\eta)$ for $SiC + Al_2O_3 + TiO_2$ / diathermic oil tri-hybrid nanofluid for various amplitudes of the Casson parameter P_1.

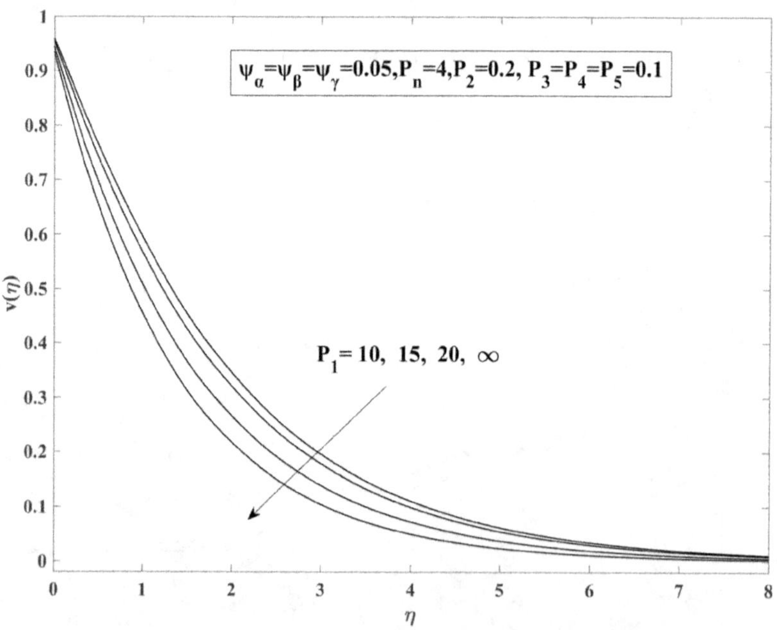

Figure 13.5 Tangential velocity $v(\eta)$ for $SiC + Al_2O_3 + TiO_2$ / diathermic oil tri-hybrid nanofluid for various amplitudes of the Casson parameter P_1.

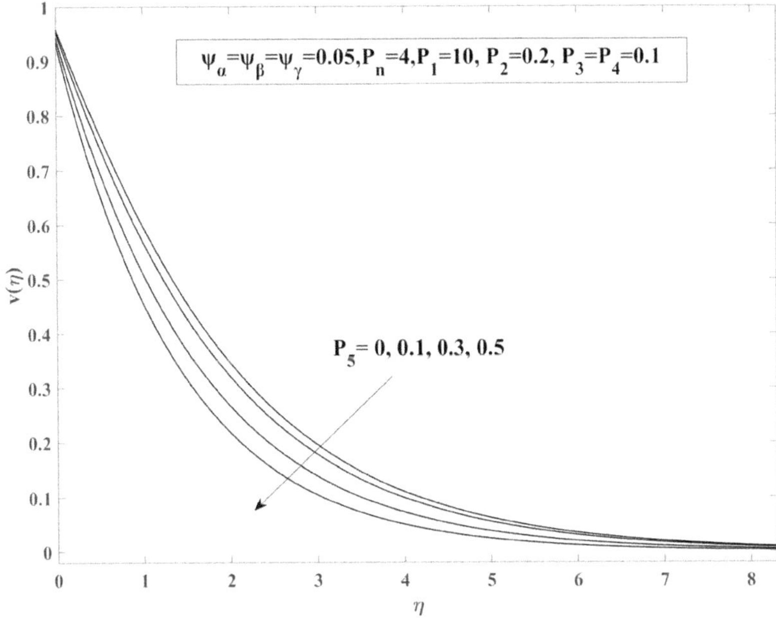

Figure 13.6 Tangential velocity $v(\eta)$ for $SiC + Al_2O_3 + TiO_2$ / diathermic oil tri-hybrid nanofluid for various amplitudes of the slip parameter P_5.

Figure 13.6 portrays that tangential velocity $v(\eta)$ decelerates for upsurging amplitudes of the slip parameter P_5. As the P_5's amplitudes augment, the flow retardation also strengthens too, so $v(\eta)$ decelerates.

Figures 13.7–13.9 are sketched the repercussions of Prandtl number P_n, SiC nanoparticle's volume fraction ψ_γ, and changeable thermal conductivity parameter P_3 upon the temperature $w(\eta)$.

Figure 13.7 elaborates that the temperature $w(\eta)$ lessens for the uplifting amplitudes of the Prandtl number P_n. Tri-hybrid nanofluid's thermal diffusivity drops down for mounting amplitudes of P_n and correspondingly temperature $w(\eta)$ reduces.

Figure 13.8 explains that the SiC nanoparticle's volume fraction ψ_γ curtails the temperature $w(\eta)$. SiC nanoparticles have fabulous thermal conductivity, thus by mixing SiC nanoparticles, the thermal conductivity of the tri-hybrid nanofluid's thermal conductivity upgrades, which fastens the heat transpiration process and, correspondingly, lessens temperature $w(\eta)$.

Figure 13.9 expounds that the temperature $w(\eta)$ scales up for the rising amplitudes of the changeable thermal conductivity parameter P_3. As the P_3 's' amplitudes uplift, the energy boundary layer lengthens resulting in an amplification in temperature $w(\eta)$.

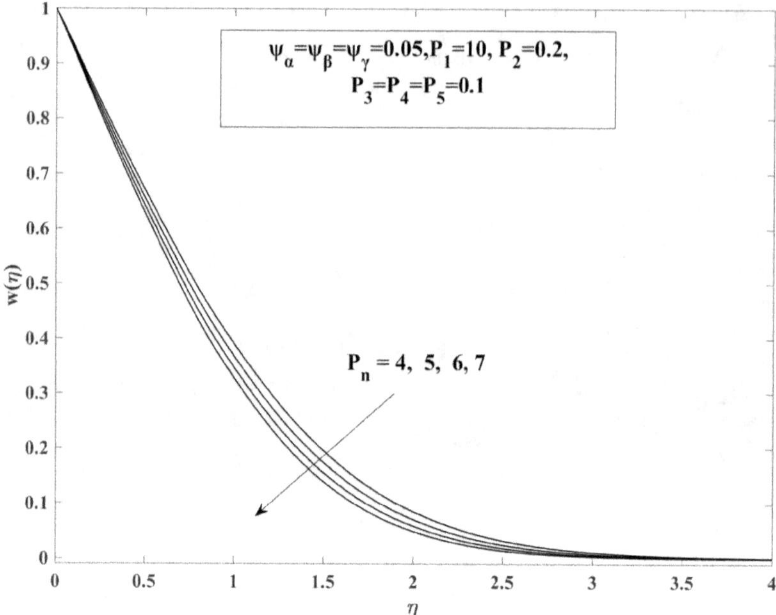

Figure 13.7 Temperature profile $w(\eta)$ for $SiC + Al_2O_3 + TiO_2$ / diathermic oil tri-hybrid nanofluid for various amplitudes of the Prandtl number P_n.

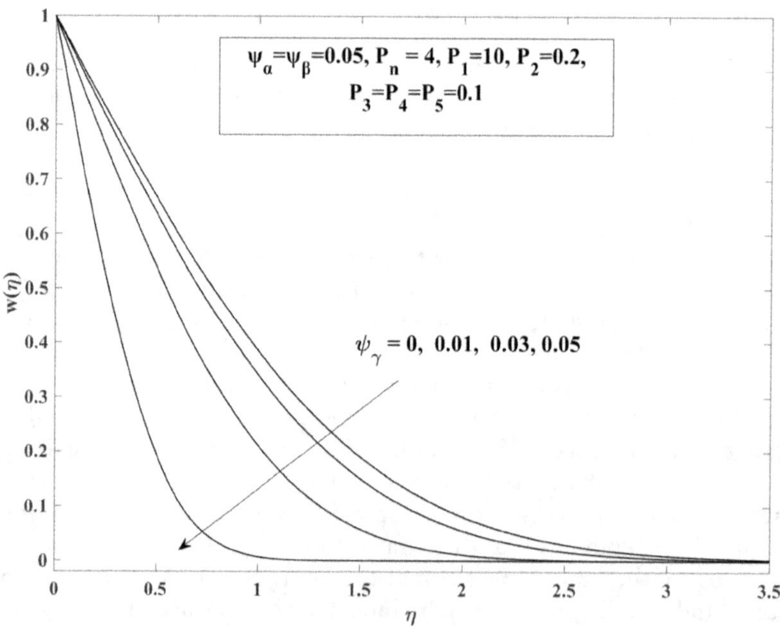

Figure 13.8 Temperature profile $w(\eta)$ for $SiC + Al_2O_3 + TiO_2$ / diathermic oil tri-hybrid nanofluid for various amplitudes of the volume fractions of SiC (Silicon carbide) nanoparticles.

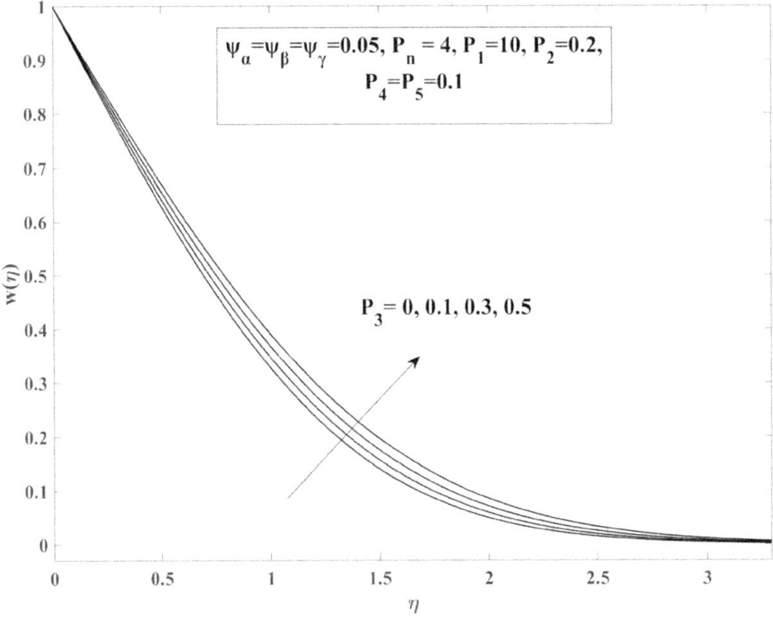

Figure 13.9 Temperature profile $w(\eta)$ for $SiC + Al_2O_3 + TiO_2$ / diathermic oil tri-hybrid nanofluid for various amplitudes of the changeable thermal conductivity parameter P_3.

Table 13.3 Computation of $Ck_1Re^{0.5}$ and $Nk_1Re^{-0.5}$ by taking $P_n = 4$, $\psi_\alpha = \psi_\beta = 0.05$, $P_2 = 0.2$, $P_5 = 0.1$

ψ_γ	P_1	P_3	P_4	$Ck_1Re^{0.5}$	$Nk_1Re^{-0.5}$
0	10	0.1	0.1	0.482632	0.602314
0.02	10	0.1	0.1	0.484437	0.634177
0.05	10	0.1	0.1	0.500321	0.661785
0.05	15	0.1	0.1	0.497545	0.682643
0.05	∞	0.1	0.1	0.483287	0.709584
0.05	10	0.3	0.1	0.483394	0.705278
0.05	10	0.5	0.1	0.485179	0.701974
0.05	10	0.5	0.2	0.485179	0.735462
0.05	10	0.5	0.3	0.485179	0.761668

Table 13.3 elucidates the effect of SiC nanoparticle's volume fraction (ψ_γ), Casson (P_1), changeable thermal conductivity (P_3), and radiation (P_4) parameters on skin friction $Ck_1Re^{0.5}$, and Nusselt number $Nk_1Re^{-0.5}$. This table discloses that higher amplitudes of ψ_γ and P_3 increase $Ck_1Re^{0.5}$, while higher amplitudes of P_1 reduce $Ck_1Re^{0.5}$. Besides, mounting amplitudes of P_4

has no worthy impact on $Ck_1Re^{0.5}$. Further, higher amplitudes of ψ_γ, P_1, and P_4 increase $Nk_1Re^{-0.5}$ while higher amplitudes of P_3 reduce $Nk_1Re^{-0.5}$.

13.5 CONCLUSION

Here 3D, MHD flow of diathermic oil tri-hybrid nanofluids through a spinning disk under slip condition is inspected. The famous Casson shear thinning fluid model is also employed to delineate the non-Newtonian fluid characteristic. Three different nanoparticles, namely TiO_2, Al_2O_3, and SiC nanoparticles are considered in this study. The novel aspects of thermal radiation, magnetic field, and changeable thermal conductivity are also taken into account. The influence of embedded parameters on the Casson tri-hybrid nanofluid flow and heat and mass transference features is discussed elaborately. The major outcomes of this inspection are:

- Velocities $u'(\eta)$ and $v(\eta)$ face retardation for mounting amplitudes of the magnetic parameter P_2.
- Velocities $u'(\eta)$ and $v(\eta)$ decelerate for larger amplitudes of the Casson parameter P_1.
- Temperature profile $w(\eta)$ depletes for higher amplitudes of P_n and ψ_γ.
- Temperature profile $w(\eta)$ rises for sturdy amplitudes of the radiation parameter P_3 .
- Higher amplitudes of P_1 and P_4 escalates the heat transference rate.

NOMENCLATURE

χ^*: Stephen Boltzmann constant
$\mathfrak{w}$: Angular speed
$\varkappa$: Thermal conductivity
τ^*: Mean absorption constant
σ: Electrical conductivity
A_1: Magnetic coefficient
$\mathcal{M}$: Dynamic viscosity
m: Slip coefficient
$\wp$: Density
P_1: Casson parameter
c_p: Heat capacity
P_n: Prandtl number
η: Dimensionless similarity variable
a: Temperature
P_2: Magnetic parameter

Re: Reynold's number
u′, *v*: Dimensionless velocity profiles
P_3: Changeable thermal conductivity parameter
w: Dimensionless temperature
ϑ: Kinematic viscosity
P_5: Slip parameter
P_4: Radiation parameter

REFERENCES

Acharya, N., Maity, S., & Kundu, P. K. (2022). Entropy generation optimization of unsteady radiative hybrid nanofluid flow over a slippery spinning disk. *Proceedings of the Institution of Mechanical Engineers, Part C: Journal of Mechanical Engineering Science*, 236(11), 6007–6024.

Adnan, & Ashraf, W. (2022). Thermal efficiency in hybrid (Al_2O_3-CuO/H_2O) and ternary hybrid nanofluids (Al_2O_3-CuO-Cu/H_2O) by considering the novel effects of imposed magnetic field and convective heat condition. *Waves in Random and Complex Media*, 1–16. doi:10.3389/fchem.2022.960369

Ahmed, W., Kazi, S.N., Chowdhury, Z.Z., Johan, M.R.B., Mehmood, S., Soudagar, M.E.M., Mujtaba, M.A., Gul, M., & Ahmad, M.S. (2021). Heat transfer growth of sonochemically synthesized novel mixed metal oxide ZnO+ Al_2O_3+ TiO_2/DW based ternary hybrid nanofluids in a square flow conduit. *Renewable and Sustainable Energy Reviews*, 145, 111025.

Arifuzzaman, S. M., Khan, M. S., Mehedi, M. F. U., Rana, B. M. J., & Ahmmed, S. F. (2018). Chemically reactive and naturally convective high speed MHD fluid flow through an oscillatory vertical porous plate with heat and radiation absorption effect. *Engineering Science and Technology, an International Journal*, 21(2), 215–228.

Asjad, M. I., Sarwar, N., Ali, B., Hussain, S., Sitthiwirattham, T., & Reunsumrit, J. (2021). Impact of bioconvection and chemical reaction on MHD nanofluid flow due to exponential stretching sheet. *Symmetry*, 13(12), 2334.

Al Nuwairan, M., Hafeez, A., Khalid, A., & Syed, A. (2022). Heat generation/absorption effects on radiative stagnation point flow of Maxwell nanofluid by a rotating disk influenced by activation energy. *Case Studies in Thermal Engineering*, 35, 102047. doi:10.1016/j.csite.2022.102047

Alharbi, K.A.M., Ahmed, A.E.S., Ould Sidi, M., Ahammad, N.A., Mohamed, A., El-Shorbagy, M.A., Bilal, M., & Marzouki, R. (2022). Computational valuation of Darcy ternary-hybrid nanofluid flow across an extending cylinder with induction effects. *Micromachines*, 13(4), 588.

Ali, L., Ali, B., &Ghori, M. B. (2022a). Melting effect on Cattaneo-Christov and thermal radiation features for aligned MHD nanofluid flow comprising microorganisms to leading edge: FEM approach. *Computers & Mathematics with Applications*, 109, 260–269.

Ali, R., Shahzad, A., Saher, K., Elahi, Z., & Abbas, T. (2022b). The thin film flow of Al_2O_3 nanofluid particle over an unsteady stretching surface. *Case Studies in Thermal Engineering*, 29, 101695. doi:10.1016/j.csite.2021.101695

Ali, B., Duraihem, F. Z., Jubair, S., Alqahtani, H., & Yagoob, B. (2023). Analysis of interparticle spacing and nanoparticle radius on the radiative alumina based nanofluid flow subject to irregular heat source/sink over a spinning disk. *Materials Today Communications*, 36, 106729.

Basha, H. T., & Sivaraj, R. (2022). Stability analysis of casson nanofluid flow over an extending/contracting wedge and stagnation point. *Journal of Applied and Computational Mechanics*, 8(2), 566–579.

Bejawada, S. G., & Nandeppanavar, M. M. (2023). Effect of thermal radiation on magnetohydrodynamics heat transfer micropolar fluid flow over a vertical moving porous plate. *Experimental and Computational Multiphase Flow*, 5(2), 149–158.

Cao, W., Animasaun, I.L., Yook, S.J., Oladipupo, V.A., & Ji, X. (2022). Simulation of the dynamics of colloidal mixture of water with various nanoparticles at different levels of partial slip: Ternary-hybrid nanofluid. *International Communications in Heat and Mass Transfer*, 135, 106069.

Das, P., Sarifuddin, S., Rana, J., & Mandal, P.K. (2021). Solute dispersion in transient Casson fluid flow through stenotic tube with exchange between phases. *Physics of Fluids*, 33(6), 061907.

Devi, S. S. U., & Mabood, F. (2020). Entropy anatomization on Marangoni Maxwell fluid over a rotating disk with nonlinear radiative flux and Arrhenius activation energy. *International Communications in Heat and Mass Transfer*, 118, 104857. doi:10.1016/j.icheatmasstransfer.2020.104857

Gouran, S., Mohsenian, S., &Ghasemi, S. E. (2022). Theoretical analysis on MHD nanofluid flow between two concentric cylinders using efficient computational techniques. *Alexandria Engineering Journal*, 61(4), 3237–3248.

Gul, H., Ramzan, M., Nisar, K. S., Mohamed, R. N., & Ghazwani, H. A. S. (2022). Performance-based comparison of Yamada-Ota and Hamilton-Crosser hybrid nanofluid flow models with magnetic dipole impact past a stretched surface. *Scientific Reports*, 12(1), 29. doi:10.1038/s41598-021-04019-8

Hamid, M., M. Usman, Z. H. Khan, R. Ahmad, & Wang, W. (2019). Dual solutions and stability analysis of flow and heat transfer of Casson fluid over a stretching sheet. *Physics Letters A*, 383(20), 2400–2408.

Hayat, T., Rashid, M., Imtiaz, M., & Alsaedi, A. (2015). Magnetohydrodynamic (MHD) flow of Cu-water nanofluid due to a rotating disk with partial slip. *AIP Advances*, 5 (6), 067169. doi:10.1063/1.4923380

Ijaz, M., Ayub, M., & Khan, H. (2019). Entropy generation and activation energy mechanism in nonlinear radiative flow of Sisko nanofluid: rotating disk. *Heliyon*, 5(6), e01863. doi:10.1016/j.heliyon.2019.e01863

Jalili, B., Sadighi, S., Jalili, P., & Ganji, D. D. (2022). Numerical analysis of MHD nanofluid flow and heat transfer in a circular porous medium containing a Cassini oval under the influence of the Lorentz and buoyancy forces. *Heat Transfer*, 51(7), 6122–6138.

Jamshed, W. (2021). Numerical investigation of MHD impact on Maxwell nanofluid. *International Communications in Heat and Mass Transfer*, 120, 104973.

Jawad, M., Saeed, A., Khan, A., & Islam, S. (2021). MHD bioconvection Darcy-Forchheimer flow of Casson nanofluid over a rotating disk with entropy optimization. *Heat Transfer*, 50(3), 2168–2196. doi:10.1002/htj.21973

Jyothi, A. M., Naveen Kumar, R., Punith Gowda, R. J., Veeranna, Y., & Prasannakumara, B. C. (2021). Impact of activation energy and gyrotactic microorganisms on flow of Casson hybrid nanofluid over a rotating moving disk. *Heat Transfer*, 50(6), 5380–5399. doi:10.1002/htj.22129

Khan, M. I., Waqas, H., Khan, S. U., Imran, M., Chu, Y. M., Abbasi, A., & Kadry, S. (2021). Slip flow of micropolar nanofluid over a porous rotating disk with motile microorganisms, nonlinear thermal radiation and activation energy. *International Communications in Heat and Mass Transfer*, 122, 105161. doi:10.1016/j.icheatmasstransfer.2021.105161

Kodi, R., ObulesuMopuri, S. S., & Venkateswaraju, K. (2022). Investigation of MHD Casson fluid flow past a vertical porous plate under the influence of thermal diffusion and chemical reaction. *Heat Transfer* 5(1), 377–394.

Kotha, G., Kolipaula, V. R., Venkata Subba Rao, M., Penki, S., & Chamkha, A. J. (2020). Internal heat generation on bioconvection of an MHD nanofluid flow due to gyrotactic microorganisms. *The European Physical Journal Plus*, 135, 1–19.

Krishna, M. V., Ahammad, N. A., & Chamkha, A. J. (2021). Radiative MHD flow of Casson hybrid nanofluid over an infinite exponentially accelerated vertical porous surface. *Case Studies in Thermal Engineering*, 27, 101229.

Kumar, R., Bhattacharyya, A., Seth, G. S., & Chamkha, A. J. (2021). Transportation of magnetite nanofluid flow and heat transfer over a rotating porous disk with Arrhenius activation energy: Fourth order Noumerov's method. *Chinese Journal of Physics*, 69, 172–185. doi:10.1016/j.cjph.2020.11.018

Loganathan, P., & Deepa, K., 2019. Electromagnetic and radiative Casson fluid flow over a permeable vertical Riga-plate. *Journal of Theoretical and Applied Mechanics*, 57(1), 987–998.

Lund, L. A., Omar, Z., and Khan, I. 2019. Analysis of dual solution for MHD flow of Williamson fluid with slippage. *Heliyon*, 5(3).

Mabood, F., Rauf, A., Prasannakumara, B. C., Izadi, M., & Shehzad, S. A. (2021). Impacts of Stefan blowing and mass convention on flow of Maxwell nanofluid of variable thermal conductivity about a rotating disk. *Chinese Journal of Physics*, 71, 260–272. doi:10.1016/j.cjph.2021.03.003

Malik, M.Y., Khan, M., Salahuddin, T., & Khan, I., 2016. Variable viscosity and MHD flow in Casson fluid with Cattaneo-Christov heat flux model: Using Keller box method. *Engineering Science and Technology, an International Journal*, 19(4), pp.1985–1992.

Majeed, A.H., Mahmood, R., Shahzad, H., Pasha, A.A., Raizah, Z.A., Hosham, H.A., Reddy, D.S.K., & Hafeez, M.B. (2023). Heat and mass transfer characteristics in MHD Casson fluid flow over a cylinder in a wavy channel: Higher-order FEM computations. *Case Studies in Thermal Engineering*, 42, 102730.

Manjunatha, G., Rajashekhar, C., Vaidya, H., Prasad, K. V., Makinde, O. D., & Viharika, J. U. (2020). Impact of variable transport properties and slip effects on MHD Jeffrey fluid flow through channel. *Arabian Journal for Science and Engineering*, 45, 417–428.

Megahed, A. M., Reddy, M. G., & Abbas, W. (2021). Modeling of MHD fluid flow over an unsteady stretching sheet with thermal radiation, variable fluid properties and heat flux. *Mathematics and Computers in Simulation*, 185, 583–593.

Mustafa, M. (2017). MHD nanofluid flow over a rotating disk with partial slip effects: Buongiorno model. *International Journal of Heat and Mass Transfer*, 108, 1910–1916. doi:10.1016/j.ijheatmasstransfer.2017.01.064

Nandeppanavar, M. M., Vaishali, S., Kemparaju, M. C., & Raveendra, N. (2020). Theoretical analysis of thermal characteristics of casson nano fluid flow past an exponential stretching sheet in Darcy porous media. *Case Studies in Thermal Engineering*, 21, 100717.

Naqvi, S. M. R. S., Muhammad, T., & Asma, M. (2020). Hydromagnetic flow of Casson nanofluid over a porous stretching cylinder with Newtonian heat and mass conditions. *Physica A: Statistical Mechanics and its Applications*, 550, 123988.

Nasir, S., Sirisubtawee, S., Juntharee, P., Berrouk, A.S., Mukhtar, S., & Gul, T. (2022). Heat transport study of ternary hybrid nanofluid flow under magnetic dipole together with nonlinear thermal radiation. *Applied Nanoscience*, 12(9), 2777–2788.

Obalalu, A. M., Adebayo, L. L., Colak, I., Ajala, A. O., & Wahaab, F. A. (2022). Entropy generation minimization on electromagnetohydrodynamic radiative Casson nanofluid flow over a melting Riga plate. *Heat Transfer*, 51(5), 3951–3978.

Palanisamy, R., Parthipan, G., &Palani, S. (2021). Study of synthesis, characterization and thermo physical properties of Al_2O_3-SiO_2-TiO_2/H_2O based tri-hybrid nanofluid. *Digest Journal of Nanomaterials & Biostructures (DJNB)*, 16(3).

Patel, H., Mittal, A., & Nagar, T. (2023). Fractional order simulation for unsteady MHD nanofluid flow in porous medium with Soret and heat generation effects. *Heat Transfer*, 52(1), 563–584.

Paul, A., Sarma, N., & Patgiri, B. (2023). Thermal and mass transfer analysis of Casson-Maxwell hybrid nanofluids through an unsteady horizontal cylinder with variable thermal conductivity and Arrhenius activation energy. *Numerical Heat Transfer, Part A: Applications*, 1–26. doi:10.1080/10407782.2023.2297000

Paul, A., Sarma, N. & Patgiri, B. (2024). Numerical assessment of MHD Thermo-mass flow of casson ternary hybrid nanofluid around an exponentially stretching cylinder. *BioNanoScience*. doi:10.1007/s12668-024-01306-2

Prameela, M., Gangadhar, K., & Reddy, G.J. (2022). MHD free convective non-Newtonian Casson fluid flow over an oscillating vertical plate. *Partial Differential Equations in Applied Mathematics*, 5, 100366.

Prasad, K. V., Vaidya, H., Makinde, O. D., & Setty, B. S. (2019). MHD mixed convective flow of Casson nanofluid over a slender rotating disk with source/sink and partial slip effects. *In Defect and Diffusion Forum*, 392, 92–122. doi:10.4028/www.scientific.net/DDF.392.92

Priam, S.S., & Nasrin, R. (2022). Numerical appraisal of time-dependent peristaltic duct flow using Casson fluid. *International Journal of Mechanical Sciences*, 233, 107676.

Ramasekhar, G., & Reddy, P. B. A (2022). Entropy generation on EMHD Darcy-Forchheimer flow of Carreau hybrid nanofluid over a permeable rotating disk with radiation and heat generation: Homotopy perturbation solution. *Proceedings of the Institution of Mechanical Engineers, Part E: Journal of Process Mechanical Engineering*, 09544089221116575. doi:10.1177/09544089221116575

Reddy, P. N., Verma, V., Kumar, A., & Awasthi, M. (2023). CFD Simulation and Thermal Performance Optimization of Flow in a Channel with Multiple Baffles. *Journal of Heat and Mass Transfer Research*, 10(2), 257–268. doi: 10.22075/JHMTR.2023.31108.1458.

Rehman, K. U., Malik, M. Y., Zahri, M., & Tahir, M. (2018). Numerical analysis of MHD Casson Navier's slip nanofluid flow yield by rigid rotating disk. *Results in Physics*, 8,744-751. doi:10.1016/j.rinp.2018.01.017

Sajid, T., Ayub, A., Shah, S. Z. H., Jamshed, W., Eid, M. R., El Din, E. S. M. T., ... & Hussain, S. M. (2022). Trace of chemical reactions accompanied with arrhenius energy on ternary hybridity nanofluid past a wedge. *Symmetry*, 14(9), 1850.

Sarada, K., Gamaoun, F., Abdulrahman, A., Paramesh, S.O., Kumar, R., Prasanna, G.D., & Gowda, R.P. (2022). Impact of exponential form of internal heat generation on water-based ternary hybrid nanofluid flow by capitalizing non-Fourier heat flux model. *Case Studies in Thermal Engineering*, 38, 102332.

Saraswathy, M., Prakash, D., Muthtamilselvan, M., & Al Mdallal, Q. M. (2022). Arrhenius energy on asymmetric flow and heat transfer of micropolar fluids with variable properties: A sensitivity approach. *Alexandria Engineering Journal*, 61(12), 12329–12352.

Sulochana, C., & Poornima, M. (2019). Unsteady MHD Casson fluid flow through vertical plate in the presence of Hall current. *SN Applied Sciences*, 1, 1–14.

Sundar, L.S., Chandra Mouli, K.V., Said, Z., & Sousa, A.C. (2021). Heat transfer and second law analysis of ethylene glycol-based ternary hybrid nanofluid under laminar flow. *Journal of Thermal Science and Engineering Applications*, 13(5), 051021.

Suresh Kumar, Y., Hussain, S., Raghunath, K., Ali, F., Guedri, K., Eldin, S. M., & Khan, M. I. (2023). Numerical analysis of magnetohydrodynamics Casson nanofluid flow with activation energy, Hall current and thermal radiation. *Scientific Reports*, 13(1), 4021.

Tassaddiq, A., Khan, S., Bilal, M., Gul, T., Mukhtar, S., Shah, Z., & Bonyah, E. (2020). Heat and mass transfer together with hybrid nanofluid flow over a rotating disk. *AIP Advances*, 10(5), 055317. doi:10.1063/5.0010181

Thakur, A., & Sood, S. (2023). Tri-Hybrid Nanofluid Flow Towards Convectively Heated Stretching Riga Plate with Variable Thickness. *Journal of Nanofluids*, 12(4), 1129–1140.

Umavathi, J. C., Prakasha, D. G., Alanazi, Y. M., Lashin, M. M., Al-Mubaddel, F. S., Kumar, R., & Gowda, R. P. (2023). Magnetohydrodynamic squeezing Casson nanofluid flow between parallel convectively heated disks. *International Journal of Modern Physics B*, 37(04), 2350031.

Waqas, H., Naseem, R., Muhammad, T., & Farooq, U. (2021). Bioconvection flow of Casson nanofluid by rotating disk with motile microorganisms. *Journal of Materials Research and Technology*, 13, 2392–2407. 10.1016/j.jmrt.2021.05.092

Zhang, X. H., A. Algehyne, E., G. Alshehri, M., Bilal, M., Khan, M. A., & Muhammad, T. (2021). The parametric study of hybrid nanofluid flow with heat transition characteristics over a fluctuating spinning disk. *Plos One*, 16(8), e0254457.

Chapter 14

Brinkman equation in nanofluids

Generalized stream function solution

Deepak Kumar Maurya and Satya Deo

14.1 INTRODUCTION

A substance involving the void spaces or pores, and in normal circumstances when fluid penetrates through these pores, is called a porous medium (Nield and Bejan, 2006). Maurya and Lata (2022) presented the analytical stream function solution of the Darcy equation in parabolic coordinates, parabolic cylindrical coordinates, and bipolar cylindrical coordinates. The Brinkman equation takes into consideration the effects of porous media on fluid flow and is a modified version of the Darcy equation, proposed by Brinkman (1947). It is extensively employed in the investigation of fluid flow in the porous material of high permeability. Mathematical expression of the Brinkman equation is

$$\nabla p = -\frac{\mu}{k}v + \mu_e \nabla^2 v. \tag{14.1}$$

Qin and Kaloni (1988) implemented the Cartesian tensor to analyze the Brinkman equation and evaluated the drag force exerted by a porous sphere in an unbounded medium. Pop and Cheng (1992) used the Brinkman model to determine the steady-state solution of an incompressible viscous fluid flow around a circular cylinder immersed in a porous medium. In the limiting case of a porous prolate spheroidal coordinates, the general stream function solution of Brinkman equation in spherical polar coordinates is investigated by Zlatanovski (1999). Deo and Maurya (2019) reported the generalized stream function solution of Brinkman equation in cylindrical polar coordinates. Maurya and Deo (2020) reported the stream function solution of Brinkman equation in parabolic cylindrical coordinates. Deo et al. (2016) investigated a general stream function solution to the Brinkman equation in cylindrical polar coordinates. Auriault (2009) discussed on the domain of validity and applicability of Brinkman's equation.

In the presence of a magnetic field, Deo et al. (2020) examined the Stokesian flow of a micropolar fluid in a cylindrical tube around an impermeable core coated in a porous layer. Uniform magnetic field is applied in a

 DOI: 10.1201/9781003494454-14

direction perpendicular to the axis of the cylinder, which is taken along the path of the fluid flow. Using four well-known cell models, Deo et al. (2021) studied the influence of magnetic field on the hydrodynamic permeability of a biporous membrane with respect to the flow of micropolar liquid. Using Nowacki's method, the governing equations of micropolar liquid are represented in a modified form. For the micropolar liquid, Khanukaeva and Deo (2019) examined the problem of an infinite flow past an infinite porous cylinder perpendicular to its axis and demonstrated that the usual Stokes paradox cannot be observed for a porous body in either a polar or non-polar liquid. The Stokes flow of a micropolar fluid through a porous cylinder was examined by Maurya et al. (2021), who also compared the flow patterns for two types of boundary value problems. The impact of a magnetic field on a Newtonian fluid that is positioned between two porous cylindrical pipes that are filled with micropolar fluids is discussed by Maurya and Deo (2022). Deo and Maurya (2022) investigated how a magnetic field affected the flow of a Newtonian fluid through a rectangular porous channel that was positioned between two non-Newtonian fluid layers. Maurya et al. (2023) studied the Stokesian flow of an axisymmetric, incompressible couple stress fluid through a porous medium enclosing a solid sphere in the presence of a uniform magnetic field. Selvi et al. (2023) investigated the flow of an incompressible couple stress fluid in a Reiner–Rivlin liquid with a permeable media.

Fluids with nanoparticles suspended in a base fluid are known as nanofluids. Metal or metallic oxide nanoparticles are dissolved into specific base fluids to create nanofluids (Hatami and Jing, 2020). Nanofluids differ totally from their separated phases in terms of their thermophysical characteristics since they exist in two distinct phases: liquid and solid. The thermal conductivity and heat transfer characteristics of the base fluid can be greatly improved by adding nanoparticles to it. The governing equations of nanofluids consist momentum equation, heat equation, and concentration equation (Ali et al., 2023). From the concentrated suspensions, the differential approach was used to get the generalized Brinkman solution by Solyaev et al. (2020). Phase change materials have been employed by Han et al. (2008) as nanoparticles in nanofluids to increase the fluids' specific heat and effective thermal conductivity at the same time. The use of nanofluid enhanced freezing may also increase the flexibility of traditional cryosurgery by manipulating the size, shape, image, and direction of ice ball production. Direct solar collectors based on nanofluids are solar thermal collectors in which solar energy is scattered and absorbed by nanoparticles suspended in a liquid medium, reported by Prasad et al. (2017).

The extended expression on the stream function solution of the Brinkman equation in cylindrical polar coordinates is the basis of the current investigation (Deo and Maurya, 2019). A standard relationship between modified Bessel's function and hypergeometric functions is expressed.

Analysis is carried out on the Wronskian of the modified Bessel's functions of the first and second kinds. Additionally, we have also verified that the stream function solution can potentially be utilized to deduce the stream function solution of the Brinkman equation in the cylindrical polar coordinates, reported by Pop and Cheng (1992) and Deo et al. (2016). In this chapter, we have also reported the importance of solution of Brinkman equation in nanofluids as well as employment of nanofluids in various fields of science, engineering, and technology.

14.2 MATHEMATICAL FORMULATION

The cylindrical polar coordinates (r,θ,z) and the position vector of a moving fluid particle in the Cartesian coordinates (x,y,z) are connected by the following relations:

$$x = r\cos\theta, \qquad y = r\sin\theta, \qquad z = z, \tag{14.2}$$

where $0 \leq r < \infty, 0 \leq \theta < 2\pi$ and $-\infty < z < \infty$.

The Darcy number (Da) is the order of magnitude of $(\mu_e / \mu)(k / \mathrm{L}^2)$, which is the ratio of the term $(\mu_e \nabla^2 \mathbf{v})$ in the Brinkman equation to the term $(\mu / \mathrm{k})\mathbf{v}$. If dynamic viscosity and effective viscosity are taken to be equal for the sake of simplicity, the Darcy number is (k / L^2). Typically, Da is substantially less than unity; nevertheless, there are certain exceptional situations when 'Da' equals around 8. The Eq. (14.1) can be transformed into a non-dimensional form by incorporating the following dimensionless parameters to investigate the Brinkman equation as a mathematical expression:

$$\tilde{\mathbf{v}} = \frac{\mathbf{v}}{U}, \tilde{r} = \frac{r}{L}, \tilde{p} = \frac{pL}{\mu U} \text{ and } \eta = \frac{k}{L^2}. \tag{14.3}$$

Regarding the purpose of the present investigation, we will now disregard the mathematical symbol tilde (~). Consequently, non-dimensional expressions of the gradient and Laplacian operators are, respectively,

$$\nabla = \hat{r}\frac{\partial}{\partial r} + \frac{\hat{\theta}}{r}\frac{\partial}{\partial \theta} \text{ and } \nabla^2 = \frac{\partial^2}{\partial r^2} + \frac{1}{r}\frac{\partial}{\partial r} + \frac{1}{r^2}\frac{\partial}{\partial \theta r^2} + \frac{\partial^2}{\partial z^2}.$$

Substituting these values in Eq. (14.1), we get a reduced version of the Brinkman equation in the non-dimensional form as:

$$\nabla p = (\nabla^2 - \alpha^2)\mathbf{v}, \text{where, } \alpha^2 = \frac{1}{\eta}. \tag{14.4}$$

Assuming that the non-dimensional space coordinates (r,θ,z) have absolutely no impact on the dynamic coefficient of viscosity μ, effective viscosity μ_e of fluid flow through porous media, and permeability parameter of the porous medium α^2. Therefore, parameter α^2 will be space independent. Through the utilization of the curl operator on both sides of Eq. (14.4), one can obtain

$$(\nabla^2 - \alpha^2)\nabla \times \mathbf{v} = \mathbf{0}. \tag{14.5}$$

Considering the fluid velocity vector in the two-dimensional fluid flow that represents the originating point of the seepage velocity component relevant to the porous medium,

$$\mathbf{v} = v_r(r,\theta), v(r,\theta), 0]. \tag{14.6}$$

To fulfill the mathematical argument of the equation of continuity for an incompressible viscous fluid flow field, a scalar valued function $\psi(r,\theta)$, known as Lagrange's stream function, is proposed in terms of the seepage velocity for two-dimensional fluid motion. For trying to accomplish this assumption, we may incorporate velocity components,

$$v_r = -\frac{1}{r}\frac{\partial \psi}{\partial \theta} \quad \text{and} \quad v_\theta = \frac{\partial \psi}{\partial r}. \tag{14.7}$$

By implementing the curl operator on the seepage velocity, the expression of the vorticity vector $\boldsymbol{\omega}(r,\theta)$ can be determined. It is possible to characterize the vorticity vector with a stream function using the Eq. (14.7) as follows:

$$\boldsymbol{\omega}(r,\theta) = \nabla \times \mathbf{v} = \hat{k}\nabla^2\psi. \tag{14.8}$$

Substituting the value of $\nabla \times \mathbf{v}$ in the Eq. (14.5), we obtain:

$$\nabla^2(\nabla^2 - \alpha^2)\psi(r,\theta) = 0. \tag{14.9}$$

14.3 GENERALIZED STREAM FUNCTION SOLUTION

To determine the generalized stream function solution of Eq. (14.9), we may substitute:

$$(\nabla^2 - \alpha^2)\psi(r,\theta) = \Phi(r,\theta), \tag{14.10}$$

then, Eq. (14.9) will reduce to

$$\nabla^2\Phi(r,\theta) = 0. \tag{14.11}$$

The Eq. (14.11) is solved by utilizing the separation of variables strategy, we obtain:

$$\Phi(r,\theta)=\sum_{n=0}^{\infty}\left[A_n^{(1)}r^n+B_n^{(1)}r^{-n}\right]_{\sin n\theta}^{\cos n\theta}. \tag{14.12}$$

Therefore, Eq. (14.10) assumes the form as:

$$\left(\nabla^2-\alpha^2\right)\psi(r,\theta)=\Phi(r,\theta)=\sum_{n=0}^{\infty}\left[A_n^{(1)}r^n+B_n^{(1)}r^{-n}\right]_{\sin n\theta}^{\cos n\theta}. \tag{14.13}$$

Now, we shall apply the method of separation of variables by assuming that

$$\psi(r,\theta)=\sum_{n=0}^{\infty}[\chi(r)]_{\sin n\theta}^{\cos n\theta}. \tag{14.14}$$

Substituting the value of $\psi(r,\theta)$ from Eq. (14.14) into the Eq. (14.13) and equating the corresponding coefficients, we obtain:

$$\frac{d^2\chi}{dr^2}+\frac{1}{r}\frac{d\chi}{dr}-\frac{n^2\chi}{r^2}-\alpha^2\chi=[A_n^{(1)}r^n+B_n^{(1)}r^{-n}]. \tag{14.15}$$

The expression in Eq. (14.15) pertains to all integral values of n because the trigonometric functions $\cos(n\theta)$ and $\sin(n\theta)$ are linearly independent. The corresponding homogeneous equation of Eq. (14.15) is

$$r^2\frac{d^2\chi}{dr^2}+r\frac{d\chi}{dr}-[(\alpha r)^2+n^2]\chi=0. \tag{14.16}$$

The differential Eq. (14.16) is the modified Bessel differential equation of order $n\in\mathbb{Z}$ which includes two linearly independent solutions, $I_n(\alpha r)$ and $K_n(\alpha r)$. Therefore, complementary function $\chi_e(r)$ of Eq. (14.15) is

$$\chi_c(r)=[A_n^{(1)}I_n(\alpha r)+B_n^{(2)}K_n(\alpha r)]. \tag{14.17}$$

To find the particular-integral $\chi_P(r)$ of Eq. (14.15), we use the method of variation of parameters with hypothesis,

$$\chi_P(r)=\beta(r)I_n(\alpha r)+\gamma(r)K_n(\alpha r). \tag{14.18}$$

where $\beta(r)$ and $\gamma(r)$ are functions of r only that can be obtained by

$$\beta(r)=-\int\left[\frac{A_n^{(1)}r^n+B_n^{(1)}r^{-n}}{W(I_n(\alpha r),K_n(\alpha r))}\right]K_n(\alpha r)dr, \tag{14.19}$$

and $$\gamma(r) = -\int\left[\frac{A_n^{(1)}r^n + B_n^{(1)}r^{-n}}{W(I_n(\alpha r),\mathrm{K}_n(\alpha r))}\right]I_n(\alpha r)dr. \tag{14.20}$$

The value of the Wronskian for linearly independent solutions $I_n(\alpha r)$ and $K_n(\alpha r)$ is $(-1/r)$. After evaluating Eqs. (14.19) and (14.20), we obtain:

$$\beta(r) = \frac{r^n}{\alpha^2}\left(-\alpha r\, I_{n+1}(\alpha r) + \left(\frac{2}{\alpha r}\right)^n \Gamma(n+1)\right)A_n^{(1)}$$

$$+\frac{r^{-n}}{\alpha^2}\left(-\alpha r\, I_{-n+1}(\alpha r) + \left(\frac{2}{\alpha r}\right)^{-n} \Gamma(-n+1)\right)B_n^{(1)}, \tag{14.21}$$

$$\gamma(r) = \left(-\frac{r^{n+1}}{\alpha}I_{n+1}(\alpha r)\right)A_n^{(1)} + \left(\frac{2^{-n+1}\alpha^{n-2}}{\Gamma n} - \frac{2^{-n+1}}{\alpha}I_{n-1}(\alpha r)\right)B_n^{(1)}. \tag{14.22}$$

Substituting these values of $\beta(r)$ and $\gamma(r)$ in Eq. (14.18) and simplify, we get:

$$\chi_P(r) = \frac{r^n}{\alpha^2}\left(-1 + {}_0F_1\left(;n+1;\frac{\alpha^2 r^2}{4}\right)\right)A_n^{(1)}$$

$$+\frac{r^{-n}}{\alpha^2}\left(-1 + {}_0F_1\left(;-n+1;\frac{\alpha^2 r^2}{4}\right)\right)A_n^{(1)}. \tag{14.23}$$

Thus, general solution of Eq. (14.15) will be:

$$\chi(r) = \frac{r^n}{\alpha^2}\left(-1 + {}_0F_1\left(;n+1;\frac{\alpha^2 r^2}{4}\right)\right)A_n^{(1)} + \frac{r^{-n}}{\alpha^2}\left(-1 + {}_0F_1\left(;-n+1;\frac{\alpha^2 r^2}{4}\right)\right)A_n^{(1)}$$

$$+ A_n^{(2)}I_n(\alpha r) + B_n^{(2)}K_n(\alpha r). \tag{14.24}$$

Therefore, the generalized stream function solution of the Brinkman equation in the cylindrical polar coordinates can be expressed in the following form:

$$\psi(r,\theta) = \sum_{n\geq 0}\Big[A_n^{(2)}I_n(\alpha r) + B_n^{(2)}K_n(\alpha r) + r^n\left(-1 + {}_0F_1\left(;n+1;\frac{\alpha^2 r^2}{4}\right)\right)A_n^{(3)}$$

$$+ r^{-1}\left(-1 + {}_0F_1\left(;-n+1;\frac{\alpha^2 r^2}{4}\right)\right)A_n^{(3)}\Big]_{\sin n\theta}^{\cos n\theta}. \tag{14.25}$$

This generalized solution for the Brinkman equation in cylindrical polar coordinates combines modified Bessel functions $I_n(\cdot), K_n(\cdot)$ and hypergeometric function ${}_0F_1(;n;r)$. The above-mentioned stream function solution of the Brinkman equation was reported earlier by Deo and Maurya (2019).

14.4 ANALYSIS ON GENERALIZED SOLUTION

Deo et al. (2016) reported a stream function solution of the Brinkman equation in cylindrical coordinates (r,θ,z)

$$\psi(r,\theta)=\sum_{n=0}^{\infty}\left[A_nI_n(\alpha r)+B_nK_n(\alpha r)]+C_nr^n+D_nr^{-n}\right]_{\sin n\theta}^{\cos n\theta}. \tag{14.26}$$

where A_n, B_n, C_n and D_n are arbitrary parameters.

Applying a few well-defined relations, we are going to demonstrate in this section that Eq. (14.26) may be obtained utilizing Eq. (14.25). Utilizing Eq. (14.26), one can establish that the stream function is the outcome of incorporating the algebraic polynomials and modified Bessel functions. Consequently, the stream function solution obtained in Eq. (14.25) consisting of an algebraic function, a hypergeometric function, and modified Bessel's functions of the first and second kinds. We require certain connections between modified Bessel's function and hypergeometric functions to generate Eq. (14.26). The hypergeometric function ${}_0F_1(;n+1;\alpha^2r^2/4)$ is expressed in the following form:

$${}_0F_1\left(;n+1;\frac{\alpha^2r^2}{4}\right)=\sum_{n=0}^{\infty}\left[\frac{\Gamma(n+1)}{\Gamma(m+1)\Gamma(m+n+1)}\left(\frac{\alpha^2r^2}{4}\right)^m\right]. \tag{14.27}$$

The modified Bessel function and the hypergeometric function can be connected using the relation:

$${}_0F_1\left(;n+1;\frac{\alpha^2r^2}{4}\right)=\left(\frac{2}{\alpha r}\right)^n\Gamma(n+1)I_n(\alpha r). \tag{14.28}$$

From Eqs. (14.25) and (14.28), we get:

$$\psi(r,\theta)=\sum_{n\geq 0}\left[A_n^{(2)}I_n(\alpha r)+B_n^{(2)}K_n(\alpha r)+r^n\left(-1+\left(\frac{2}{\alpha r}\right)^n\Gamma(n+1)I_n(\alpha r)\right)A_n^{(3)}\right.$$
$$\left.+r^{-n}\left(-1+\left(\frac{2}{\alpha r}\right)^{-n}\Gamma(-n+1)I_{-n}(\alpha r)\right)A_n^{(3)}\right]_{\sin n\theta}^{\cos n\theta}. \tag{14.29}$$

Since the modified Bessel's functions $I_n(\alpha r)$ and $I_{-n}(\alpha r)$ are dependent for all $n \in \mathbb{Z}$, therefore, these functions can be expressed as:

$$I_{-n}(\alpha r) = \rho(n) I_n(\alpha r). \tag{14.30}$$

By adjusting the value of $I_{-n}(\alpha r)$ from Eq. (14.30) into Eq. (14.29), we may derive the explicit form of the stream function solution to the Brinkman equation as follows:

$$\psi(r,\theta) = \sum_{n\geq 0} \left[C_n^{(1)} I_n(\alpha r) + D_n^{(1)} K_n(\alpha r) + C_n^{(2)} r^n + D_n^{(2)} r^{-n} \right]_{\sin n\theta}^{\cos n\theta}, \tag{14.31}$$

where arbitrary constants $C_n^{(i)}, D_n^{(i)}$, $i = 1,2$ are related with constants $A_n^{(j)}, B_n^{(j)}$, $j = 2,3$ by the following relations:

$$C_n^{(1)} = A_n^{(2)} + A_n^{(3)} \left(\frac{2}{\alpha}\right)^n \Gamma(n+1) + B_n^{(3)} \left(\frac{2}{\alpha}\right)^{-n} \Gamma(-n+1)\rho(n),$$

$$D_n^{(1)} = B_n^{(2)}, C_n^{(2)} = -A_n^{(3)}, D_n^{(2)} = -B_n^{(3)}.$$

Deo et al. (2016) established that the stream function solution of Eq. (14.25) is the generalized form of Eq. (14.26), which is the stream function solution of the Brinkman equation expressed in the cylindrical polar coordinates. From the generalized stream function solution indicated in Eq. (14.25), the stream function solution of the Brinkman equation can be found as a specific case ($n=1$), which was reported earlier by Pop and Cheng (1992).

14.4.1 Alternative Approach for Generalized Solution

Furthermore, it is possible to solve the Brinkman equation in the Laplacian operator form described in Eq. (14.9) by making the following assumption:

$$\nabla^2 \psi(r,\theta) = H(r,\theta), \tag{14.32}$$

then, the differential equation (14.9) will reduce to

$$(\nabla^2 - \alpha^2) \mathrm{H}(r,\theta) = 0. \tag{14.33}$$

Applying the method of variation of parameters and the technique of separation of variables to solving Eqs. (14.32) and (14.33), we obtain an explicit expression of the stream function solution which is identical to the obtained solution of Eq. (14.26), which requires the arrangement of arbitrary constants.

14.5 APPLICATIONS OF BRINKMAN EQUATION IN NANOFLUIDS

The flow of hybrid dusty nanofluids across porous media with mass transpiration and heat transfer has been studied using the Brinkman type equation by Sneha et al. (2021). According to their study, the fundamental similarity equations accept two phases for both stretching and shrinking surfaces. Selvi et al. (2024) employed the cell model technique to study the axisymmetric flow of a steady incompressible Reiner–Rivlin liquid sphere enclosed by a spherical porous envelope. To study the hydrodynamics through porous media, isotropic permeability is taken into consideration and the Brinkman-extended Darcy model is implemented. Applications of the Brinkman equation can be found in several domains, such as biological fluxes associated with biofilms, blood clots, and flagellar motion in gels. It has also been used to simulate fluid flow in nanofluids and anisotropic porous media as studied by Karageorghis et al. (2021). The COMSOL Multiphysics software uses the Brinkman equation to calculate the fluid velocity and pressure fields of single-phase flow in porous media under laminar flow conditions, reported by Ouadefli *et al.* (2022). Kim et al. (2007) reported several applications of Reiner–Rivlin nanofluids that contribute to a variety of industrial heat transfer procedures and have attracted more research interests, in recent years. These nanofluids are also used in pharmaceutical applications, with some including nanodrugs. Hard drives, understanding turbine systems, jet engines, optical sensors, heating and cooling components are examples of products that have applications for these individuals.

14.6 APPLICATIONS OF NANOFLUIDS IN DIVERSE FIELD

To assess the potential sustainability of nanofluids in nuclear applications, several researchers enhanced the efficiency of any heat-removal limited water-cooled nuclear system. Applications include primary coolant for pressurized water reactors, backup safety systems, accelerator targets, plasma divertors, and many more. Using nanofluids instead of water coats fuel rods with nanoparticles like alumina, avoiding bubble formation and enhancing critical heat flux substantially. Nanofluids can be utilized as a coolant in emergency cooling systems, reducing overheating, and improving powerplant safety. The usage of nanofluids in power plants raises concerns about the unpredictable number of nanoparticles carried away by boiling vapor. A further concern is the need for additional safety procedures when disposing of nanofluids. Nanofluid coolants are doubtful to be used in boiling water reactors due to problems with erosion and contamination resulting from nanoparticles in the turbine and condenser. Engine oils, automatic transmission fluids, coolants, lubricants, and other synthetic high-temperature heat transfer fluids used in traditional vehicle thermal systems such as radiators, engines, heating, ventilation, and air-conditioning have poor heat transfer qualities. Nanofluids with added nanoparticles may provide great thermal conductivity. Using nanofluids as coolants can reduce radiator size and improve location. The improved effectiveness allows for smaller coolant pumps and higher engine temperatures, resulting in more horsepower.

14.7 CONCLUSION AND FUTURE SCOPE

The method of variation of parameters is employed to examine the generalized stream function solution of the Brinkman equation in the cylindrical polar coordinates (r,θ,z) Numerous algebraic functions, modified Bessel's function of the first and second kinds, and hypergeometric functions are incorporated to generate the specified generalized stream function. The reported analytical solutions can be used to investigate real life problems, such as fluid flow through a swarm of fibrous cylindrical particles and in the process of extraction of oils/minerals through porous cylindrical pipes, etc. Nanofluids can be used as a coolant in emergency cooling systems, which reduces overheating and increases power plant safety. The use of nanofluids in power plants raises issues about the unpredictable number of nanoparticles transported by boiling vapor. Products with uses for these people include hard drives, knowledge of turbine systems, jet engines, optical sensors, heating and cooling parts, and hard drives.

ACKNOWLEDGMENT

To carry out this research work, Deepak Kumar Maurya is grateful to the Council of Science and Technology, Uttar Pradesh, India, for the Research Project (Ref. No. CST/D-1517).

NOMENCLATURE

k	Permeability of porous medium
v	Fluid velocity
ω	Vorticity vector
p	Fluid pressure
μ_e	Effective viscosity
μ	Dynamic viscosity of fluid
∇	Gradient operator

∇^2	Laplacian operator
ψ	Stream function
Da	Darcy number
L	Characteristic length
n^2	Separation constant
$I_n(\cdot)$	Modified Bessel function of first kind of order n
$K_n(\cdot)$	Modified Bessel function of second kind of order n
${}_0F_1(;n,r)$	Hypergeometric function
$W(I_n(\cdot), K_n(\cdot))$	Wronskian of $I_n(\cdot)$ and $K_n(\cdot)$
$\rho(n)$	Scalar valued function
v_r, v_θ	Velocity components
$\hat{r}, \hat{\theta}$	Unit vectors
h_1, h_2, h_3	Scale factors
(r, θ, z)	Cylindrical polar coordinates
(x, y, z)	Cartesian coordinates
$A_n^{(i)}, B_n^{(i)}, C_n^{(i)}, D_n^{(i)}$	Arbitrary constants

REFERENCES

Ali, H.M., Hassan, A. and Wahab, A. (2023). *Nanofluids for Heat Exchangers.* Springer, 2023.

Auriault, J.L. (2009). On the domain of validity of Brinkman's equation, *Transp. Porous Media*, vol. 79, pp. 215–223.

Brinkman, H.C. (1947). A calculation of viscous force exerted by a flowing fluid on a dense swarm of particles, *Appl. Sci. Res.*, vol. A1, pp. 27–34.

Deo, S., Ansari, I.A. and Srivastava, B.G. (2016). On the general stream function solution of Brinkman equation in the cylindrical polar coordinates, *Adv. Theo. Appl. Mech.*, vol. 9, no. 1, pp. 21–30.

Deo, S. and Maurya, D.K. (2019). Generalized stream function solution of the Brinkman equation in the cylindrical polar coordinates, *Spec. Topics Rev. Porous Med.*, vol. 10, no. 5, pp. 421–428.

Deo, S., Maurya, D.K. and Filippov, A.N. (2020). Influence of magnetic field on micropolar fluid flow in a cylindrical tube enclosing an impermeable core coated with porous layer, *Colloid J.*, vol. 82, no. 6, pp. 649–660.

Deo, S., Maurya, D.K. and Filippov, A.N. (2021). Effect of magnetic field on hydrodynamic permeability of biporous membrane relative to micropolar liquid flow, *Colloid J.*, vol. 83, pp. 662–675.

Deo, S. and Maurya, D.K. (2022). Investigation of MHD effects on micropolar-Newtonian fluid flow through composite porous channel, *Microfluid. Nanofluid.* vol. 26, Article no. 64.

Han, Z.H., Cao, F.Y. and Yang, B. (2008). Synthesis and thermal characterization of phase-changeable indium/polyalphaolefin nanofluids, *Appl. Phy. Lett.*, vol. 92, no. 24, Article no. 243104.

Hatami, M. and Jing, D. (2020). *Nanofluids: Mathematical, Numerical, and Experimental Analysis*, Academic Press.

Karageorghis, A., Lesnic, D. and Marin, L. (2021). The method of fundamental solutions for Brinkman flows. Part I. exterior domains, *J. Eng. Math.*, vol. 126, Article no. 10.

Khanukaeva, D.Y. and Deo, S. (2019). On the Stokes paradox in a micropolar liquid, *Colloid J.*, vol. 81, pp. 395–400.

Kim, S.J., Bang, I.C., Buongiorno, J. and Hu, L.W. (2007). Surface wettability change during pool boiling of nanofluids and its effect on critical heat flux, *Int. J. Heat Mass Transf.*, vol. 50, no. 19–20, pp. 4105–4116.

Maurya, D.K. and Deo, S. (2020). Stream function solution of the Brinkman equation in parabolic cylindrical coordinates, *Int. J. Appl. Comput. Math.*, vol. 6, Article 167.

Maurya, D.K., Deo, S. and Khanukaeva, D. (2021). Analysis of Stokes flow of micropolar fluid through a porous cylinder, *Math. Meth. Appl. Sci.*, vol. 44, pp. 6647–6665.

Maurya, D.K. and Deo, S. (2022). Effect of magnetic field on Newtonian fluid sandwiched between non-Newtonian fluids through porous cylindrical shells, *Spec. Top. Rev. Porous Media*, vol. 13, pp. 75–92.

Maurya, P.K., Deo, S. and Maurya, D.K. (2023). Couple stress fluid flow enclosing a solid sphere in a porous medium: Effect of magnetic field, *Phy. Fluids*, vol. 35, Article no. 072006, 2023.

Maurya, D.K. and Lata, S. (2022). Analytical solution of the Darcy equation in parabolic, bi-polar cylindrical and parabolic cylindrical coordinates, *J. Progress. Sci.*, vol. 13, no. 1–2, pp. 20–27.

Nield, D.A. and Bejan, A. (2006). *Convection in Porous Media*, Springer.

Ouadefli, L.E., Akkad, A.E., Moutea, O.E., Moustabchır, H., Elkhalfi, A., Scutaru, M.L. and Muntean, R. (2022). Numerical simulation for Brinkman system with varied permeability tensor, *Mathematics*, vol. 10, no. 18, Article no. 3242

Pop, I. and Cheng, P. (1992). Flow past a circular cylinder embedded in a porous medium based on the Brinkman model, *Int. J. Eng. Sci.*, vol. 30, no. 2, pp. 257–262.

Prasad, A.R., Singh, S. and Nagar, H. (2017). A review on nanofluids: Properties and applications, *Int. J. Adv. Res. Innov. Ideas Educ.*, vol. 3, no. 3, pp. 3185–3209.

Qin, Y. and Kaloni, P.N. (1988). A Cartesian-tensor solution of the Brinkman equation, *J. Eng. Math.*, vol. 22, pp. 177–188.

Selvi, R., Maurya, D.K. and Shukla, P. (2023). Analytical solution of a couple stress fluid saturated in a porous medium through a Reiner–Rivlin liquid sphere, *Phys. Fluids*, vol. 35, Article no. 073106.

Selvi, R., Maurya, D.K., Shukla, P. and Chamkha, A.J. (2024). Analysis of a Reiner–Rivlin liquid sphere enveloped by a permeable layer, *Phys. Fluids*, vol. 36, no. 1.

Sneha, K.N., Mahabaleshwar, U.S., Bennacer, R. and Ganaoui, M.E.L. (2021). Darcy Brinkman equations for hybrid dusty nanofluid flow with heat transfer and mass transpiration, *Computation*, vol. 9, no. 11, Article no. 118.

Solyaev, Y.O., Lurie, S.A. and Semenov, N.A. (2020). Generalized Einstein's and Brinkman's solutions for the effective viscosity of nanofluids, *J. Appl. Phys.*, vol. 128, Article no. 035102.

Zlatanovski, T. (1999). Axisymmetric creeping flow past a porous prolate spheroidal particle using the Brinkman model, *Quart. J. Mech. Appl. Math.*, vol. 52, no. 1, pp. 111–126.

Chapter 15

Future trends and emerging research in nanofluids for aerospace applications

Milad Heidari, Sivasakthivel Thangavel, Khulood Al Ghafri, and Ashwani Kumar

15.1 INTRODUCTION

Nanofluids are engineered colloidal suspensions comprising a base fluid, typically a conventional liquid coolant, and nanoparticles with diameters in the nanometer range, often less than 100 nm. The fluid's thermal and heat transport capabilities are improved by the dispersion of these nanoparticles, which can be carbon-based compounds or metallic oxides. Improved thermal conductivity, convective heat transfer, and rheological behavior are the outcomes of the small size and high surface area of the nanoparticles (Awasthi et al. 2024). This makes nanofluids a promising solution for advanced heat dissipation and thermal management applications in a variety of fields, including aerospace, electronics, and energy systems (Dewangan et al. 2023; Kundu et al. 2023).

In the aerospace industry, where strict operating conditions necessitate precise temperature control to maintain the dependability, effectiveness, and safety of diverse systems, thermal management is essential. Thermal management is essential in aerospace applications to dissipate the high heat produced during atmospheric re-entry, avionics functioning, and propulsion. The influence that efficient heat management has on the overall functionality, durability, and structural integrity of aeronautical systems and components highlights how important it is (Lv et al. 2022; Jiang et al. 2022).

The significant heat produced by propulsion devices, including rocket motors and jet engines, is one of the main challenges. To avoid overheating, which could reduce engine efficiency and result in catastrophic failures, effective cooling measures are crucial. Moreover, heat is produced during operation by avionic systems' sensors, communication devices, and electronic parts. Critical aerospace components may perform worse, be less reliable, or have a shorter lifespan if these systems have insufficient heat control.

Furthermore, aerodynamic heating causes spacecraft to encounter extremely elevated temperatures during atmospheric re-entry. For a vehicle to descend safely and under control, its thermal protection systems (TPSs)

 DOI: 10.1201/9781003494454-15

must adequately shelter it from the extreme heat. Managing thermal loads becomes crucial for creating re-entry vehicles that can tolerate these harsh circumstances.

Thermal management is critical to aircraft because it keeps components from degrading, keeps operating temperatures at ideal levels, and protects the structural integrity of aerospace systems. By guaranteeing spacecraft reliability in the most hostile settings, addressing these thermal problems not only improves the overall efficiency and safety of aerospace vehicles but also opens new horizons for exploration. The success and durability of aerospace activities depend increasingly on efficient thermal management as technology develops and aerospace missions get more ambitious (Lv et al. 2022).

Because of the distinct behaviors of nanoparticles at the nanoscale, nanofluids show great promise for improving thermal characteristics. These tiny particles, which are frequently metallic or ceramic in origin, give the base fluid exceptional thermal conductivity and heat transfer capabilities. Utilizing the increased surface area and fluid–nanoparticle interfacial interactions to promote effective thermal transport and improved heat dissipation is crucial (Verma et al. 2024a, 2024b).

The enhanced thermal conductivity that nanofluids can provide is especially useful in situations when traditional fluids are inadequate. In sectors including electronics cooling, energy systems, and aircraft, there is a critical need for sophisticated heat management. An inventive solution to these problems is provided by nanofluids, which promise increased dependability and efficiency. This overview of nanofluids highlights their importance as a state-of-the-art approach to next-generation thermal management, providing a way to meet the changing demands of contemporary technologies that demand accurate temperature control and effective heat dissipation. As research in nanofluids progresses, their potential applications continue to expand, making them a promising area of exploration for engineers and scientists seeking to optimize thermal performance in diverse industrial sectors (Barai et al. 2023).

The exploration of nanofluids in aerospace high-performance thermal management is motivated by the pressing need for advanced solutions to address the escalating thermal challenges faced by aerospace systems. Conventional cooling techniques are unable to manage the higher temperatures and power densities produced by contemporary avionics and propulsion systems. Because of their remarkable thermal characteristics, nanofluids present a viable solution to these constraints. Their ability to dramatically improve heat transfer efficiency and thermal conductivity presents a game-changing strategy for heat dissipation in aerospace components, guaranteeing peak performance and prolonging the life of vital systems (Martins Obalalu et al. 2023).

Because of the critical weight and space limits in the aircraft industry, the small size and effective thermal management offered by nanofluids are particularly enticing. Aerospace vehicles can reduce their overall weight by using more lightweight and compact thermal control systems, which can be designed in response to the possibility of increased heat dissipation. Furthermore, the investigation of nanofluids is consistent with the industry's goal of increased sustainability and economy because better temperature control can boost propulsion system performance while lowering fuel consumption and minimizing environmental effects.

As aerospace missions become more ambitious, involving longer durations and extreme operating conditions, the motivation to harness the benefits of nanofluids intensifies. As engineers and researchers explore the world of nanofluids, they are looking for dependable, lightweight, and energy-efficient thermal management solutions in the hopes of revolutionary breakthroughs in aerospace technology.

15.2 FUNDAMENTALS OF NANOFLUIDS

15.2.1 Characteristics of nanofluids

Nanofluids represent a category of advanced heat transfer fluids meticulously engineered by dispersing nanoparticles with dimensions on the nanometer scale within a base fluid. These nanoparticles, which are usually non-metallic or metallic, have special thermal and rheological characteristics that improve the nanofluid's overall performance. Compared to conventional heat transfer fluids, these particles' increased surface area due to their nanoscale size promotes better thermal conductivity and convective heat transfer efficiency (Ahmadizadeh et al. 2024; Chitt et al. 2024; Prajapati et al. 2011).

The characteristics of nanofluids are defined by their nanoparticle composition, size, and concentration. Various metallic oxides, such as alumina, copper oxide, or titanium dioxide, are commonly employed due to their favorable thermal properties. The nanoparticles, often ranging from 1 to 100 nm, contribute to augmented thermal conductivity while preserving fluidic properties. Total thermal enhancement is determined by the concentration of nanoparticles in the base fluid, an important aspect of nanofluid formulation. The most effective concentrations are ascertained by considering the application, guaranteeing harmony between enhanced heat transmission and negligible influence on fluid viscosity. Nanofluids exhibit non-linear enhancements in thermal conductivity, often surpassing the predictions of classical models, and their rheological behavior remains a subject of ongoing research. The definition and properties of nanofluids highlight its promise as a ground-breaking class of heat transfer fluids that may be customized to meet the specific needs of a variety of industries, such as energy, electronics, and aerospace, which require sophisticated thermal management capabilities (Manuel et al. 2023).

15.2.2 How nanoparticles influence thermal conductivity, viscosity, and heat transfer

The influence of nanoparticles on thermal conductivity, viscosity, and heat transfer in nanofluids is a complex interplay of nanoscale physics, fluid dynamics, and material science. Understanding these effects is critical for harnessing the full potential of nanofluids in various applications, such as aerospace thermal management, electronics cooling, and energy systems.

Thermal Conductivity Enhancement: By utilizing the concepts of increased surface area and phonon scattering, nanoparticles dramatically improve thermal conductivity. Heat-carrying phonons in a base fluid run against particles in the form of nanoparticles. As a result of this scattering effect, which impedes phonons' normal progression, thermal resistance rises, and thermal conductivity rises as well. In addition to enabling closer packing within the fluid, the smaller size of the nanoparticles opens new channels for effective heat transfer. Nanofluids are superior heat conductors than their base fluids because of additional quantum phenomena at the nanoscale that lead to increased thermal conductivity (Reddy et al. 2023).

Viscosity Changes: Viscosity is impacted when nanoparticles are added because it changes the rheological behavior of nanofluids. Crucial factors include the kind, concentration, and form of the nanoparticles. Although certain nanoparticles behave in a Newtonian manner, others introduce non-Newtonian traits that impact the physics of flow. Because of steric hindrance and increasing particle–particle interactions, viscosity typically tends to increase with higher nanoparticle concentrations. The surface chemistry and dispersion techniques, however, can lessen this impact. Improved dispersion is made possible by engineered coatings on nanoparticles, which also minimize agglomeration and lower viscosity elevation. For practical applications, it is imperative to optimize the performance of nanofluids by striking a balance between increased thermal characteristics and controllable viscosity.

Heat Transfer Mechanisms: Using conduction, convection, and radiation mechanisms, nanofluids improve heat transfer. Enhanced thermal conductivity facilitates effective heat movement by directly improving heat conduction within the fluid. Convection, the transfer of heat by fluid motion, is facilitated by the enhanced fluidic characteristics and greater heat-carrying ability of nanoparticles. Improved heat conduction and viscosity are two of the distinct behaviors of nanofluids that increase the efficiency of convective heat transfer. Moreover, heat transport from radiation can be impacted by nanoparticles in nanofluids. The heat transfer coefficient rises overall because of effective radiation being encouraged by the high surface area of nanoparticles (Bhattad et al. 2023).

Nanoparticle Size and Shape: The influence of nanoparticles on thermal characteristics is mostly determined by their size and form. Because smaller nanoparticles have a bigger surface area than volume, they typically show more noticeable improvements in heat conductivity. Because of their simplicity, spherical nanoparticles are frequently employed; nevertheless, anisotropic forms, like nanotubes or platelets, may provide special benefits in particular applications. Because the orientation and arrangement of nanoparticles influence the overall thermal behavior, precise control is essential during the synthesis of nanofluids.

Stability and Agglomeration: Nanoparticle stability within the fluid is paramount for sustained thermal enhancement. Agglomeration, the undesired clustering of nanoparticles, can negate the positive effects on thermal properties. The addition of stabilizing chemicals or the use of surfactants are examples of surface alterations that are essential for preventing agglomeration and guaranteeing the long-term stability of nanofluids.

15.3 NANOFLUIDS SYNTHESIS AND CHARACTERIZATION

Nanofluids synthesis involves methods like one-step chemical processes or physical techniques, ensuring precise control over nanoparticle properties. Characterization utilizes techniques such as TEM, XRD, and DLS to assess size, distribution, and stability, crucial for optimizing nanofluid performance in diverse applications (Tables 15.1 and 15.2).

The importance of nanoparticle size, shape, and concentration in thermal performance lies at the core of tailoring nanofluids for specific applications, particularly in the realm of advanced thermal management. These parameters intricately influence the heat transfer characteristics of nanofluids,

Table 15.1 Synthesis methods of nanofluids

Synthesis methods	*Characteristics/considerations*
One-step chemical method	– Direct synthesis in a single step using chemical reactions. – Requires precise control of reaction parameters. – Limited scalability for large-scale production.
Two-step chemical method	– Involves initial nanoparticle synthesis followed by dispersion. – Allows better control over nanoparticle properties. – Offers flexibility in choosing base fluid and stabilizers.
Two-step chemical method	– Includes methods like ball milling, laser ablation, and others. – Mechanical or thermal energy used to break down bulk materials. – High energy consumption; careful control of parameters crucial.
Physical methods	– Includes methods like ball milling, laser ablation, and others. – Mechanical or thermal energy used to break down bulk materials. – High energy consumption; careful control of parameters crucial.
Sol-gel method	– Formation of nanoparticles in a liquid precursor (sol). – Gelation to form a solid matrix with embedded nanoparticle – Applicable for a variety of materials and shapes.

Table 15.2 Characterization techniques of nanofluids

Characterization techniques	*Parameters/properties assessed*
TEM (Transmission electron microscopy)	– Nanoparticle size, shape, and distribution. – Confirming dispersion within the base fluid.
XRD (X-ray diffraction)	– Crystal structure and phase identification of nanoparticles. – Assessing purity and potential agglomeration
DLS (Dynamic light scattering)	– Hydrodynamic size and stability of nanoparticles in solution. – Detecting any agglomeration or changes over time.
Zeta potential measurement	– Surface charge of nanoparticles, influencing stability. – Predicting long-term dispersion behavior

playing a pivotal role in determining their efficacy and applicability across diverse industries.

Nanoparticle Size: The size of the nanoparticles is a significant element affecting the thermal performance of nanofluids. As the size of nanoparticles decreases, the ratio of surface area to volume increases considerably. This procedure results in increased thermal conductivity, which significantly enhances the interaction between the nanoparticles and the base fluid. Smaller nanoparticles have more accessible phonon scattering sites, which increase heat transmission efficiency. Finding the right balance is crucial, though, since too-small nanoparticles may exhibit more surface scattering and defeat the advantages of greater heat conductivity.

Nanoparticle Shape: Thermal performance becomes even more difficult due to the geometry of the nanoparticles. Anisotropic forms, such as nanotubes, or platelets, can impart unique thermal properties, while spherical nanoparticles are often used due to their simplicity and ease of manufacturing. Thermal anisotropy and conductivity are affected in numerous ways depending on how anisotropic nanoparticles are arranged and oriented concerning one another. When directional heat transfer is crucial, taking advantage of the advantages of various forms can be quite helpful.

Nanoparticle Concentration: One aspect that needs to be carefully considered is the concentration of nanoparticles in the base fluid. The thermal benefits might not be significant enough at lower concentrations to outweigh the viscosity rise and processing difficulties. Higher concentrations, however, may cause agglomeration, change rheological behavior, and lessen the benefit of improving thermal conductivity. Reaching the ideal concentration is essential to retaining the nanofluid's usability for industrial applications while also optimizing heat transfer improvement. The synergistic influence of these parameters on thermal performance becomes particularly evident in applications like aerospace thermal management, where stringent weight and space constraints necessitate precise control over nanofluid properties. In such scenarios, understanding how variations in nanoparticle size, shape, and concentration impact thermal conductivity and heat transfer efficiency becomes imperative for designing efficient cooling systems for propulsion and avionics (Pierre et al. 2022).

15.4 THERMAL MANAGEMENT CHALLENGES IN AEROSPACE

Aerospace applications confront a myriad of unique thermal challenges due to the extreme operating conditions and stringent performance demands inherent in this field. Sophisticated thermal management techniques are required to guarantee the longevity, safety, and effectiveness of aerospace systems and components due to these problems, which arise from avionics, atmospheric re-entry, and propulsion systems.

One of the primary thermal issues in aerospace is the extreme heat produced by propulsion systems, like rocket motors and jet engines. Extreme temperatures are encountered during combustion in these systems, which places significant thermal demands on engine parts. Effective cooling systems are essential for reducing overheating and stopping material deterioration. To preserve structural integrity and improve engine performance overall, sophisticated cooling solutions are necessary for turbine blades, combustion chambers, and nozzles, which are subjected to extreme thermal loads. A special thermal issue in aerospace propulsion is striking a careful balance between avoiding excessive thermal strains and reaching ideal combustion temperatures for propulsion efficiency.

Aerospace vehicles need complex avionic systems for monitoring, communication, and navigation. These avionics produce a lot of heat during operation, thus effective thermal management techniques are required. Sensors, CPUs, and communication devices are examples of electronic components that function best at a range of temperatures. Poor heat control can result in deteriorating performance, fewer reliable components, and in severe situations, catastrophic failures. The task at hand involves effectively dispersing heat from closely spaced electronic systems while taking weight constraints into account and guaranteeing the dependability of crucial avionic components in a variety of flying scenarios.

During atmospheric re-entry, spacecraft encounters extreme temperatures due to aerodynamic heating. Temperatures can reach several thousand degrees Celsius because of the friction created by the quick drop through Earth's atmosphere. Systems for preventing thermal damage become essential for protecting the spacecraft from this extreme heat. The difficulty lies in creating structures and materials that can endure these harsh environments while maintaining aerodynamic stability. To avoid structural collapse and preserve the integrity of the spacecraft's heat shield, thermal protection materials, such as ablative shields and heat-resistant ceramics, must efficiently drain heat.

Applications in the aerospace industry are extremely sensitive to weight and space limitations. Any heat management system needs to be lightweight because extra weight reduces performance and fuel efficiency. The size and complexity of thermal systems are also limited by space limits within aircraft and spacecraft. The task of creating small-scale, extremely effective thermal solutions that satisfy the stringent specifications of aircraft applications falls to engineers (Pan, Lu et al. 2023).

The integration of nanofluids and other sophisticated materials, together with other advancements in materials science, present potential and problems for aerospace thermal control. Although these materials provide improved heat dissipation and thermal conductivity, compatibility, stability, and safety must be carefully considered when integrating them. The difficult part of the task is managing the intricate interactions between new materials and current aerospace system architectures to ensure a smooth integration without sacrificing overall reliability.

15.4.1 Traditional cooling methods and their limitations

The overview of traditional cooling methods in the context of aerospace and other high-performance applications is crucial for understanding the evolution toward more advanced thermal management strategies. Traditional cooling methods encompass various techniques such as conduction, convection, and radiation, each with its own set of advantages and limitations.

Traditional cooling often relies on conduction as a primary method to transfer heat away from high-temperature regions. This entails distributing and absorbing heat away from the source by using conductive materials, such as metal heat spreaders or sinks. Conduction has limits even if it works well for modest heat dissipation, particularly in circumstances with high heat fluxes. Conduction-based cooling is less effective at reducing heat loads in aerospace propulsion and avionics systems because it becomes less effective as the temperature difference across the material grows.

Convective cooling utilizes the transfer of heat through the movement of fluids, either natural convection or forced convection with the aid of fans or pumps. Even though convective cooling is a common technique in aerospace applications, particularly for cooling electronic components, it can be difficult to get the best possible heat dissipation. Dependence on surrounding air or other cooling media presents drawbacks in situations when convective heat transfer rates are impeded by elevated temperatures or confined spaces. Furthermore, the heat exchange surface area affects convective cooling efficiency, making it difficult to sustain effective cooling in small, densely packed systems (Wang et al. 2022).

Thermal radiation is emitted in the process of radiative cooling to remove heat from a surface. Because of the elevated temperatures encountered in aircraft applications, radiative cooling becomes important, particularly during atmospheric re-entry. Nevertheless, the dependence of conventional radiative cooling on emissivity, surface shape, and ambient conditions poses certain constraints. It is nevertheless difficult to achieve ideal radiative heat transfer under various circumstances and material limitations, which reduces the cooling method's overall efficiency.

Traditional phase-change cooling methods, such as liquid cooling or refrigeration, rely on the absorption of heat during the phase transition from liquid to vapor. These techniques, while useful in some situations, have drawbacks due to the requirement for large, complicated infrastructure, as well as weight issues with aeronautical systems. Additional complications arise when managing phase-change processes in microgravity settings or during quick movements (Mohamadkhani 2023; Zhou et al. 2022).

Limitations of Traditional Cooling in Aerospace: Using conventional cooling techniques on aircraft systems has built-in drawbacks. Bulky cooling systems are not practical in aviation or spacecraft due to weight and space limits. Furthermore, more effective and compact solutions are needed because of the rising power density and thermal loads associated with contemporary avionics and propulsion. The drawbacks of conventional cooling techniques are most noticeable in high-temperature settings where there is a greater demand for efficient heat dissipation than can be satisfied by conventional means. A movement toward more sophisticated heat management strategies, like the use of phase-change materials, active cooling methods, and nanofluids, is emerging in response to these constraints. These advancements are intended to address the inadequacies of conventional cooling techniques and satisfy the changing needs of aerospace applications, where weight, efficiency, and dependability are critical factors. The search for improved thermal management techniques is essential to expanding the capabilities of aerospace systems and guaranteeing their peak performance in demanding environments.

15.4.2 Need for advanced thermal management solutions

The imperative for advanced thermal management solutions has become increasingly pronounced across various industries, driven by the relentless pursuit of enhanced performance, efficiency, and reliability in complex systems. This emphasis is particularly critical in sectors such as aerospace, electronics, and energy, where the demand for compact, lightweight, and high-power systems continually challenges the capabilities of traditional thermal management approaches.

Within the aerospace sector, the harsh operating conditions that propulsion systems, avionics, and spacecraft encounter across several mission stages highlight the necessity for sophisticated thermal management technologies. Sophisticated cooling solutions are necessary to prevent component damage and maintain optimal engine efficiency due to the high heat generated by jet engines and rocket motors. Additionally, the need for efficient heat management technologies grows as aircraft missions become more ambitious, entailing longer durations and intricate maneuvers. To preserve the structural integrity of spacecraft under these demanding conditions and to resist the severe temperatures encountered during atmospheric re-entry, innovative technologies are essential.

Modern technologies are powered by electrical gadgets that are becoming smaller and have ever-higher power densities. This tendency presents a serious threat to conventional cooling techniques. For electronically packed components to disperse heat effectively, advanced thermal management technologies are essential. The constraints of traditional air and liquid cooling methods have forced the semiconductor industry to investigate innovative materials including phase-change materials and nanofluids. In addition to ensuring peak device performance, efficient electronics cooling increases the lifespan of electronic parts and lowers the possibility of thermally induced failures (Lv et al. 2022).

As energy systems strive for greater sustainability and efficiency, improved thermal management is becoming increasingly important. To optimize energy conversion efficiency, power generation—whether from conventional sources or newly developed technologies like concentrated solar power—needs efficient heat dissipation. Thermal management systems are essential for maximizing power plant efficiency, enhancing electrical component dependability, and reducing energy loss. The need for creative thermal solutions to handle the difficulties presented by intermittent energy generation and distributed energy systems grows as the energy landscape shifts toward renewable sources and smart grid technology.

Thermal management has expanded into new areas, thanks to the development of sophisticated materials like nanomaterials and metamaterials. Engineered colloidal suspensions with improved thermal properties, known as nanofluids, are becoming increasingly well-known for their capacity to improve heat transmission in a variety of applications. Phase-change material integration is a promising avenue for small-scale and effective thermal energy storage. These resources highlight the revolutionary potential of innovative thermal management technologies, giving engineers the means to overcome the constraints of conventional techniques and create systems that perform beyond expectations.

The demand for sophisticated thermal management solutions is in line with the overarching objective of resource conservation at a time when sustainability is becoming increasingly important. Effective thermal control lowers energy usage, lessens the impact on the environment, and increases the overall sustainability of technological systems. The creation and application of innovative thermal solutions become essential to accomplishing these sustainability goals as industries work to comply with strict environmental requirements and embrace greener practices.

15.5 APPLICATIONS OF NANOFLUIDS IN AEROSPACE

The potential of nanofluids to address crucial thermal management issues in avionics, re-entry vehicles, and propulsion systems has drawn a lot of interest to their applications in aerospace. Because of their special thermal characteristics, colloidal suspensions of nanoparticles in base fluids, or nanofluids are attractive options for improving the efficiency of heat transmission in aeronautical applications (Riehl and Murshed 2022).

Nanofluids present a tempting option in propulsion systems, where engines produce extremely elevated temperatures during combustion. Nanofluids can increase the coolant's thermal conductivity by mixing nanoparticles—such as metal oxides or carbon-based materials—into the coolant. This prevents overheating and improves the overall performance of the propulsion system by increasing the efficiency with which heat is dissipated from engine components. The tough conditions seen in aerospace propulsion are comparable to the high-temperature temperatures that nanofluids can tolerate.

For avionics and electronic components, nanofluids provide an avenue to tackle the challenges associated with heat dissipation. Electronic systems operate at elevated temperatures; therefore, maintaining component dependability requires effective cooling. Because of their improved heat transfer properties and thermal conductivity, nanofluids can be used as efficient coolants. Their special qualities allow for effective heat extraction from electronic parts, guaranteeing steady functioning even in highly powered and closely spaced avionic systems.

Aerodynamic heating causes spacecraft to experience extremely elevated temperatures during atmospheric re-entry. Because of their capacity to improve thermal conductivity, nanofluids present a viable option for creating efficient TPSs. The use of nanofluids in heat shields or ablative materials can strengthen their resistance to elevated temperatures, protecting the spacecraft's structural integrity and safety during re-entry.

Aerospace applications of nanofluids highlight how versatile these tailored fluids are in managing a range of thermal issues. The goal of ongoing research is to optimize nanofluids for aeronautical applications by investigating novel formulations and considering variables including fluid compatibility, concentrations, and types of nanoparticles. Nanofluids present a promising new frontier in the search for sophisticated thermal management solutions as the aerospace industry endeavors to create more dependable and efficient systems.

15.5.1 How nanofluids can be applied in aerospace systems

The use of nanofluids in aerospace systems presents a revolutionary solution to the complex thermal issues that avionics, atmospheric re-entry, and propulsion systems face. Nanofluids show great promise for improving thermal protection, reducing overheating, and increasing heat transfer efficiency. Notwithstanding the difficulties, continued study and cooperative efforts can open the door for the use of nanofluids as a pillar in the development of aerospace thermal management. Because of their revolutionary effect, nanofluids will play a significant role in determining the direction of aircraft technology in the future.

Nanofluids in Propulsion Systems: Aerospace applications of nanofluids highlight how versatile these tailored fluids are in managing a range of thermal issues. The goal of ongoing research is to optimize nanofluids for aeronautical applications by investigating novel formulations and considering variables including fluid compatibility, concentrations, and types of nanoparticles. Nanofluids present a promising new frontier in the search for sophisticated thermal management solutions as the aerospace industry endeavors to create more dependable and efficient systems.

Avionics Cooling with Nanofluids: For avionics—which include sensors, computers, and communication devices—to function dependably during flight, effective cooling is essential. Because of their superior thermal characteristics, nanofluids provide a revolutionary response to the problems brought on by the shrinking and rising power densities of electronic components. Because of their improved thermal conductivity, nanofluids can dissipate heat more efficiently, preventing overheating and extending the lifespan and dependability of avionic systems. This use is especially important since the aerospace industry is moving toward electrical gadgets that are more compact and potent.

Thermal Protection during Atmospheric Re-entry: Atmospheric re-entry exposes spacecraft to extreme temperatures due to aerodynamic heating. Ensuring the structural integrity of spacecraft during this phase requires the design of efficient TPSs. When added to ablative materials or used in heat shields, nanofluids have the potential to enhance TPS functionality.

15.5.2 Advantages and limitations of using nanofluids in aerospace thermal management

Table 15.3 provides a detailed overview of both the advantages and limitations associated with the utilization of nanofluids in aerospace thermal management, covering various criteria relevant to propulsion systems, avionics cooling, thermal protection during atmospheric re-entry, material compatibility, weight considerations, and the current state of research and development in the field (Table 15.4).

Table 15.3 Advantages and limitations of using nanofluids in aerospace thermal management

Criteria	*Advantages*
Thermal conductivity enhancement	– Significantly higher thermal conductivity. – Potential for nanoparticle agglomeration. – Enhanced heat transfer efficiency. – Challenges in maintaining stability over time – Improved thermal performance compared to conventional coolants.
Heat dissipation in propulsion systems	– Mitigation of overheating in propulsion systems – Increased engine efficiency and performance. – Potential for weight reduction in cooling systems due to improved thermal properties.
Avionics cooling	– Enhanced cooling of high-power electronic components – Improved reliability and lifespan of avionic systems. – Efficient heat dissipation in compact and densely packed electronic systems.
Thermal protection in atmospheric re-entry	– Improved heat dissipation during atmospheric re-entry. – Potential for enhancing thermal protection systems. – Optimized performance of ablative materials.
Material compatibility	– Compatibility with a variety of base fluids. – Flexibility to tailor nanofluids for specific aerospace applications. – Potential for integration with existing cooling systems.
Weight considerations	– Potential for weight reduction due to improved thermal properties. – Especially relevant in weight-sensitive aerospace applications. – Potential for compact and lightweight thermal management solutions.

Table 15.4 Limitations of using nanofluids in aerospace thermal management

Criteria	*Limitations*
Thermal conductivity enhancement (Kumar et al. 2023a)	– Potential for nanoparticle agglomeration – Challenges in maintaining stability over time –Viscosity changes may impact pump power requirements
Heat dissipation in propulsion systems	– Limited studies on long-term effects in dynamic systems – Compatibility and stability challenges with engine materials
Avionics cooling	– Potential challenges in maintaining nanofluid stability in confined spaces – Selection of suitable nanofluids for electronic components may require specific considerations
Thermal protection in atmospheric re-entry	– Challenges in formulating nanofluids for compatibility with ablative materials – Limited understanding of nanofluid behavior in extreme atmospheric conditions
Material compatibility	– Limited understanding of long-term effects on materials
Weight considerations	– Need for additional studies on nanofluid impact on system weight and balance

15.6 HEAT TRANSFER ENHANCEMENT MECHANISMS

Aerospace applications demand rigorous thermal management strategies to address the extreme conditions experienced by propulsion systems, avionics, and spacecraft during flight and re-entry. Nanofluids, engineered colloidal suspensions of nanoparticles in base fluids, have emerged as a promising solution to enhance heat transfer efficiency in these demanding environments. This in-depth analysis explores the underlying mechanisms and scientific principles governing how nanofluids augment heat transfer in aerospace applications.

Nanofluid Composition and Thermal Conductivity: Intentionally adding nanoparticles to conventional coolants is the fundamental method of improving heat transmission using nanofluids. The surface area-to-volume ratio is significantly increased by these nanoparticles, which are usually much smaller than the base fluid (1–100 nm). Improved heat conductivity is the outcome of better interactions between the fluid molecules and the nanoparticles made possible by the increased surface area. One crucial factor influencing how well nanofluids transport heat is their thermal conductivity. Because they are effective heat transporters, nanoparticles aid in phonon scattering in the fluid. This procedure increases the nanofluid's total thermal conductivity while interfering with the phonons' normal evolution. This phenomenon is highly dependent on the size and material characteristics of nanoparticles, with smaller particles typically displaying more noticeable increases.

Brownian Motion and Dispersion Stability: The dispersion and thermal behavior of nanoparticles in nanofluids are influenced by their dynamic Brownian motion. A stable colloidal suspension is maintained, and nanoparticle agglomeration is inhibited by Brownian motion, which is fueled by collisions with fluid molecules. This stability is essential for long-term augmentation of heat transfer since aggregated nanoparticles could lower the effective surface area and prevent increases in thermal conductivity. To manage Brownian motion and guarantee the stability of nanofluids throughout time, the right stabilizing agents or surfactants must be used. In aerospace applications, where long-duration flights or exposure to fluctuating climatic conditions need dependable and stable heat management solutions, proper dispersion is especially important.

Impact on Convective Heat Transfer: Convective heat transfer is essential for removing heat from crucial components in aerospace applications. Convective heat transfer efficiency is significantly increased by nanofluids, which affects forced as well as natural convection. Improved convective heat transfer coefficients result from nanofluids' higher thermal conductivity, which facilitates more efficient heat transmission within the fluid. Additionally, changing the rheological characteristics and viscosity of nanofluids affects fluid dynamics and encourages more effective convective heat transmission. Because of the way that nanofluids behave differently in various flow conditions, such as turbulent or laminar flows, it is possible to customize heat transfer properties to meet the demands of aerospace systems.

Nanofluid-Fluid Interaction and Phase Change: How nanofluids and base fluid interact is one of the many factors affecting heat transfer in aeronautical applications. Fluid characteristics are changed by nanoparticles, which have an impact on hydrodynamic and thermophysical behaviors. This interaction becomes especially important in phase-change processes that are frequently used in aerospace thermal control, such as condensation or boiling (Meena et al. 2022). In comparison to conventional coolants, nanofluids have higher boiling heat transfer coefficients, which makes them useful in applications where phase-change processes are common. The nucleation sites that nanoparticles give help to increase heat transfer by promoting more effective bubble formation and departure during boiling. Furthermore, enhanced critical heat flow in nanofluids permits better thermal control in situations where heat dissipation is essential (Mohammed and Bencs 2023).

15.7 CHALLENGES AND FUTURE DIRECTIONS

15.7.1 Current challenges and limitations in implementing nanofluids

Nanofluids have attracted a lot of attention for their potential to improve heat transfer in a variety of industrial applications, including thermal management in aircraft. However, there are restrictions and difficulties associated with integrating nanofluids into real-world systems, which calls for careful consideration of several variables. This scholarly and scientific investigation explores the present obstacles and constraints related to the use of nanofluids, highlighting areas in need of additional study and advancement (Murshed 2023).

The tendency of nanoparticles to aggregate and form clusters that reduce their effective surface area and, thus, their capacity to promote heat transmission, is one of the main implementation issues of nanofluids. Several variables, including surface chemistry, size, and concentration of nanoparticles, affect this aggregation. Ensuring the stability and constant performance of nanofluids in real-world applications requires the development of efficient ways to prevent or mitigate nanoparticle aggregation (Shafiq and Abletis 2023; Alktranee and Bencs 2023).

The long-term stability of nanofluids is still a major concern. Over time, nanoparticles may settle or clump together, changing the characteristics of the fluid. Nanofluid dependability can be jeopardized by stability problems, especially in situations where operating times are long, like space missions. Comprehending the enduring characteristics of nanofluids and devising methods to uphold their steadiness is essential for their efficacious integration in practical situations (Schjerve 2023).

The interaction between nanofluids and the materials comprising the cooling systems or structural components is a complex and critical aspect. There could be problems with compatibility, which could cause material deterioration or corrosion. As materials in aerospace applications need to endure severe environmental conditions, it is critical to make sure nanofluids work well with current materials. Finding appropriate nanofluid formulations that preserve the structural integrity of materials frequently utilized in aircraft systems should be the main goal of research efforts.

While laboratory-scale studies have demonstrated the heat transfer enhancement potential of nanofluids, scaling up production for large-scale industrial applications remains a challenge. One crucial factor to consider is how economical it is to produce nanofluids in quantities necessary for industrial and aerospace applications. Since cost constraints are inherent in actual applications, developing scalable and economically viable methods for nanofluid production is essential to their widespread acceptance.

The lack of standardized testing protocols for nanofluids poses a challenge in assessing and comparing their performance across different studies and applications. Standardization is necessary to create standardized procedures for assessing the thermal conductivity, stability, and compatibility of nanofluids. The creation of widely recognized testing procedures would make it easier to compare results accurately and offer a solid foundation for the application of nanofluids in a variety of industries.

The environmental and safety aspects of nanofluids warrant careful consideration. Concerns include the possible toxicity of some nanoparticles as well as the effects of their manufacture and disposal on the environment. Research should concentrate on creating ecologically acceptable formulations and comprehending the ecological effects of using nanofluids. Furthermore, the responsible use of nanofluids in industrial settings depends on the protection of workers engaged in the manufacturing and handling of nanofluids (Mondal et al. 2023; Kumar et al. 2023b).

In the aerospace sector, implementing nanofluids faces additional challenges related to the unique operating conditions and stringent requirements. The weight and space constraints in aircraft and spacecraft demand nanofluid solutions that are not only effective in enhancing heat transfer but also practical in terms of system weight and volume. Furthermore, the potential impact of nanofluids on the reliability of critical aerospace components, such as avionics and propulsion systems, necessitates thorough investigation.

15.7.2 Ongoing research and future directions in the field

Ongoing research in the field of nanofluids for heat transfer enhancement continues to explore novel formulations, advanced synthesis techniques, and application-specific optimizations. In the realm of aerospace thermal management, researchers are delving into the intricacies of nanofluid behavior under extreme conditions, aiming to bridge the gap between laboratory-scale experiments and real-world applications.

Addressing the issues with nanoparticle aggregation and long-term stability is one line of current research. To stabilize nanofluids, researchers are investigating surface modification methods. They are using functionalized nanoparticles to reduce the tendency for agglomeration. Research is being done to find better stabilizers and dispersants that can sustain nanofluid stability for extended periods, which is important for the dependability of aircraft systems.

One of the main goals of nanofluid research continues to be the hunt for scalable and affordable synthesis techniques. Scholars are investigating novel methodologies, like microfluidic devices and continuous flow methods, to facilitate large-scale production without jeopardizing the economic feasibility of nanofluids. The aerospace industry, where there is a need for effective heat management solutions, requires the scalability of nanofluid synthesis to be produced at industrial levels.

Attention is also being drawn to developments in testing procedures and nanofluid characterization. The goal of standardization is to create widely recognized protocols for evaluating the properties of nanofluids, guaranteeing reporting uniformity and enabling significant cross-study comparisons. Scientists are working to improve methods for determining thermal conductivity, stability, and compatibility to create trustworthy standards for nanofluid performance under various aircraft circumstances.

Developing formulations specifically for avionics cooling or propulsion systems is one of the future objectives in the aerospace domain's nanofluid research. Researchers are investigating the integration of nanofluids into multifunctional materials for increased performance, as a route for tailoring nanofluids to meet the specific thermal difficulties of aircraft conditions.

Collaborative efforts between academia, industry, and research institutions are crucial for advancing the field. Teams of researchers with diverse backgrounds in materials science, fluid dynamics, and aeronautical engineering are working together to better understand the intricacies of nanofluid behavior and maximize its application in thermal control. Computational modeling and simulations are integrated with experimental studies to provide insights into the behavior of nanofluids under various situations.

15.8 CONCLUSION

This chapter provides a thorough examination of nanofluids as an innovative solution for advanced thermal management, particularly in aerospace applications. Facing escalating thermal challenges in propulsion systems, avionics, and spacecraft, the study highlights nanofluids' promising attributes, such as enhanced thermal conductivity and improved convective heat transfer. These characteristics make nanofluids well-suited for the stringent requirements of the aerospace industry, where precise temperature control is crucial amidst weight, space constraints, and extreme operating conditions. This chapter delves into the fundamentals of nanofluids, detailing their characteristics, synthesis methods, and key parameters influencing thermal performance. It explores the complex interplay of nanoparticle size, shape, and concentration affecting thermal conductivity and heat transfer efficiency. Additionally, it critically examines traditional cooling methods and emphasizes the need for advanced thermal management solutions in aerospace. Meticulous discussions cover challenges in implementing nanofluids in practical aerospace systems, including nanoparticle aggregation, long-term stability, material compatibility, scalability, and the lack of standardized testing protocols. Despite these challenges, ongoing research points to promising avenues for overcoming limitations through surface modification methods, improved stabilizers, scalable synthesis techniques, and standardized testing procedures.

This chapter underscores the urgency of advancing thermal management solutions in aerospace, driven by the growing complexity of systems and the need for compact, lightweight, and energy-efficient technologies. Nanofluids are positioned as a revolutionary approach to overcome the limitations of traditional cooling methods, with potential transformative breakthroughs in aerospace technology. Collaborative efforts between academia, industry, and research institutions are highlighted as crucial for pushing the boundaries of thermal management technologies. The integration of computational modeling and experimental studies, along with interdisciplinary research teams, promises to unlock new insights into nanofluid behavior under diverse conditions. In conclusion, this chapter emphasizes the pivotal role of nanofluids in next-generation thermal management, offering a promising avenue for optimizing thermal performance in aerospace and other high-performance industrial sectors.

REFERENCES

A. Bhattad, B. Nageswara Rao, V. Atgur, N. R. Banapurmath, A. M. Sajjan, C. Vadlamudi, S. Krishnappa., T. M. Yunus Khan, N. H. Ayachit (2023). Studies on evaluation of the thermal conductivity of alumina titania hybrid suspension nanofluids for enhanced heat transfer applications. *ACS Omega*, 8, 24176–24184. doi:10.1021/acsomega.2c07513.

A. K. Dewangan, S. Quadir Moinuddin, M. Cheepu, S. K. Sajjan, A. Kumar (2023). Thermal energy storage: opportunities, challenges and future scope. In: *Thermal Energy Systems: Design, Computational Techniques and Applications*; Kumar, A.; Singh, V. P.; Meena, C. S.; Dutt, N.; Editors; CRC Press, Boca Raton, FL, pp. 17–28. doi.:10.1201/9781003395768-2.

A. Kumar, A. Kumar, A. Kumar (2023b) *Laser Based Technologies for Sustainable Manufacturing*. CRC Press, Boca Raton, FL. doi:10.1201/9781003402398.

A. Kumar, V.P. Singh, C.S. Meena, N. Dutt (2023a). *Thermal Energy Systems: Design, Computational Techniques and Applications*. CRC Press, Boca Raton, FL. doi:10.1201/9781003395768.

A. Kundu, A. Kumar, N. Dutt, C. S. Meena, V. P. Singh (2023). Introduction to thermal energy resources and their smart applications. In: *Thermal Energy Systems: Design, Computational Techniques and Applications*; Kumar, A.; Singh, V. P.; Meena, C. S.; Dutt, N.; Editors; CRC Press, Boca Raton, FL, pp. 1–15. doi:10.1201/9781003395768-1.

A. Martins Obalalu, H. Ahmad, S. O. Salawu, O. A. Olayemi, C. Odetunde, A. O. Ajala, A. S. Abdulraheem (2023). Improvement of mechanical energy using thermal efficiency of hybrid nanofluid on solar aircraft wings: an application of renewable, sustainable energy. *Waves in Random and Complex Media*, 1–30. doi:10.1080/17455030.2023.2184642

A. Shafiq, J. Abletis (2023). Nanoparticle aggregation effect on nonlinear convective nanofluid flow over a stretched surface with linear and exponential heat source/sink. *International Journal of Thermofluids*, 19, 100355–100355. doi:10.1016/j.ijft.2023.100355

C. S. Meena, A. Kumar, S. Roy, A. Cannavale, A. Ghosh (2022). Review on boiling heat transfer enhancement techniques. *Energies*, 15, 5759. doi:10.3390/en15155759.

F. Jiang, S. Zhao, L. Wang, H. Guan, J. Liu (2022). Energy harvesting and thermal management system in aerospace. *Frontiers in Materials*, doi:10.3389/fmats.2022.907858

J. Manuel, S. Telmo, M. Gabriela Chango (2023). Recent progress on preparation, properties, and applications of nanofluids. *Latin American Applied Research*, 53(2), 77–88. doi:10.52292/j.laar.2023.620

L. Zhou, J. Rada, Y. Tian, Y. Han, Z. Lai, M. F, McCabe, Q. Gan (2022). Radiative cooling for energy sustainability: Materials, systems, and applications. *Physical Review Materials*, 6(9). doi:10.1103/physrevmaterials.6.090201

M. Ahmadizadeh, M. Heidari, S. Thangavel, E.A. Naamani, M. Khashehchi, V. Verma, A. Kumar (2024). Technological advancements in sustainable and renewable solar energy systems. In: *Highly Efficient Thermal Renewable Energy Systems: Design, Design, Optimization and Applications*; V. Verma, S. Thangavel, N. Dutt, A. Kumar, R. Weerasinghe Editors; CRC Press, Boca Raton, FL, pp. 23–39

M. Chitt, S. Thangavel, V. Verma, A. Kumar (2024). Green hydrogen productions: methods, designs and smart applications. In: *Highly Efficient Thermal Renewable Energy Systems: Design, Design, Optimization and Applications*; V. Verma, S. Thangavel, N. Dutt, A. Kumar, R. Weerasinghe Editors; CRC Press, Boca Raton, FL, pp. 261–276.

M. H. R. Alktranee, P. Bencs (2023). Factors affecting nanofluids' behaviour: A review. *International Review of Applied Sciences and Engineering*, 14(2), 241–255. doi:10.1556/1848.2022.00531

M. K. Awasthi, A. Kumar, N. Dutt, S. Singh (Editors) (2024). *Computational Fluid Flow and Heat Transfer: Advances, Design, Control and Applications*. CRC Press, Boca Raton, FL.

M. Mohamadkhani, P. Bencs (2023). Radiative cooling surfaces: principles, performance evaluation and applications. *Future Technology*, 2(3), 17–23. doi:10.55670/fpll.futech.2.3.4.

N. Schjerve (2023). Review of the stability of nanofluids. doi:10.5772/intechopen.107154

P. Lu, T. Yang, T. Gao, C. Xia, F. Yu (2023). Application of new materials in aerospace thermal management. *Advances in Engineering Technology Research*, 4(1), 50–50. doi:10.56028/aetr.4.1.50.2023

P. N. Reddy, V. Verma, A. Kumar, M. Awasthi (2023). CFD simulation and thermal performance optimization of flow in a channel with multiple baffles. *Journal of Heat and Mass Transfer Research*, 10(2), pp. 257–268. doi:10.22075/JHMTR.2023.31108.1458.

R. M. Barai, D. Kumar, A. V. Wankhade (2023). Heat transfer performance of nanofluids in heat exchanger: a review. *Journal of Thermal Engineering*, 9(1), 86–106. doi:10.18186/thermal.1243398

R. R. Riehl and S. M. S. Murshed (2022). In: *Fundamentals and Transport Properties of Nanofluids*. The Royal Society of Chemistry, pp. 399–417.

R. Wang, X. Fan, D. Li, R. Qu, L. Li, T. Zou (2022). convective heat transfer characteristics on end-winding of stator immersed oil-cooled. *Electrical Machines for Aerospace Applications*, 8, 4265–4278. doi:10.1109/TTE.2022.3186800

S. M. S. Murshed (2023). Heat transfer and fluids properties of nanofluids. *Nanomaterials*, 13(7), 1182–1182. doi:10.3390/nano13071182

S. Mondal, B. Bhattacharjee, S. Dey, S. Bhowmick (2023). Impacts of nanofluids and nanomaterials on environment and human health: a review. *Nanoscience and Nanotechnology*, doi:10.2174/2210681213666230601103342

S. Pierre, L. Svetlana U. Y. E. Milián, M. Nemś, A. Nemś (2022). Application of nanofluids in improving the performance of double-pipe heat exchangers-a critical review. *Materials*, 15(19), 6879–6879. doi:10.3390/ma15196879.

V. Verma, S. Thangavel, N. Dutt, A. Kumar, R. (2024a). *Weerasinghe Editors, Highly Efficient Thermal Renewable Energy Systems: Design, Design, Optimization and Applications*. CRC Press, Boca Raton.

V. Verma, S. Thangavel, N. Dutt, A. Kumar, R. Weerasinghe (2024b). Recent development of thermal energy storage: solar, geothermal and hydrogen energy. In: *Highly Efficient Thermal Renewable Energy Systems: Design, Design, Optimization and Applications*; V. Verma, S. Thangavel, N. Dutt, A. Kumar, R. Weerasinghe Editors; CRC Press, Boca Raton, FL, pp. 1–22.

Y. Lv, G.-P. Zhang, Q. Wang, W. Chu (2022). Thermal management technologies used for high heat flux automobiles and aircraft: a review. *Energies*, 15(21):8316–8316. doi:10.3390/en15218316

Y.K. Prajapati, P.K. Gupta, A. Kumar (2011). Free surface undulation and air entrainment in a rectangular tank. *International Journal of Applied Engineering Research*, 6(4), 409–419.

Chapter 16

Evolution of green concrete in the building industry

A sustainable approach for nanomaterials: development and application

Aditi Bonde, Pratibha S. Agrawal, Mayur Chaware, Shweta Dode, Srijan Gupta, and Fasaha Ahmad

16.1 INTRODUCTION TO GREEN CONCRETE

In the phase of climate change and the growing need for eco-conscious construction, the development and application of green concrete stands as a beacon of innovation and responsibility. Traditional concrete, while a cornerstone of modern construction, poses environmental challenges, contributing significantly to carbon emissions (Shen et al. 2021; Winkler et al. 2023). The demand for environmentally friendly alternatives has intensified as we strive for more sustainable living. Green concrete, a term encompassing various eco-friendly formulations, emerges as a promising solution. This chapter explores the evolution of green concrete, its key components, and its diverse applications in sustainable construction. The choice of construction materials plays a pivotal role in mitigating environmental impact. Green concrete exemplifies a paradigm shift toward sustainable building practices, addressing both immediate construction needs and long-term environmental goals (Boobalan et al. 2022; Sureshkumar, Kumar, and Ravikanth 2019). The global market for sustainable building materials has been on the rise, driven by increased awareness of environmental issues and a growing commitment to sustainable practices in construction. There has been a noticeable increase in the adoption of renewable and recycled materials in construction with a focus on sustainable practices. The Leadership in Energy and Environmental Design (LEED) certification has influenced the use of sustainable materials. Projects seeking LEED certification often prioritize material with low environmental impact (Iqbal et al. 2023). The integration of energy-efficient technologies such as smart windows, energy-efficient HVAC systems, and solar panels has become more common in sustainable construction. Many countries have implemented or strengthened regulations encouraging or requiring the use of sustainable building materials. This has further propelled the market for eco-friendly construction materials.

DOI: 10.1201/9781003494454-16

Green concrete, also known as sustainable or environmentally friendly concrete, represents a significant advancement in the construction industry. Here are key points about green concrete incorporates recycled materials, or aggregates, reducing the demand for traditional raw materials and decreasing environmental impact. The benefits and restrictions of green concrete are depicted in Figure 16.1. The production of green concrete often involves lower carbon emissions compared to conventional concrete. This reduction contributes to mitigating the environmental footprint of construction projects. Green concrete can exhibit improved durability and performance characteristics, leading to longer-lasting structures (Marceau and VanGeem 2003). This longevity contributes to sustainability by reducing the need for frequent repair or replacement. Industrial byproduct and waste material in green concrete enhances resource efficiency, as it utilizes materials that might otherwise end up as waste. The construction industry has shown an increasing interest in green concrete as a response to global sustainability goals. Builders, architects, and developers are recognizing its potential to contribute to environmentally responsible construction practices. Government and regulatory bodies in various regions are promoting the use of sustainable building materials, including green concrete through incentives, certifications, and regulations. Academic institutions and research centers are actively involved in studying and advancing green concrete technologies, to make it more sustainable toward environmental benefit (Askar, Al-Kamaki, and Hassan 2023; Ghinaya and Masek 2021). As the construction sector continues to prioritize environmental sustainability, green concrete is positioned as a valuable solution offering a balance between structural performance and reduced ecological bearing. This chapter goals to delve into the advanced journey of green concrete,

Benefits:

- Plans for blending in advance mean easier maintenance, improved consistency, and easier cleanup.
- Decrease in shrinkage and creep.
- They use recycled and local materials in their concrete.
- Compared to conventional cement, green cement hydrates at a fundamentally lower intensity.

This leads to a decrease in temperature rise in large cement pours, which is especially advantageous for green cement. It also reduces material waste and opens up new uses for waste materials in this day and age.

Restrictions:

- Use of hardened steel results in an increase in support costs.
- The lifespan of green cement structures is comparatively shorter than that of conventional structures.
- The part pressure of green cement differs slightly from that of regular cement.
- If the totals used are not freed from consumption driving specialists, erosion in the steel bars may also occur.

Figure 16.1 Benefits and restrictions of sustainable construction (Osial et al. 2022).

examining its composition, manufacturing processes, and highlighting its versatile applications. By the end, readers will gain insights into the tangible benefits and challenges associated with integrating green concrete into contemporary construction projects. We will explore the evolution of green concrete, analyze its components, and manufacturing techniques, and showcase real-world applications. Additionally, the environmental impact and potential challenges paving the way for a comprehensive understanding of this sustainable building material are also discussed. As we embark on this exploration of green concrete, let's uncover the strides made in sustainable construction and the possibilities that lie ahead. The following sections will unravel the fascinating journey of green concrete, from its inception to its current status as a cornerstone of eco-conscious architecture.

This chapter also discusses the environmental benefits of green concrete such as reduced footprint and decreased reliance on non-renewable resources concrete. Furthermore, it explores the practical application of green concrete in sustainable building projects. Case studies and examples showcase successful implementations, emphasizing the economic feasibility and long-term advantages of choosing green concrete over conventional options. The potential challenges and further direction in the widespread adoption of green concrete are also briefly addressed. The development and application of green concrete represent a significant advancement in sustainable construction practices. In the development phase, researchers and engineers focused on incorporating recycled materials such as fly ash, slag, a recycled aggregate into the concrete mix. Additionally, alternative binders such as geo polymers or calcium sulfoaluminate cements are explored to replace or supplement traditional Portland cement which is a major contributor to carbon emission in concrete production (Golewski 2021; M. Liu et al. 2023). Optimizing the mix design and production processes is another crucial aspect of green concrete development. Researchers work to enhance the performance and durability of green concrete while maintaining or improving its structural properties. This involves fine-tuning proportions, curing methods, and other factors to ensure that green concrete meets or exceeds the standards set by conventional concrete.

16.2 PROPERTIES AND CHARACTERISTICS OF NANOMATERIALS IN GREEN CONCRETE

16.2.1 Compressive strength

One important characteristic of freshly mixed, uncured concrete is its compressive strength (CS), which is influenced by various mix design factors such as the ratio of cement to water. For maximum strength development, proper curing—which involves controlling moisture, temperature, and time—is crucial because improper curing can reduce durability (Mater et al. 2023; Murad et al. 2020). The hydration process, where water reacts with cement particles, plays a crucial role, with CS closely tied to the degree of hydration in the early stages. Cylindrical specimens are put through compressive loading procedures until they fail. Temperature and time are two examples of variables that can affect strength. Regular testing and quality control procedures are essential for consistent and predictable CS. When designing and evaluating concrete structures, engineers consider both the long-term and early-age strength characteristics. They emphasize the significance of comprehending and optimizing these factors for optimal concrete performance.

16.2.2 Splitting tensile strength

The splitting tensile strength of green concrete, referring to freshly mixed and uncured concrete, is a critical mechanical property that assesses its resistance to tensile stresses. Understanding the behavior of concrete under bending or tensile forces requires an understanding of its tensile strength, which is influenced by various factors including mix design, curing conditions, and the age of the concrete. Tensile strength development requires proper curing, which includes controlling temperature and moisture content; improper curing can lead to lower values (Bamigboye et al. 2022; Oyebisi et al. 2020; Sldozian, Hamad, and Al-Rawe 2023). During testing, cylindrical or prismatic specimens are loaded axially or diametrically until they break. It is interesting to note the correlation between splitting tensile strength and CS, where higher tensile strength is frequently correlated with higher CS. The tensile strength of green concrete, which may have lower values in its early stage of maturity, is influenced by the maturity of the concrete, which is influenced by temperature and time during the hydration process. In order to ensure structural integrity and durability, engineers take these factors into account when designing concrete mixes that exhibit adequate splitting tensile strength.

16.2.3 Elastic modulus

Several factors affect green concrete's elastic modulus, which measures its stiffness and capacity to deform under stress. The ratios of cement, water, aggregates, and admixtures in the mix design are crucial; a higher elastic modulus is typically the consequence of a lower water-to-cement ratio. Variations in stiffness are also influenced by the type of aggregates that are used. The development of the elastic modulus depends on the right curing conditions, and an insufficient curing process can affect the material's stiffness. Additionally, the age of the concrete affects the elastic modulus, which tends to increase as the hydration process progresses. The elastic modulus is measured using a variety of testing techniques, such as compression tests, tension tests, or ultrasonic pulse velocity. The outcomes vary depending

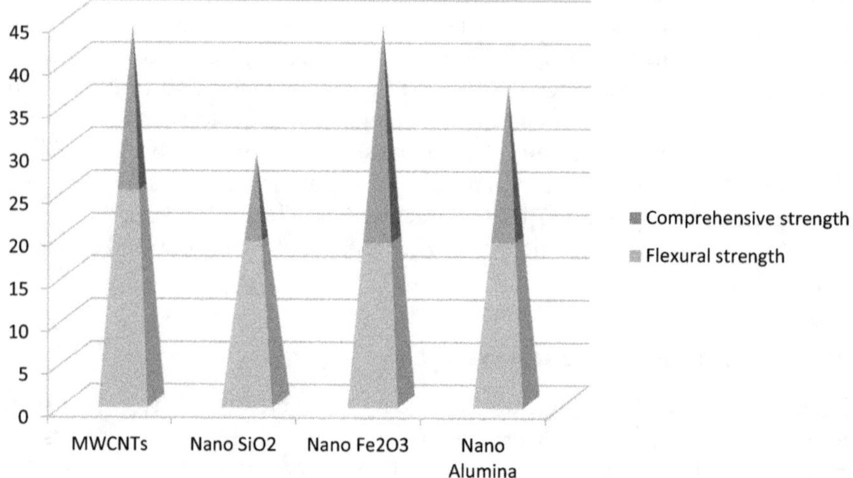

Figure 16.2 Effect of nanoparticles on the strength of concrete.

on the technique utilized (Fang et al. 2023; Mohammed and Hama 2022). In order to design concrete mixes with the appropriate elastic modulus—a crucial parameter in determining the deformation characteristics of structures and guiding structural analysis and design choices—engineers take these factors into account (Zhang et al. 2023). While green concrete might have a lower elastic modulus than fully cured concrete, this is a property that changes as the concrete ages.

Using nanotechnology, new or improved building materials can be created. The majority of nanoparticles that are added to cement and concrete either quicken the process of hardening, seal the surface of the concrete, or enhance its characteristics. Figure 16.2 shows how adding certain nanoparticles to the concrete mix can result in an increase in strength. Multi-walled carbon nanotubes have a major impact on the concrete mix's flexural strength, while nano-ferrous oxide greatly boosts its CS.

16.2.4 Resistance against acid, water, and sunlight

The ability of green concrete—that is, freshly mixed, uncured concrete—to withstand acid, water, and sunlight is essential for determining how long-lasting it is. Although concrete typically has a strong water resistance, which acts as a barrier to keep moisture out, its acid resistance may be a cause for concern (Hashmi et al. 2022; Khabaz 2023). Because green concrete is still developing, it might be more susceptible to acid attack. Acid resistance can be increased by including additives in the mix design, such as fly ash or silica fume. Sufficient curing and sealing techniques are necessary to enhance water resistance in the crucial initial phases. Concrete's appearance can be impacted by sunlight exposure, particularly ultraviolet radiation, which can cause surface discoloration or "solar bleaching." While concrete's structural integrity is usually not greatly affected by exposure to sunlight, protective coatings or pigments could be taken into consideration for outdoor applications to reduce aesthetic changes and improve overall resistance (Coppola et al. 2022). To ensure that green concrete structures are resilient to environmental conditions over time, engineers and construction professionals need to consider these factors during the design and construction phase.

16.2.5 Sound insulation enhancement

According to the World Health Organization (WHO), noise pollution is any excessive or undesired sound that may be harmful to both human health and the environment. Sound barriers can be used to reduce noise entering the building from the outside. For sound insulation, a variety of acoustic materials are used in building construction. Selecting the right material is crucial to ensuring the residential building's interior acoustic comfort (Azhdarpour, Nikoudel, and Taheri 2016; Mutiari et al. 2023). Concrete's capacity to absorb sound energy is determined by two factors: its noise reduction value (NRC) and its sound absorption coefficient. According to recent studies, concrete that contains plastic as a filler material has a higher absorption capacity than regular concrete. It also has an NRC value that is 57% higher for plastic or lightweight concrete than for regular concrete.

The privately developed acoustic protection estimation tools can be used to estimate acoustic protection. In any case, the modified concrete and poly squares have less CS due to the reused polymer materials. Furthermore, the concrete/poly squares modified with waste polymers can be used as advancement materials due to their unique qualities, such as assurance of sound and warmth and the necessary capability of the waste polymer materials. Since this understanding permits and stores a significant amount of plastic waste on Earth, the use of plastics has increased steadily. For instance, research indicates that substituting 75% of coarse and fine aggregates with waste polyvinyl chloride (PVC) plastic enables the creation of a composite material with intriguing acoustic insulation properties. The characteristics that PVC adds to concrete qualify the materials for use in medical facilities. Such concrete can also be used as an admixture in noisy locations, like heavily trafficked road areas close to residential areas, to provide a longitudinal sound-proof barrier and shield the surrounding area from noise pollution. However, using this plastic will also cause the density to drop by up to 19%, which will alter the concrete elements' mechanical performance (Daud, Selamat, and Rivai 2013; Syamsiyah et al. 2020; Warnock 1998).

Furthermore, since plastic waste is not biodegradable, repurposing these kinds of structures offers us a potential solution to raise the concrete's quality. That is why recycled plastic wastes like polyethene terephthalate (PET), polypropylene (PP), and low- and high-density polyethene (LDPE and HDPE, respectively) are being added as a component to solid concrete to enhance its qualities. Dehydol LS-12, a non-ionic surfactant, is used to regulate the rise in plastic material's wetting capacity. The interfacial weight and contact edge decrease to 31–32 m N/m and 65°–68°, respectively, at the optimal centralization of different phases of the basic micelle fixation (CMC), freely (Thiam, Fall, and Diarra 2021; Tulashie et al. 2022). When compared to mortar experiments, the properties of cement are demonstrated and settled. Porosity rises as plastic volume increases, but compressive quality, mass thickness, warm conductivity, and an isolating, unyielding nature were still achieved. The solid models with a volume of 30% LDPE filaments had the least warm conductivity; the remaining models were made of PET, PP, and HDPE separately. Ultimately, the process of pretreatment employed to prepare the aggregates resulted in a brittle bond between the concrete adhesive and the elastic material; consequently, it improved the cement's ability to withstand vibration and sound absorption. The properties that provide protection are not significantly affected by the effects of freezing and defrosting. It was discovered that all of the protection characteristics significantly improved at high frequencies. Additionally, it demonstrates that elastic cement can be used outside of building envelopes to absorb sound through multi-story urbanized buildings; full-scale testing in the vicinity is necessary.

16.2.6 Thermal insulation improvement

Lightweight concrete which is thermally insulated and made from recycled plastic may significantly increase thermal efficiency in the construction of buildings by substituting an affordable material for traditional aggregates. When plastic is added to concrete, the material's thermal conductivity drops. Plastic aggregates in concrete have a major impact on thermal insulation. Numerous studies conducted in the past demonstrate that thermal insulation is opposite to thermal opposition. It was observed that the thermal conductivity was increased by 49% when filler materials and plastic waste made up 100% of the natural aggregate, in contrast to the natural lightweight aggregate concrete. The findings imply that concrete may find use in lightweight applications that call for high insulation qualities and low density. Furthermore, it was observed that concrete blocks containing 50% plastic had a markedly lower thermal conductivity. The study demonstrated the use of PET plastics to replace coarse aggregate in concrete blocks by examining their thermal conductivity (Deraman et al. 2021; Horma et al. 2022). According to the findings, adding 5% PET reduces thermal conductivity by 36% and volumetric density by 20%.

Cement's thermal conductivity is primarily affected by substituting waste plastic totals for average totals. A few analyses have shown that cement types with wasted total plastic expansion have less thermal conductivity. It was discovered that, generally speaking, the thermal conductivity was related to the dry thickness. There were two main explanations given for the decrease in thermal conductivity: (1) The thermal conductivity of waste plastic aggregates is approximately ten times lower than that of silica sand (0.28–0.3 W/mK and 1.3–1.4 W/mK, respectively), resulting in a reduction in composite conductivity. (2) The total amount of waste plastic leads to an increase in air-void substance, which further reduces thermal conductivity. In this sense, concrete that combines waste and recycled aggregates has significantly better protective qualities than concrete that only contains aggregate has the impact of plastic waste on concrete's thermal conductivity when it is used as aggregate. Waste based on PVC, HDPE, and PP were utilized (Mohanraj, Malathy, and Ravisankar 2022). The same volume of plastic aggregates was used to replace varying percentages (25%, 50%, and 75%) of the natural aggregates. The concrete containing 75% PVC was found to have the lowest thermal conductivity. Additionally, a significant decrease in thermal conductivity was observed as the amount of PET plastic waste aggregate increased. For structural insulation, lightweight concrete members can be made of concrete that contains a high percentage of PET plastic waste aggregate (between 40% and 50% volumetric replacement of natural coarse aggregate).

It was investigated whether it would be possible to produce lightweight concrete (1,500 kg/m^3) using aggregate made entirely of recycled plastic. Conversely, the control concrete's thermal conductivity was 1.7 W/mK, compared to 1.1–0.5 W/mK in the recycled plastic aggregate. An experimental investigation was conducted on the characteristics of concrete that substitutes natural aggregates with up to 50% plastic waste. Such material presents a potential solution from the perspectives of both plastic waste disposal and improvement of building energy efficiency, according to the obtained experimental results. In summary, it was said that the material's thickness, surface area, and temperature slope during the continuous movement of heat waves are all important factors in thermal insulation. At 25°C, the conductivities of PP, HDPE, and LDPE are 0.35 W/(m·K), 0.23 W/(m·K), and 0.43 W/(m·K), respectively. In contrast, the conductivities of conventional cement were typically three times higher, ranging from 1.34 to 2.92 W/(m·K) (Stevulova et al. 2021; Záleská et al. 2019). Thus, we can rely on plastic filaments to assist us in reducing the thermal conductivity of cement. In this way, the addition of plastic reduces thermal conductivity, allowing us to use this kind of concrete in locations that are both desirable and appropriate.

16.3 INCORPORATING NANOTECHNOLOGY INTO THE GREEN CONCRETE MIXTURE

Nanomaterials derived from silica improve mechanical properties, decrease permeability problems, and improve particle packing. It improves the strength and durability of concrete by forming C-S-H gel, which solves calcium leaching issues in aquatic environments. In addition to its hydrophilic and self-cleaning qualities, Nano-TiO_2 strengthens concrete by quickening the hydration of cement. It can give concrete structures durability, self-cleaning capabilities, and antibacterial qualities (Xu et al. 2021). As a filler, nano-calcium carbonate, or nano-$CaCO_3$, adds strength and quickens the rate of hydration. Although it helps to reduce early-age shrinkage and improves the physical and chemical properties of concrete, it can have an impact on flowability. Zinc oxide, also known as nanoZnO, has antifungal, antibacterial, and anticorrosive qualities. In addition to increasing mechanical strength and decreasing harmful pores in concrete structures, nanoZnO can also function as a nanofiller (Nayak et al. 2022). Iron oxide (Nano-Fe_2O_3), though it affects the setting time of the concrete, increases concrete strength, decreases water permeability, and can function as a nanofiller to decrease harmful pores. While aluminum oxide (Nano-Al_2O_3) increases CS, it may also make concrete less workable. In concrete specimens, it recovers pore structures and strengthens gel formation (Abdalla et al. 2022; Du et al. 2023). When added in small amounts, carbon nanotubes (CNTs), cylindrical structures with high thermal conductivity, improve the mechanical properties of concrete.

A life cycle assessment (LCA) of CNTs has been obtained from "Environmental Assessment of Single-Walled Carbon Nanotube Processes" in accordance with the inputs required by the model. Out of the three different CNT manufacturing processes examined in(Healy, Dahlben, and Isaacs 2008), the LCA of CNT using the High-pressure carbon monoxide (HiPco) process has been selected because it has the least detrimental effect on the environment. This finding supports the goal of the study, which is to determine whether adding nano-waste particles to concrete instead of cement—the most expensive and traditional ingredient—will have a sustainable impact.

To further support the inputs in the SDSS model, the LCA of rice husk ash (RHA), a pozzolanic material blended with Portland cement (PC) to create green concrete, was derived from earlier studies (Garas, Sayed, and Bakhoum 2021; Zerbino, Giaccio, and Isaia 2011). Cement and aggregate LCAs were gathered from research projects by (Bakhoum et al. 2017) and (Healy et al. 2008). The information needed for each component is added to the model's Life Cycle Inventory for all used materials, CO_2, SO_x, NO_x, particulates, embodied energy, raw material consumption, and solid waste. However, RHA is thought to emit no CO_2, SO_x, NO_x. Additionally, there isn't the energy and raw material consumption shown nor is there the release of particulates and solid wastes. As an agricultural waste, RHA uses very little energy in its natural state, and very little energy is required during the incineration process. The price of cement and aggregate is determined by the local market price in Egypt at the time of writing. CNTs were purchased from nearby vendors for a fixed price, while RHA was free because it was disposed of waste that contributed to land pollution. Acidification and air pollution are two methods used by the SDSS model to quantify pollution. Disability-Adjusted Life Years are a common metric used to measure air pollution.

The study offered a quantitative evaluation of the sustainability impact of using two different kinds of nano-waste particles in place of some of the cement in the green concrete mixture. By basing decisions about the assessment and selection of nanomaterials in green concrete mixes on sustainable environmental criteria, it is meant to be a step toward sustainable construction. The primary advantages include cutting back on the amount of cement used and the energy it requires, as well as enhancing environmental quality by lowering air pollution, acidification, and solid waste. There was no discernible difference in the final scores for the same manufacturing phase between the three mixes' sustainability factors assessed using different weighing techniques. Nonetheless, the concrete mix containing RHA demonstrated the best sustainability outcomes across the board. This is because RHA, being an agricultural residue of a natural material, fell into category zero of the life cycle. Replacing 5% of the cement in the mix with RHA cement helped to reduce energy use and carbon emissions during the cement manufacturing process. Although CNT's substitution proportion was small, its high cost and high energy consumption during production resulted in a negative impact on the evaluation's final outcome. Further research on the LCA of nano-waste particles derived from natural sources is advised in order to optimize the amount of cement substituted in the concrete mix while maintaining its mechanical properties and improving sustainability.

Figure 16.3 shows a thorough analysis of every material in light of the ten sustainability factors alone. Despite the fact that CNT replaced a portion of the cement content, its sustainability scores are almost identical

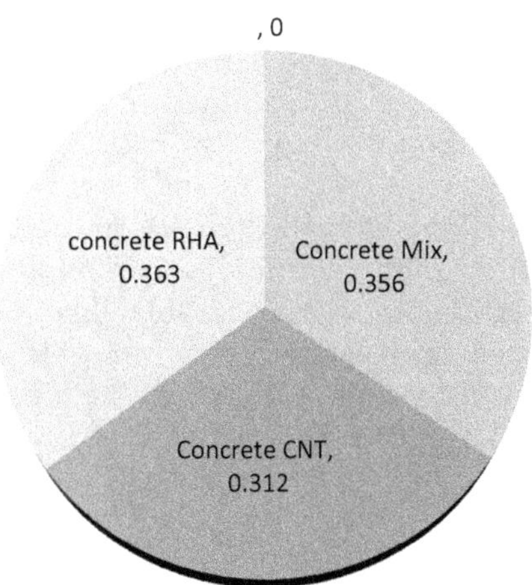

Figure 16.3 Sustainable evaluation of materials.

to those of the control mix because only 0.1% of the cement content was replaced, which had almost no impact on the score outcome. The mix containing CNT had the lowest energy consumption sustainability score. This is because, based on its LCA, the manufacturing of CNTs uses a lot of energy. The energy score of the RHA concrete mix is higher than that of the concrete-CNT and control concrete mixes by 3.63 and 0.03, respectively. This is explained by the fact that RHA is thought to have zero energy consumption, so adding it to the concrete mix reduced the amount of cement and, in turn, the energy consumption of the mixture. Furthermore, the concrete-control mix's score is 3.6 points higher than that of the concrete-CNT mix. This is because, based on its LCA, the manufacturing of CNTs uses a lot of energy.

16.4 SUSTAINABLE CONSTRUCTION PRACTICES WITH GREEN CONCRETE

Sustainable building material plays a crucial role in fostering environmentally responsible construction practices. By opting for these materials, construction projects can contribute to resource conservation, energy efficiency, and overall environmental sustainability. The choice of building materials aligns with the growing global emphasis on green construction practices and the need for more resilient, eco-friendly structures. Sustainable building materials are essential for environmentally friendly construction. These organizations such as the USGBC (U.S. Green Building Council), promote such practices through certifications like LEED. BMRA (Building materials reuse Association) and building green offer extensive databases of sustainable building materials facilitating informed choices during the design and construction phases. They may include details on materials' environmental impact, recyclability, and potential for reuse. Choosing FSC (Forest Stewardship Council)-certified wood supports sustainable forestry practices and biodiversity (Cerutti et al. 2017; Malovrh et al. 2019). This certification evaluates products based on their entire lifecycle, considering factors such as materials health impact recyclability and renewable energy use. Tools like EC3 help quantify the embodied carbon in building materials, aiding architects and builders in selecting low-carbon alternatives. Reducing embodies carbon is crucial for overall sustainability in construction. Local environmental or building departments may provide guidelines or incentives for sustainable construction. Compliance with government standards can contribute to achieving a more sustainable and eco-friendly construction project. Architectural platforms like ArchDaily or Dezeen often feature articles, case studies, and interviews related to sustainable architecture and innovative building material (Barry 2018). Following these platforms can keep you updated on the latest trends and advancements in sustainable construction. By combining insights from these resources, builders,

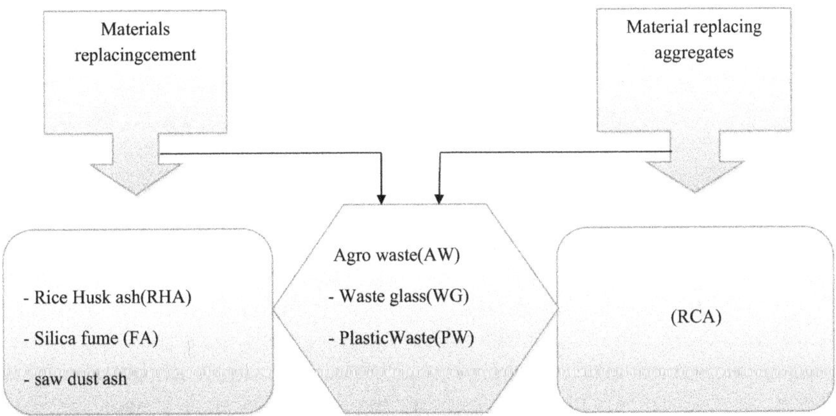

Figure 16.4 The most common byproduct used as substitutes.

and architects can make well-informed decisions to create environmentally sustainable structures.

Figure 16.4 clarifies the methodology of this review for the categorization system of the reviewed byproducts used as concrete ingredients which are explained below as follows:

16.4.1 Plastic waste

Because of its well-known benefits, concrete is the most widely used, durable, stronger, and economically sustainable material. It is made up of just three basic ingredients: cement, water, and coarse fine aggregate. However, concrete has significant drawbacks as well. For example, its tensile strength is much lower, leading to a straightforward failure from tensile stresses (Ispir, Ates, and Ilki 2022). Researchers investigated more sophisticated fiber-reinforced material to deal with those drawbacks by increasing the mechanical properties of the material (Tahir, Sbahieh, and Al-Ghamdi 2022). As an additional illustration, manufacturers of concrete emphasize on recycling of waste aggregates to control the loss of quarries and, consequently, lessen the adverse environmental effects. These creative approaches to concrete improvement, however, ignore the issue of plastic waste and recycling. As concrete has a longer service life than plastic, plastics may be added to concrete to lower the amount of waste that is not recycled and lessen the effect on ecosystems. At present, 4.4 billion tons of concrete are produced per annum worldwide. Then, employing recycled plastic waste in concrete could be a lot of promise for using plastic waste to make concrete which can result in economically efficient and sustainable development in the building sector.

Numerous plastic types have applications in the building sector and have the potential to be recycled. According to a study, after recycling, plastic waste may find use in a variety of applications depending on its characteristics. Thus, plastic lumber, tables, chairs, and other furniture can be made from HDPE, a material that is reasonably stiff and hard. Bricks and blocks made of LDPE, a flexible material, may be produced using it. Because of its flexibility and hardness, PP has potential uses as aggregate in asphalt mixtures. When compared to traditionally used materials, PET is incredibly inexpensive to recycle and reuse because it is a waste and requires expensive storage. PET-based cement materials do not exhibit any weaker strength characteristics when compared to traditional-based cement materials. PET waste itself possesses extremely low absorbability which is reflected in its composite cements. It also shows greater resistance to biocorrosion (Abu-Saleem et al. 2021; Rezvan et al. 2023). Buildings with better energy efficiency require less maintenance because plastic wastes can be used to produce insulation materials. This kind of waste is available practically anywhere locally. When they are used instead of natural raw materials, which sometimes have excavation sites far from the locations of their applications, it reflects in decreased transportation cost (Tota-Maharaj, Adeleke, and Nounu 2022). This method also lowers other transportation-related pollutants and CO_2 emissions.

Plastic fibers-blended concrete can be more sustainable and have better structural strength providing more durability. Concrete can be reinforced with synthetic materials to increase its ductility because polymers do not oxidize like the traditional steel reinforcement does. Plastic microfibers are already employed in a variety of settings, including tunnel linings and pavements. However, when the elements produced have significant structural performance, these fibers are used in combination with steel fibers. Microfibers are another way that plastic fibers are used in concrete. These polymer-produced fibers, which are shorter than 54 mm, improve the fiber-reinforced concrete's durability rather than its structural performance (Akbar and Liew 2020; Signorini and Volpini 2021). Waste plastic could be used for this latter purpose because waste polymers have a lower mechanical performance than high-density polymers, which are typically used to produce microfibers (Araya-Letelier et al. 2019).

The use of plastic as a coarse or fine aggregate is highly beneficial for the environment. According to a study, concrete can benefit from using up to 50% of the polyethylene terephthalate (PET) in place of sand. In general, recycled PET can be used, at specific replacement rates, in the production of environmentally friendly concrete. Because such plastic has unique properties that set it apart from other aggregates, it can be employed in places where less strength is required. This reduces the self-weight of concrete in structures (Aocharoen and Chotickai 2021; ramalingam, Sundarakannan, and Vasudevan 2022). Therefore, by combining such plastic waste with concrete to meet sustainable building requirements and improve thermal properties, reduce the use of natural resources, avoid pollution, avoid consuming waste, and save energy, the concrete industry's sustainability may be increased.

Plastic is incorporated into concrete at sites using a variety of techniques. Its effectiveness is also impacted by the adoption of these techniques. Mixing concrete with plastic as aggregate is the simplest way. Reactors that can produce batches for various uses can do so with this easy method. Another method is by heating the plastic, adding it as a coarse aggregate, and then after cooling, plastic will adhere firmly. In addition to filling some of the concrete's pores, plastic can also enhance the material's response in terms of thermal and acoustic conductivity. Therefore, it is necessary to understand the engineering properties and relationship of the added plastic material within concrete if large amounts of plastic are used in its production. According to studies, these plastic aggregates outperform mineral aggregates in terms of thermal and acoustic insulation, construction quality, and energy conservation. They also mark a commendable advancement in sustainable building practices. Plastic aggregates are a great way to make low-CO_2 concrete that also aims to cut down on the amount of structural strength needed and in turn minimize transportation expenses. The majority of reviews of the literature have concentrated on how adding plastic wastes affects the mechanical qualities of concrete, with less attention paid to the material's physical attributes, like its ability to insulate against heat and sound (Yilmaz, 2021).

16.4.2 Fly ash

FA is now necessary for civil engineering projects due to its benefits to the environment and economy. The amount of free lime in the ash limits how much cement FA can substitute. In addition to its chemical composition, the phase composition, amount of glassy phase, the temperature at which lignite or coal burns, specific surface area, etc., all influence how reactive FA is. FA is a pozzolanic substance. Pozzolanic materials are those that, when exposed to lime and water, form insoluble cementitious compounds but have little-to-no cementing action when left alone. FA, or pulverized fuel ash, is a byproduct of coal-fired power plants that is added as a mineral to cement or pulverized fuel ash, is a byproduct of coal-fired power plants that is added as a mineral to cement. Particulate coal is driven into the burning zone of the furnace, where its combustible elements (mostly carbon, hydrogen, and oxygen) ignite at approximately 1,500°C (2,700°F). In this temperature range, inflammable minerals like pyrite, feldspar, quartz, calcite, gypsum, and the minerals in the clay melt to create tiny liquid droplets. The droplets carried by the flue gases from the burning zone quickly cool to form small, spherical, glassy particles. Mechanical and electrical precipitators or baghouses are used to extract solid particles from flue gases (Naenudon et al. 2022; Niyazbekova et al. 2023; S. Singh et al. 2023). The ash particles that travel with the flue gases as they "fly" out of the furnace are referred to as FAs. The properties of FA are influenced by a multitude of factors, including the type of coal utilized, the burning environment, the collection system, etc.

Historically, the percentage of FA by mass of the cementitious material component in concrete has ranged from 15% to 25%, depending on the application, the properties of FA, the constraints of the specification, the climate, and the location. It was reported that in a few unique instances, concrete with up to 80% FA had been placed successfully (Mondal et al. 2023). Class F and Class C FAs are the two types that ASTM states are used in concrete. The class F FA is produced as a byproduct of burning bituminous coal. Class F FA has a high iron, silica, and alumina content but a low calcium content. It's a glassy substance that requires cement or lime activation. FA that is produced by burning lignite and sub-bituminous coal falls into class C. It contains more calcium than class FFA. Concrete with class CFA gains strength significantly faster than concrete with class FFA. In addition to changing the properties of the concrete in both its fresh and hardened states, FA is a cost-effective way to improve workability, strength, and drying shrinkage in concrete. Additionally, using FA in concrete solves the problem of storing and disposing of FA, an industrial byproduct. The FA is complex because it contains over 188 different mineral groups, whereas the coal only contains over 316 different mineral groups. Conversely, all FA contain significant levels of silicon dioxide as well as oxides of calcium, iron, and aluminum.

FA's reactivity is influenced by its mineralogical composition and particle properties. FA needs to have specific surfaces or granulometry that is similar to PC in order to prevent variations in concrete properties like workability. China, the world's biggest producer of FA, uses 70% of the 600 million tons of FA it produces each year (Nie et al. 2014). Of the 226 million tons of FA produced in India, 83% are used there. The requirements for FA classification vary by nation (Bhatt et al. 2019). In Japan, the primary consideration is performance; in Europe, fineness and ignition loss are considered; in China and the USA, the primary consideration is CaO content. The total SiO_2, Al_2O_3, and Fe_2O_3 content of FA used in concrete must be more than 70% by mass, per EN 450–1. For siliceous and calcareous pulverized fuel, at least 70% and 50%, respectively, of SiO_2 plus Al_2O_3 plus Fe_2O_3 are needed. As per IS 3812 (Part 1) ashes FA's higher free lime content was found to be the cause of both earlier setting and higher CS, especially at younger ages. A high concentration of free lime in FA reduces the significance of the CS improvement. Concrete that contains a high alkali content in gel products may swell and expand due to an alkali-silica reaction. Concrete that exhibits expanding behavior due to a high SO_3 level may become unstable over time. The consistency of blended cement increases when FA is substituted; more water is needed to get the same consistency as ordinary Portland cement (OPC). In one study, adding FA to 30% and 50% of OPC resulted in an increase in standard consistency of 1.6% and 3.1%, respectively, due to the blended mix's improved fineness. A 30% substitution of calcareous FA resulted in a 2.6% increase in consistency, according to reports. Similar observations were made in blended cement with the FA from the precipitator of the thermal station and with the hydrated FA, as the FA replacements increased the water demand for the normal consistency (Z. Liu et al. 2023). The consistency was tested using a 20% FA replacement, either with or without superplasticizer (SP). This study shows that the admixed paste, with or without SP, has a higher consistency than the OPC paste. Furthermore, the consistency of the paste was lower with SP than it was without. Up to 35% FA replacement, blended cement paste maintained the same normal consistency as plain PC paste, according to experimental studies on mortar incorporating class FFA. Furthermore, they found that because of the spherical shapes of the FA particles, which reduced the interparticle frictional force and the particle surface-to-volume ratio, lowering the water requirement, FA paste had a deeper penetration depth than PC paste. Other studies also found that when the FA level increased, the water demand decreased (Shao et al. 2020). The w/c ratio required for normal consistency decreased as the FA concentration increased, which is consistent with the glassy spherical particles found in FA.

16.4.3 Rice husk ash

RHA is a valuable pozzolanic material due to its high silica content. Pozzolans are compounds that increase the strength and durability of concrete by combining with calcium hydroxide, which is found in cement. The strength and durability of concrete are enhanced when RHA is added to mixtures because it reacts with calcium hydroxide to produce more calcium silicate hydrate (C–S–H) gel. RHA can also lessen the heat produced when concrete hydrates, lowering the possibility of thermal cracking. One byproduct of the rice milling business is RHA. Reusing what would otherwise be deemed waste in the production of concrete lowers environmental pollution and landfill accumulation. By incorporating RHA into concrete, less conventional cement is used, which has a large carbon footprint during production. By lowering the amount of cement required in concrete mixes, RHA utilization contributes to a reduction in the total CO_2 emissions related to the production of concrete. RHA-containing concrete is more durable, especially in harsh environments like coastal areas, as it demonstrates better resistance to chemical attacks from materials like sulfates and chlorides. RHA can improve concrete's workability, making it simpler to handle and install during construction. Although the initial cost of implementing RHA may differ, using it frequently results in long-term costs. savings as a result of longer-lasting structures, less maintenance, and better concrete performance.

16.4.4 Saw dust ash

Sawdust ash is an important pozzolanic material because of its high silica and other reactive compound content. It interacts with cement's calcium hydroxide, much like other pozzolans do, to create new cementitious compounds that improve the strength and longevity of concrete (Dulipalla 2018).

Sawdust ash has a variety of compositions, but amorphous silica is usually present, which adds to its pozzolanic qualities. Additionally, by filling in gaps in the concrete matrix, its fine particles increase strength and density overall. A byproduct of burning or processing wood is sawdust ash. By recycling this byproduct, lowering waste accumulation, and minimizing the environmental impact of disposal, integrating sawdust (SDA) into concrete production provides a sustainable solution. The total amount of cement in concrete is reduced by adding SDA in place of some of the cement. This decrease in cement consumption supports environmentally friendly building methods by lowering CO_2 emissions associated with cement production. Concrete formulations gain strength and durability when sawdust ash is added. The concrete's mechanical qualities are improved by the additional cementitious compounds formed as a result of the pozzolanic reaction between SDA and calcium hydroxide (Otieno, Kabubo, and Gariy 2023). Sawdust ash helps to improve density and reduce pore size in the microstructure of concrete. As a result, there is less permeability in the concrete, which increases its durability by preventing the entry of dangerous materials like water, chloride ions, and sulfates. Although the initial cost may differ, using sawdust ash can result in long-term cost savings because it improves the performance of concrete, requires less maintenance, and extends the lifespan of structures (Fapohunda et al. 2021). When applied correctly, sawdust ash can increase the cohesiveness and workability of concrete mixtures, making handling and placement during construction easier.

16.4.5 Palm oil fuel ash

Because of its beneficial qualities, palm oil fuel ash (POFA), a byproduct of burning palm oil waste, is an important component in the creation of green concrete. In the presence of moisture, its natural pozzolanic activity—which is typified by reactive silica and alumina content—enables a chemical reaction with calcium hydroxide. The C–S–H gel produced by this reaction significantly increases the strength and durability of the concrete. Beyond its technical advantages, using POFA lessens the environmental effect of producing concrete by reducing the use of conventional cement, a method known for emitting large amounts of carbon dioxide (Santhosh, Subhani, and Bahurudeen 2022; Zeyad et al. 2022). By repurposing the residue from the production of palm oil, the integration of POFA also embodies a sustainable approach to waste management and reduces environmental pollution. Furthermore, adding POFA to concrete formulations can enhance a number of characteristics like workability, durability, and resistance to chemical deterioration, which may increase the useful life of built structures. Adopting POFA reduces costs by taking the place of some of the cement, which usually accounts for a significant amount of the costs associated with producing concrete. To guarantee the intended results in the creation of green concrete, however, rigorous testing, and quality control procedures are required. The effectiveness of POFA depends on a number of factors, including its quality, processing methods, and the particular composition of the concrete mix (Choong et al. 2022).

16.4.6 Agro-waste

The productivity of agriculture has tripled in the past 50 years. The population of the planet will surpass that of Earth by 2050, and currently, 24 million tons of food are produced annually worldwide. At every stage of agricultural production, processing, and consumption, massive amounts of solid waste are generated; depending on the crop's intended use, these wastes vary greatly. These wastes are disposed of in public spaces and close to bodies of water, where they are carelessly burned outside and contaminate the land and water. Therefore, it is crucial to find uses for agricultural waste for the environment, animals, and human health. One application for solid crop waste recycling is the construction of buildings, along with its fibers, leaves, and husk. Research shows that substituting some of the sand with agricultural wastes such as rice husks, rice straw, peanut shells, and coconut shells (CNSs) can help partially meet building standards for the strength and durability of cement blocks (Charitha et al. 2021). A variety of crops, including rice, sugarcane, bamboo, coconut, palm, and sawdust, have been used to create manufactured agricultural construction materials, including RHA, Sugarcane Bagasse Ash, Bamboo Leaves Ash, Groundnut Shell, sawdust (SDA), Oil Palm Shell, Cork Waste Ash, and CNS. In the past, solid reinforcements for binding clay and sand particles together in construction projects were created by combining earth with rice husk or coconut coir. Fibrous materials aid in a building's tensile strength, whereas binding materials aid in its CS. Eggshells have also been used as building materials in the past. People have been using naturally occurring materials for construction since the Neolithic era. The 20th century saw the development of energy-intensive building materials like concrete and steel, which are now commonplace in construction. There is currently a lot of interest in the application of biomaterials in non-hazardous construction. Conventional building materials like cement and concrete are costlier because of their high thermal and electrical energy requirements (Alaneme, Olonade, and Esenogho 2023; Antony Godwin et al. 2018). Therefore, it becomes imperative to create building materials that are less expensive and energy-intensive without compromising the structural integrity of the building. Agricultural waste-derived materials have lower embodied energy and higher thermal performance, making them more environmentally friendly. This bridges the divide and creates a closed loop between the construction of buildings and the agricultural sectors.

16.4.7 Industrial waste

The residue or byproducts from industrial processes in a variety of industries, including manufacturing, production, mining, and agriculture, are referred to as industrial waste (Dey et al. 2022; Ghanim et al. 2023; Munir et al. 2023). In order to reduce risks to the environment and public health, this waste usually contains chemicals, metals, and other potentially hazardous materials that need to be handled, treated, and disposed of properly. Solid waste, liquid waste, gaseous waste, and hazardous waste are the four categories of industrial waste. Paper, plastics, metals, and other non-liquid materials are examples of solid waste materials. Liquid waste from industrial processes is frequently produced; examples include chemical effluents, manufacturing wastewater, and contaminated water from different operations. Gaseous waste is composed of Industrial waste is also a result of hazardous gas emissions from industrial processes, such as carbon monoxide or sulfur dioxide from burning fossil fuels or industrial chemicals released into the atmosphere. Hazardous waste refers to waste products that, because of their toxic, reactive, flammable, or corrosive properties, could be dangerous or present risks to the environment or human health. Examples include certain chemicals, heavy metals, and radioactive materials. Various plant-based agricultural wastes, including RHA, sawdust ash, rubber crump, POFA, coconut husk and shell, and molasses waste, are used to make green concrete (Mohanraj et al. 2022).

16.4.8 Coconut husk and shell

The utilization of coconut husks and shells has been instrumental in the advancement of green concrete, providing a range of uses that support environmentally friendly building methods. These materials are used in a variety of ways. For example, fibers from coconut husks are used in concrete compositions in place of traditional aggregates. These fibers provide an environmentally friendly substitute for heavier conventional aggregates by increasing the concrete's strength, resilience to impact, and overall density reduction. Meanwhile, finely ground CNS particles act as fillers in concrete, lowering the amount of cement required (Bello et al. 2023). This reduces carbon emissions associated with cement production, making the material more environmentally friendly, and it also lessens its negative effects on the environment by recycling agricultural waste. Furthermore, the addition of components derived from coconuts improves the thermal insulation capabilities of the concrete, guaranteeing energy efficiency in built environments. By utilizing these plentiful agricultural byproducts, incorporating coconut husks and shells into concrete formulations not only solves waste management issues but also presents a potential cost-saving opportunity (Janani et al. 2022). These resources are also easily obtainable and renewable. For the best outcomes in the production of green concrete, however, strict testing and quality assurance procedures must be followed during formulation (Mokhtar et al. 2022) including the caliber of these materials, exact processing techniques, and their intended application within the concrete mix.

16.4.9 Molasses waste

Because of its benefits, molasses waste—a byproduct of the production of sugar—is a valuable ingredient in the production of green concrete. One important benefit is that it has natural retarding qualities, which help to extend the concrete's setting time. Better handling and construction techniques are made possible by this trait, which is especially useful in construction scenarios requiring extended workability and placement periods. Moreover, adding molasses waste improves the workability of concrete without compromising its structural integrity, enabling the concrete mix to have the ideal consistency (Al-hasan and Hartantyo 2020; Jiménez, Fontes Vieira, and Colorado 2022). This feature makes placement simpler and eliminates the need for excessive water content, which otherwise might jeopardize the concrete's ultimate strength. Using molasses waste to produce concrete is an environmentally friendly way to manage industrial waste by keeping it out of the landfill and other disposal sites. This reduces pollution in the environment. Furthermore, as a readily available or even cost-free byproduct, it may be more affordable than other concrete additives or chemicals, making it a viable supplement or alternative. Accepting molasses waste in the process of making concrete helps reduce the overall carbon footprint of the concrete manufacturing process while also demonstrating environmental responsibility. However, the incorporation of molasses waste into concrete requires careful evaluation of its quality, chemical makeup, and specific function in the concrete mix. This calls for extensive testing and assessment to guarantee that the performance requirements and desired concrete properties are fulfilled.

The waste categories used are displayed in Figure 16.5. With 79.2%, both industrial and agricultural waste have the highest percentage. Nineteen

Figure 16.5 Categories of waste utilized.

completed projects have made use of both industrial and agricultural waste in the building process. On the other hand, there is comparatively less application of using municipal waste in construction projects that the respondents have taken on. It only requires 25% of the total or six projects.

16.4.10 Animal waste

One of the main animal wastes used to make green concrete is cow dung. After being cleaned and prepared, cow dung ash can partially replace cement in the creation of concrete. Because of its alkaline qualities, which enhance concrete's strength and durability, it can be used in place of or in addition to more conventional materials (B. Veera Narayana 2020; Mathur and Chhipa 2022). Furthermore, due to their unique chemical compositions and qualities, other animal wastes, such as chicken droppings or specific kinds of animal bones, have also been investigated for their possible use in concrete. To guarantee they meet quality standards without sacrificing the structural integrity of the concrete, thorough testing, processing, and treatment are necessary before they can be applied as widely as possible in concrete.

Organic matter found in poultry droppings can be processed and treated to create ash or powder that can be used in concrete. Phosphorus is abundant in poultry waste, and this element may influence the chemical composition of concrete. As a binding agent, phosphorus may increase the strength of concrete (Zhang et al. 2021). Pork manure contains organic materials that may be beneficial as a concrete additive if they are cleaned up and processed. It contains potassium, phosphorus, and nitrogen, all of which, under the right circumstances, may give concrete beneficial chemical qualities. Concrete needs minerals like calcium and phosphorus, which are found in fish waste, especially in the form of bones and scales. It may be possible to add these minerals to concrete mixes by processing fish waste, which could improve the concrete's characteristics. When turned into bone meal, animal bones provide an abundant supply of calcium and phosphorus. Bone meal has the potential to strengthen and prolong concrete when added as a mineral additive when ground finely.

16.4.11 Municipal waste

Glass that has been recycled is frequently found in municipal garbage. Glass cullet is created by crushing the glass into tiny, fine particles. In concrete mixes, glass cullet can partially replace conventional aggregates. Recycling is encouraged and the need for natural aggregates is decreased when leftover glass is used in concrete (Ahmad et al. 2022). Fly ash is a byproduct of burning coal in power plants and is frequently regarded as a kind of garbage that comes from cities. After being processed to remove impurities, fly ash is added to concrete as an additional cementitious material. By reducing the amount of cement required, its incorporation improves the strength and durability of the concrete while reducing the carbon footprint associated with cement production. Certain municipal plastic waste types can be processed into small particles or shreds. In order to address the issue of plastic waste and possibly enhance certain concrete properties, the processed plastic particles can be used in concrete mixes in part place of conventional aggregates. Slag is a byproduct of making steel and can be found in garbage from cities. Like fly ash, slag strengthens and preserves concrete while offering an environmentally friendly way to handle industrial waste. Recycled aggregates can be obtained from demolished buildings, construction waste, and other sources included in municipal waste management programs. In concrete mixes, recycled aggregates can take the place of conventional aggregates, protecting natural resources and lessening the environmental effect of aggregate extraction. Rubber is crushed into tiny pieces, or crumb rubber, from used tires and other sources (Ji and Sun 2023). In some concrete applications, crumb rubber can be used in part place of aggregates, adding to the material's flexibility and impact resistance. The ash produced by burning municipal solid waste is gathered and treated. Processed ash can be added to concrete as an additive to improve strength and durability.

16.4.12 Concrete debris

Concrete debris, acquired from destroyed buildings or construction debris, goes through a careful processing stage. This procedure entails breaking down the waste into smaller pieces so that it can be recycled and used as aggregate in new concrete mixes. By reducing reliance on virgin aggregates like crushed stone or gravel, this integration helps to conserve natural resources. By reducing the need to extract new materials, the use of recycled aggregate made from concrete debris helps the construction industry practice responsible resource management and lessens its negative environmental effects (Gyawali 2022). Reclaimed concrete waste is a strong foundation material for many building needs, such as walkways and road bases. When compacted, it provides strength and longevity, serving as a stable base for upcoming building projects. Concrete waste that has been finely ground can be used as an additional cementitious material when making concrete. When used in part place of cement, it increases the durability and strength of the resulting concrete structures. In line with sustainable waste management principles, the addition of concrete debris to green concrete formulations allows for the significant diversion of waste material away from landfills. Using recycled concrete waste is economical because it eliminates the need to buy new aggregates and lowers disposal costs at the same time. Reusing concrete waste makes a significant contribution to reducing the environmental impact of the extraction, processing, and transportation of virgin materials, which in turn promotes an environmentally conscious construction sector.

16.5 ENVIRONMENTAL IMPACTS AND LCA

In order to create development materials that have an impact on the climate, geopolymer concrete, also known as green cement, is necessary. A mixture of 25%–100% modern waste and an inorganic polymer is used in its production. These are various benefits of using green cement.

- **Longer Life Span:** Compared to concrete made mostly of Portland concrete, green concrete significantly increases strength more quickly and shrinks more slowly. Because green cement can withstand temperatures of up to 2,400°C, structures made with it have a better chance of surviving a fire (Öhrn Sagrelius et al. 2022). Additionally, it offers stronger protection against consumption, which is important given the effect that pollution has on the climate (corrosive rains significantly shorten the lifespan of conventional building materials). All of those components add up to a structure that will last a lot longer than one built with regular cement. Similar substantial combinations can also be found in ancient Roman designs. This material was also used in the 1950s and 1960s in the Ukraine (Hashmi et al. 2022). Those Ukrainian structures are still in place more than 40 years after the incident. If structures don't need to be altered, less construction material is needed, and the impact on the environment is lessened.
- **Uses Industrial Waste:** Green significantly uses 25%–100% fly debris instead of a combination of 100% Portland concrete. Fly debris is a side-effect of coal burning and is accumulated from the chimney stacks of modern plants (for example, power plants) that utilize coal as a power source. Fly debris is disposed of on a vast number of sections of land. The use of green cement in development will provide a way to dispose of fly waste and, ideally, free up large areas of land.
- **Reduces Energy Utilization:** You will use less energy if, during the concrete-blending process, you use more fly ash and less Portland concrete. In order to heat the materials needed to make Portland concrete to the proper temperature, massive amounts of coal or gaseous gasoline are needed. Fly ash is currently produced as a byproduct of another contemporary interaction, so using it to create green cement won't require significantly more energy. The fact that green cement is more resistant to temperature changes means that buildings made of it require less energy. This can be used by an engineer to design a green, substantial building that uses energy for heating and cooling more effectively.
- **Reduces Carbon Dioxide Emissions:** To make Portland concrete—one of the primary fixings in standard concrete—pulverized limestone, dirt, and sand are warmed to 1,450°C involving petroleum gas or coal as a fuel (Yuan et al. 2022). Five to eight percent of global carbon dioxide (CO_2) emissions can be attributed to this interaction. Up to 80% less

Table 16.1 Relative weights of sustainable factors for the material life cycle phases

	Climate change	*Pollution*	*Energy consumption*	*Resources and waste*	*Cost*	*Recyclability*	*Health and safety*	*Human satisfaction*	*Practicability*
Phase I	12.27%	9.95%	12.76%	9.53%	10.39%	11.78%	8.05%	7.74%	9.13%
Phase II	9,53%	8.05%	10.39%	12.27%	9.95%	8.39%	11.78%	7.74%	12.76%
Phase III	8.39%	8.05%	9.95%	10.39%	9.13%	12.76%	9.53%	11.78%	12.27%

CO_2 is emitted during the assembly of green large deliveries. A major contribution to the global effort to reduce emissions is the complete switch to using green cement in development.

The system's default weight, which accounts for all nine sustainability factors across all three life cycle phases of a material, was used to conduct this assessment. The overall sustainability weights of each factor are shown in Table 16.1. Cost, energy consumption, pollution, and climate change were the factors that carried the most weight during phase 1 (the manufacturing phase). Practicality ruled during the building phase, while recyclability had the highest relative weight during the demolition phase.

16.6 CHALLENGES AND FUTURE DIRECTIONS IN SUSTAINABLE GREEN CONCRETE

The main significant institutional barrier is development business practices. The quick development plans, not the life-cycle cost reserve funds from the preservation of energy and materials, are what keep the structure of business productivity in the air. Experience has shown that faster development is not always more cost-effective. Poor quality concrete generally crumbles more quickly and needs costly repairs. Owners ought to incur a high life cycle cost as a result. The current development economy needs to be fundamentally rebuilt. Another institutional barrier that is put in place by construction laws is the reduction of the use of recycled materials. Outdated codes don't specify a particular level of execution; rather, they specify the use of particular materials and blend extents for a task. For instance, all codes pertaining to substantial blends support the lowest amount of concrete substance or the highest fly debris content, which is typically 15%–25% of the cementitious material's mass. High Volume Excoriate Debris concrete combined with elite execution significantly shows that execution-based principles should take the place of prescriptive details. The lack of a thorough methodology in the planning of education and research (Nilimaa 2023; Y. Singh et al. 2023). The first step in moving from reductionistic to comprehensive development rehearsals should be to change the way that training and assessment are currently organized in the areas of significant science and innovation. Before green cement takes the place of conventional concrete as the preferred material for general development, the concrete development industry as a whole need to continue going green. The building sector understands the importance of environmental concerns and the case for green building (GB) as a practical means of achieving social, economic, and environmental objectives. The sector is starting to embrace objectives. However, the adoption of GB practices and technologies has not been smooth in many parts of the world. Many academics have written books that give a summary of the challenges, issues, or obstacles (henceforth called barriers) that keep GB from being effectively implemented in different countries (Pan et al. 2023). According to a survey of the literature, these obstacles fall into five major categories: financial concerns; information, knowledge, and awareness; government and management; technology and training; and attitude and the market. Comparable GB barrier classifications from the literature served as the main foundation for these clusters.

16.6.1 Financial concerns

Meryman and Silman 2004 state that financial concerns are believed to be the primary barrier to green construction. Cost remains one of the most significant economic factors in GB projects. Because green materials are usually more expensive than conventional materials, starting a GB project is more expensive than starting a conventional building project (Hwang and Tan 2012; Kibert, Sendzimi, and Guy 2007). As per the United States Green Building Council, the incorporation of green features and technologies may result in an initial project cost increase of 2%–7%. According to (Tramontin, Loggia, and Basciu 2010), the adoption of green technologies is probably going to increase both the initial and overall project costs. The sustainability in finances is depicted in Figure 16.6. This is due to the possibility that stakeholders will have to pay the greater capital expenses connected with putting these technologies into use. The capital costs of GB projects are estimated by Tagaza and Wilson (Wilson and Tagaza 2006) to be 1%–25% higher. Other factors that are known to impact the cost of GB projects include building certification and design complexity. It can be concluded that the primary and undeniable barrier preventing GB from surviving in the market is the high-cost premium. One important economic

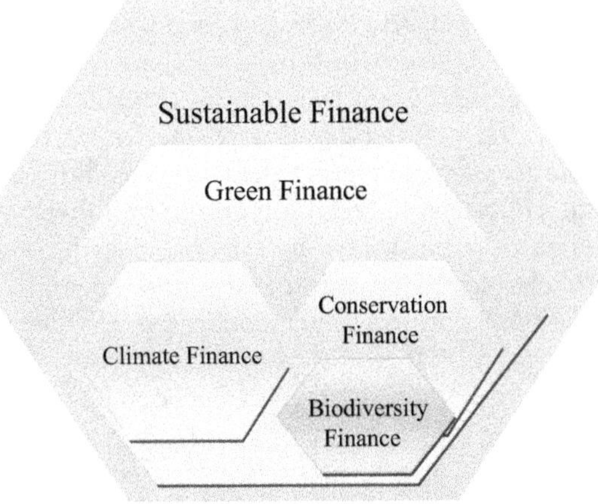

Figure 16.6 Sustainable development finance.

consideration in building projects is time. Time is important to almost all project stakeholders and is used to measure project performance, much like cost is. Green practices could cause a project's completion to be delayed, which could have detrimental effects on the economy. This is because project delays usually result in higher costs, harm to parties, and a bad effect on the reputations of corporations. GB project delays can be caused by a variety of factors, such as the time it takes for new green technologies to be approved within an organization and the time it takes to implement technologies on-site. Project delays are a significant barrier keeping stakeholders from adopting green practices. The literature goes on to address additional financial obstacles that prevent GB adoption, including high market prices, rental fees, and GBs' lengthy payback periods. Most customers (or tenants) find GBs unattractive due to the high rental fees.

16.6.2 Market mindset

The roles that stakeholders' attitudes, lifestyles, behaviors, and cultures play in the GB market are undeniable. Mainly, because they are the ones with the money, clients are in a great position to decide whether or not to adopt GB technologies. Unless the client indicates interest in doing so, the project team is unable to implement green technologies and practices. Furthermore, consumers' and clients' willingness to support GBs pushes the industry to meet more environmentally friendly standards. Taking these facts into account, any negative views that stakeholders may have could affect the GB market. One common example is when stakeholders are unwilling to give up on using traditional building methods and technologies. Researchers (Alqahtani and Zafar 2023) also concluded that a major barrier to the GB market is the reluctance to deviate from the use of conventional building methods. Many people may have doubts regarding the efficacy and quality of green technologies due to the relatively recent nature of most of them and the deeply held traditional beliefs held by many stakeholders. As a result, there is a great deal of mistrust for GB systems and goods, which may force some individuals to embrace new religions and worldviews. The capacity of present manufacturers, suppliers, and developers is determined by market demand. For instance, if people don't demand GBs, how can the developer or contractor incorporate green technologies in buildings? And without the developers' interest or request, how can the provider provide? Studies have revealed that stakeholders' lack of interest in and demand for GB technologies is impeding their adoption. This jeopardizes the dependability of the entire GB supply chain. By learning from the successes of previous, similar projects, the complexity of a new green project can be reduced. These kinds of projects validate the effectiveness of some implemented green technologies, at least in part. As a result, a number of other barriers in the market, such as the absence of comparable schemes for comparison and the lack of tried-and-true green materials and technologies in the local market, may provide stakeholders good reasons not to adopt green technologies.

16.6.3 Knowledge resources

Information is utmost essential to increase public acceptance and awareness as well as to acquire relevant knowledge. Industry is becoming more and more aware of the connection between information and GB adoption that is successful. GB technologies differ from conventional technologies and are more complex. Research has shown that managing green technologies is a laborious task for clients, developers, and contractors (Seghier et al. 2018). While some researchers think that this issue is the result of gaps in GB data, others think that issues with awareness, familiarity, knowledge, and technical expertise, as well as issues with research and education, are the root causes of this issue. Still, it's clear that these challenges are related. Insufficient data and researchers, for instance, could lead to incomplete knowledge, which could obstruct awareness, and so on. Without sufficient research, it is difficult to make a compelling and appealing sustainable model for GBs, as the USGBC pointed out. Sustainably oriented actions and increased awareness among stakeholders are contingent upon the availability of credible research and sufficient information. Ignorance is the main barrier keeping stakeholders from implementing green practices. A major barrier is the dearth of trustworthy research. All of these have a negative impact on stakeholders' knowledge and awareness because most of the people don't have awareness about the advantages of GB technologies. Convincing ignorant stakeholders that their technologies and practices are GB and that they don't know enough would be difficult. Analysis of venture capital investment trends in GB technologies, the absence of awareness campaigns or other initiatives to encourage investment in GB technologies prevents their broader adoption. There are two main reasons why stakeholders might decide against implementing GB practices: first, a lack of technical expertise or knowledge; second, a lack of familiarity with green design, materials, products, and technologies.

Cutting-edge GB technologies must be used in order for GB projects to succeed. However, because most green technologies are complex, utilizing them requires some technical consideration to maximize the prospect of reaching the anticipated sustainability objectives. Researchers (Pearce and Vanegas 2002) suggested that increasing the efficacy and efficiency of currently used construction technologies while simultaneously creating new environmentally friendly ones is critical to creating a sustainable built situation. Analyzing existing technologies is crucial to pinpointing the crucial areas in need of development. The complex and varied environmental issues and expectations make it difficult to evaluate and quantify construction activities. Moreover, figuring out the lifespans and performance of novel materials and technologies is challenging. Because there are no reliable data sources or technical standard operating procedures for GB technologies, stakeholders may be dissuaded from pursuing green construction. Many times, lack of expertise suffer the whole construction process which in turn again will be dangerous in all angles. Each and every component needs to be trained so that the whole structure will lie within the sustainability range.

For example, if an architect is unable to make informed decisions about the integration of solar panels into the design (for example, on the facade or on top of the roof), which could also cause delays to the project. Due to the inability to predict the potential performance of GB technologies, there is a strong need to formulate a new pathway that will alter the risks and drawbacks and will lead to successful and sustainable practices. This frequently reduces GB's efficiency and may lead stakeholders to resume using conventional construction methods.

16.7 CONCLUSION: TOWARD SUSTAINABLE BUILDING SOLUTION

Green substantial is available in a variety of forms, including geopolymer concrete, lightweight concrete, fly debris concrete with a high volume, self-solidified concrete, high-strength concrete, and super elite execution concrete. Due to differences in the developed needs, capabilities, and limits of the neighborhood development industry, different approaches would be adopted in each country to support green cement in development. Most people consider concrete to be a composite material made of finely ground "sand" and coarsely ground "rock" that have been joined together and reinforced with concrete glue before gradually solidifying over time. Concrete is known to be a major area of strength for extreme pressure because the pressure load is effectively conveyed by the total, especially the coarse total. These days, "green" refers to the entire climate, which is the primary component globally, rather than just variety. "Green Cement" is a type of material that is made from regular substantial waste and is regarded as environmentally friendly. These wastes include a variety of materials, including slag, waste from power plants, recycled cement, waste from mining and quarrying, waste glass, buildup from incinerators, red mud, used dirt, sawdust, combustor debris, and foundry sand. With its varied blend plan and arrangement, green cement ensures more durable construction while reducing natural effects on wastewater, CO_2 emissions, and energy consumption. Reducing cement's environmental impact is the aim of the Middle for Green Cement. New innovations are developed in order to strengthen this. Innovation encompasses all aspects of execution and considers every stage of a significant development's life cycle, including the underlying model, detail, assembly, and maintenance.

REFERENCES

Abdalla, Jamal A., Blessen Skariah Thomas, Rami A. Hawileh, Jian Yang, Bharat Bhushan Jindal, and Erandi Ariyachandra. 2022. "Influence of Nano-TiO2, Nano-Fe2O3, Nanoclay and Nano-CaCO3 on the Properties of Cement/Geopolymer Concrete." *Cleaner Materials*.

Abu-Saleem, Mahmoud, Yan Zhuge, Reza Hassanli, Mark Ellis, Md Mizanur Rahman, and Peter Levett. 2021. "Stress-Strain Behaviour and Mechanical Strengths of Concrete Incorporating Mixed Recycled Plastics." *Journal of Composites Science*. doi: 10.3390/jcs5060146.

Ahmad, Jawad, Roberto Alonso González-Lezcano, Ali Majdi, Nabil Ben Kahla, Ahmed Farouk Deifalla, and Mohammed A. El-Shorbagy. 2022. "Glass Fibers Reinforced Concrete: Overview on Mechanical, Durability and Microstructure Analysis." *Materials*. doi: 10.3390/ma15155111.

Akbar, Arslan, and K. M. Liew. 2020. "Assessing Recycling Potential of Carbon Fiber Reinforced Plastic Waste in Production of Eco-Efficient Cement-Based Materials." *Journal of Cleaner Production*. doi: 10.1016/j.jclepro.2020.123001.

Al-hasan, Susilo Abadi, and Sugeng Dwi Hartantyo. 2020. "PENGARUH LIMBAH PABRIK GULA MOLASE SEBAGAI BAHAN TAMBAH (ADMIXTURE) KUAT TEKAN BETON K-175 DENGAN MENGGUNAKAN PASIR LOKAL PASIR JOMBANG." *UKaRsT*. doi: 10.30737/ukarst.v4i1.701.

Alaneme, George Uwadiegwu, Kolawole Adisa Olonade, and Ebenezer Esenogho. 2023. "Eco-Friendly Agro-Waste Based Geopolymer-Concrete: A Systematic Review." *Discover Materials*. doi: 10.1007/s43939-023-00052-8.

Alqahtani, Fahad K., and Idrees Zafar. 2023. "Construction of Green Concrete Incorporating Fabricated Plastic Aggregate from Waste Processing." *Sustainability (Switzerland)*. doi: 10.3390/su15054114.

Antony Godwin, I., S. Nancy Deborah, I. Julius Ponraj, R. Vinslin Blessho, and C. Stephen. 2018. "Experimental Investigation on the Mechanical and Microstructural Properties of Concrete with Agro-Waste." *International Journal of Engineering and Technology(UAE)*. doi: 10.14419/ijet.v7i3.12.15858.

Aocharoen, Yanika, and Piya Chotickai. 2021. "Compressive Mechanical Properties of Cement Mortar Containing Recycled High-Density Polyethylene Aggregates: Stress–Strain Relationship." *Case Studies in Construction Materials*. doi: 10.1016/j.cscm.2021.e00752.

Araya-Letelier, Gerardo, Pablo Maturana, Miguel Carrasco, Federico Carlos Antico, and María Soledad Gómez. 2019. "Mechanical-Damage Behavior of Mortars Reinforced with Recycled Polypropylene Fibers." *Sustainability (Switzerland)*. doi: 10.3390/su11082200.

Askar, Mand Kamal, Yaman S. S. Al-Kamaki, and Ali Hassan. 2023. "Utilizing Polyethylene Terephthalate PET in Concrete: A Review." *Polymers*.

Azhdarpour, Amir Mahyar, Mohammad Reza Nikoudel, and Milad Taheri. 2016. "The Effect of Using Polyethylene Terephthalate Particles on Physical and Strength-Related Properties of Concrete; A Laboratory Evaluation." *Construction and Building Materials*. doi: 10.1016/j.conbuildmat.2016.01.056.

B. Veera Narayana, C. V. K. Chaitanya Kumar. 2020. "Study on Mechanical and Durability Properties of Concrete with Partial Replacement of Cement with Cow Dung Ash." *International Journal of Advanced Science and Technology*.

Bakhoum, E. S., G. L. Garas, M. E. Allam, and H. Ezz. 2017. "The Role of Nano-Technology in Sustainable Construction: A Case Study of Using Nano Granite Waste Particles in Cement Mortar." *Engineering Journal*. doi: 10.4186/ej.2017.21.4.217.

Bamigboye, Gideon O., Uchechi E. Okechukwu, David O. Olukanni, Daniel E. Bassey, Uchechukwu E. Okorie, Joshua Adebesin, and Kayode J. Jolayemi. 2022. "Effective Economic Combination of Waste Seashell and River Sand as Fine Aggregate in Green Concrete." *Sustainability (Switzerland)*. doi: 10.3390/su141912822.

Barry, Bronwyn. 2018. "Energy and Design Criticism: Is It Time for a New Measure of Beauty?" *Architectural Design*. doi: 10.1002/ad.2266.

Bello, Sefiu Adekunle, Nasirudeen Kolawole Raji, Maruf Yinka Kolawole, Mohammed Kayode Adebayo, Jeleel Adekunle Adebisi, Kehinde Adekunle Okunola, and Mustekeem Olanrewaju AbdulSalaam. 2023. "Eggshell Nanoparticle Reinforced Recycled Low-Density Polyethylene: A New Material for Automobile Application." *Journal of King Saud University - Engineering Sciences*. doi: 10.1016/j.jksues.2021.04.008.

Bhatt, Arpita, Sharon Priyadarshini, Aiswarya Acharath Mohanakrishnan, Arash Abri, Melanie Sattler, and Sorakrich Techapaphawit. 2019. "Physical, Chemical, and Geotechnical Properties of Coal Fly Ash: A Global Review." *Case Studies in Construction Materials*. doi: 10.1016/j.cscm.2019.e00263.

Boobalan, S. C., M. Salman Shereef, P. Saravanaboopathi, and K. Siranjeevi. 2022. "Studies on Green Concrete – A Review." *Materials Today: Proceedings*. doi: 10.1016/j.matpr.2022.04.392.

Cerutti, P. O., G. Lescuyer, L. Tacconi, R. Eba'a Atyi, E. Essiane, R. Nasi, P. P. Tab. Eckebil, and R. Tsanga. 2017. "Social Impacts of the Forest Stewardship Council Certification in the Congo Basin." *International Forestry Review*. doi: 10.17528/cifor/004487.

Charitha, V., V. S. Athira, V. Jittin, A. Bahurudeen, and P. Nanthagopalan. 2021. "Use of Different Agro-Waste Ashes in Concrete for Effective Upcycling of Locally Available Resources." *Construction and Building Materials*. doi: 10.1016/j.conbuildmat.2021.122851.

Choong, Wei Sheng, Jian Chong Chiu, Flavio Lopez-Martinez, Abdullah Alaklabi, Mariana Claudia Oliveira, Surya Dewi Puspitasari, and Julius Adebayo. 2022. "Utilization of Green Material for Concrete in Construction." *Civil and Sustainable Urban Engineering*. doi: 10.53623/csue.v2i2.116.

Coppola, Luigi, Silvia Beretta, Maria Chiara Bignozzi, Fabio Bolzoni, Andrea Brenna, Marina Cabrini, Sebastiano Candamano, Domenico Caputo, Maddalena Carsana, Raffaele Cioffi, Denny Coffetti, Francesco Colangelo, Fortunato Crea, Sabino De Gisi, Maria Vittoria Diamanti, Claudio Ferone, Patrizia Frontera, Matteo Maria Gastaldi, Claudia Labianca, Federica Lollini, Sergio Lorenzi, Stefania Manzi, Milena Marroccoli, Michele Notarnicola, Marco Ormellese, Tommaso Pastore, Mariapia Pedeferri, Andrea Petrella, Elena Redaelli, Giuseppina Roviello, Antonio Telesca, and Francesco Todaro. 2022. "The Improvement of Durability of Reinforced Concretes for Sustainable Structures: A Review on Different Approaches." *Materials*.

Daud, M. A. M., M. Z. Selamat, and A. Rivai. 2013. "Effect Pf Thermoplastic Polymer Waste (PET) in Lightweight Concrete." in *Advanced Materials Research*.

Deraman, Rafikullah, Mohd Nasrun Mohd Nawi, Mohd Norazam Yasin, Mohd Hanif Ismail, and Rami Salah Mohd Osman Mohd Ahmed. 2021. "Polyethylene Terephthalate Waste Utilisation for Production of Low Thermal Conductivity Cement Sand Bricks." *Journal of Advanced Research in Fluid Mechanics and Thermal Sciences*. doi: 10.37934/arfmts.88.3.117136.

Dey, Dhrutiman, Dodda Srinivas, Biranchi Panda, Prannoy Suraneni, and T. G. Sitharam. 2022. "Use of Industrial Waste Materials for 3D Printing of Sustainable Concrete: A Review." *Journal of Cleaner Production*.

Du, Xiaoqi, Yanlong Li, Binghui Huangfu, Zheng Si, Lingzhi Huang, Lifeng Wen, and Meiwei Ke. 2023. "Modification Mechanism of Combined Nanomaterials on High Performance Concrete and Optimization of Nanomaterial Content." *Journal of Building Engineering*. doi: 10.1016/j.jobe.2022.105648.

Dulipalla, Spurthi. 2018. "Study on Partial Replacement of Cement by Saw Dust Ash in Concrete." *International Journal for Research in Applied Science and Engineering Technology*. doi: 10.22214/ijraset.2018.2133.

Fang, Guohao, Jieting Chen, Biqin Dong, and Bing Liu. 2023. "Microstructure and Micromechanical Properties of Interfacial Transition Zone in Green Recycled Aggregate Concrete." *Journal of Building Engineering*. doi: 10.1016/j.jobe.2023.105860.

Fapohunda, C., A. Kilani, B. Adigo, L. Ajayi, B. Famodimu, O. Oladipupo, and A. Jeje. 2021. "A Review of Some Agricultural Wastes in Nigeria for Sustainability in the Production of Structural Concrete." *Nigerian Journal of Technological Development*. doi: 10.4314/njtd.v18i2.1.

Garas, Gihan, Alaa Mohamed Sayed, and Emad Shaker Hana Bakhoum. 2021. "Application of Nano Waste Particles in Concrete for Sustainable Construction: A Comparative Study." *International Journal of Sustainable Engineering*. doi: 10.1080/19397038.2021.1963004.

Ghanim, Abdulnoor A. J., Fayyaz Ur Rahman, Waqas Adil, Abdullah M. Zeyad, and Hassan M. Magbool. 2023. "Experimental Investigation of Industrial Wastes in Concrete: Mechanical and Microstructural Evaluation of Pumice Powder and Fly Ash in Concrete." *Case Studies in Construction Materials*. doi: 10.1016/j.cscm.2023.e01999.

Ghinaya, Zahra, and Alias Masek. 2021. "Eco-Friendly Concrete Innovation in Civil Engineering." *ASEAN Journal of Science and Engineering*. doi: 10.17509/ajse.v1i3.39475.

Golewski, Grzegorz Ludwik. 2021. "Green Concrete Based on Quaternary Binders with Significant Reduced of Co2 Emissions." *Energies*. doi: 10.3390/en14154558.

Gyawali, Tek Raj. 2022. "Re-Use of Concrete/Brick Debris Emerged from Big Earthquake in Recycled Concrete with Zero Residues." *Cleaner Waste Systems*. doi: 10.1016/j.clwas.2022.100007.

Hashmi, Dr. Ahmad Fuzail, M.S. Khan, M. Bilal, M. Shariq, and A. Baqi. 2022. "Green Concrete: An Eco-Friendly Alternative to the OPC Concrete." *CONSTRUCTION*. doi: 10.15282/construction.v2i2.8710.

Healy, Meagan L., Lindsay J. Dahlben, and Jacqueline A. Isaacs. 2008. "Environmental Assessment of Single-Walled Carbon Nanotube Processes." *Journal of Industrial Ecology*. doi: 10.1111/j.1530-9290.2008.00058.x.

Horma, Othmane, Mouatassim Charai, Sara El Hassani, Aboubakr El Hammouti, Mohammed A. Moussaoui, and Ahmed Mezrhab. 2022. "Thermal Performance Study of a Cement-Based Mortar Incorporating EPS Beads." *Frontiers in Built Environment*. doi: 10.3389/fbuil.2022.882942.

Hwang, Bon-Gang, and Jac See Tan. 2012. "Sustainable Project Management for Green Construction: Challenges, Impact, and Solutions." in *In World construction conference*.

Iqbal, Ashik, Ismat Jahan, Qudrati Al Wasiew, Imtiaz Ahmed Emu, and Dipta Chowdhury. 2023. "From Existing Conventional Building towards LEED Certified Green Building: Case Study in Bangladesh." *Frontiers in Built Environment*. doi: 10.3389/fbuil.2023.1194636.

Ispir, Medine, Ali Osman Ates, and Alper Ilki. 2022. "Low Strength Concrete: Stress-Strain Curve, Modulus of Elasticity and Tensile Strength." *Structures*. doi: 10.1016/j.istruc.2022.01.018.

Janani, S., P. Kulanthaivel, G. Sowndarya, H. Srivishnu, and P. G. Shanjayvel. 2022. "Study of Coconut Shell as Coarse Aggregate in Light Weight Concrete- a Review." *Materials Today: Proceedings*. doi: 10.1016/j.matpr.2022.05.329.

Ji, Yongcheng, and Qijun Sun. 2023. "Experimental and Numerical Investigation of Recycled Rubber Foam Concrete." *Alexandria Engineering Journal*. doi: 10.1016/j.aej.2023.06.057.

Jiménez, Juan E., Carlos Mauricio Fontes Vieira, and Henry A. Colorado. 2022. "Composite Soil Made of Rubber Fibers from Waste Tires, Blended Sugar Cane Molasses, and Kaolin Clay." *Sustainability (Switzerland)*. doi: 10.3390/su14042239.

Khabaz, Amjad. 2023. "Optimum Thermal Performance of Green Walls Systems and Design Requirements against Heat Transfer of Conventional External Walls of Low-Rise Concrete Buildings in Hot Regions." *Journal of Building Engineering*. doi: 10.1016/j.jobe.2023.107654.

Kibert, Charles J., Jan Sendzimi, and Bradley G. Guy. 2007. "Construction Ecology: Nature as the Basis for Green Buildings & Sustainable Construction: Green Building Design and Delivery,." *Journal of Industrial Ecology*.

Liu, Miao, Ruihan Hu, Youchao Zhang, Changqing Wang, and Zhiming Ma. 2023. "Effect of Ground Concrete Waste as Green Binder on the Micro-Macro Properties of Eco-Friendly Metakaolin-Based Geopolymer Mortar." *Journal of Building Engineering*. doi: 10.1016/j.jobe.2023.106191.

Liu, Zonghui, Jiaqi Li, Liqiang Hu, Xiaolei Zhang, Shiying Ding, and Haodong Li. 2023. "Strength and Environmental Behaviours of Municipal Solid Waste Incineration Fly Ash for Cement-Stabilised Soil." *Sustainability (Switzerland)*. doi: 10.3390/su15010364.

Malovrh, Špela Pezdevšek, Dženan Bećirović, Bruno Marić, Jelena Nedeljković, Stjepan Posavec, Nenad Petrović, and Mersudin Avdibegović. 2019. "Contribution of Forest Stewardship Council Certification to Sustainable Forest Management of State Forests in Selected Southeast European Countries." *Forests*. doi: 10.3390/f10080648.

Marceau, Medgar L., and Martha G. VanGeem. 2003. "Using Concrete to Increase LEED™ Ratings of Buildings." in *Architectural Engineering, Building Integration Solutions*.

Mater, Yasser, Mohamed Kamel, Ahmed Karam, and Emad Bakhoum. 2023. "ANN-Python Prediction Model for the Compressive Strength of Green Concrete." *Construction Innovation*. doi: 10.1108/CI-08-2021-0145.

Mathur, Mahim, and R. C. Chhipa. 2022. "Durability Assessment of Cow Dung Ash Modified Concrete Exposed in Fresh Water." *International Journal of Advanced Technology and Engineering Exploration*. doi: 10.19101/IJATEE.2021.875968.

Meryman, Helena, and Robert Silman. 2004. "Sustainable Engineering - Using Specifications to Make It Happen." *Structural Engineering International: Journal of the International Association for Bridge and Structural Engineering (IABSE)*.

Mohammed, Thanaa Khalaf, and Sheelan Mahmoud Hama. 2022. "Mechanical Properties, Impact Resistance and Bond Strength of Green Concrete Incorporating Waste Glass Powder and Waste Fine Plastic Aggregate." *Innovative Infrastructure Solutions*. doi: 10.1007/s41062-021-00652-4.

Mohanraj, Erode Krishnasamy, Ramesh Malathy, and Kanjikovil Loganathan Ravisankar. 2022. "Utilization of Industrial Waste Materials in Concrete-Filled Steel Tubular Columns." *Revista Materia*. doi: 10.1590/S1517-707620220002.1388.

Mokhtar, Mardiha, Ain Fakhirah, Abdul Halim, Hizati Mahmod, Nurin Nabila, and Mohd Rashdan. 2022. "Use of Rice Husk and Coconut Shell Ash in Concrete Production." *Multidisciplinary Applied Research and Innovation*.

Mondal, Sukanta K., Carrie Clinton, Hongyan Ma, Aditya Kumar, and Monday U. Okoronkwo. 2023. "Effect of Class C and Class F Fly Ash on Early-Age and Mature-Age Properties of Calcium Sulfoaluminate Cement Paste." *Sustainability (Switzerland)*. doi: 10.3390/su15032501.

Munir, Qaisar, Mariam Abdulkareem, Mika Horttanainen, and Timo Kärki. 2023. "A Comparative Cradle-to-Gate Life Cycle Assessment of Geopolymer Concrete Produced from Industrial Side Streams in Comparison with Traditional Concrete." *Science of the Total Environment*. doi: 10.1016/j.scitotenv.2022.161230.

Murad, Yasmin, Rana Imam, Husam Abu Hajar, Dua'a Habeh, Abdullah Hammad, and Zaid Shawash. 2020. "Predictive Compressive Strength Models for Green Concrete." *International Journal of Structural Integrity*. doi: 10.1108/IJSI-05-2019-0044.

Mutiari, Dhani, Nur Rahmawati Syamsiyah, Yayi Arsandric, Suharyani, Muhammad Ali Rofik, and Saidah Aliyatul Himmah. 2023. "The Acoustic Comfort in the House Made of Plastic Waste." *Civil Engineering and Architecture*. doi: 10.13189/cea.2023.110119.

Naenudon, Sakchai, Anousit Vilaivong, Yuwadee Zaetang, Weerachart Tangchirapat, Ampol Wongsa, Vanchai Sata, and Prinya Chindaprasirt. 2022. "High Flexural Strength Lightweight Fly Ash Geopolymer Mortar Containing Waste FIber Cement." *Case Studies in Construction Materials*. doi: 10.1016/j.cscm.2022.e01121.

Nayak, Chittaranjan B., Pratik P. Taware, Umesh T. Jagadale, Nitin A. Jadhav, and Samadhan G. Morkhade. 2022. "Effect of SiO2 and ZnO Nano-Composites on Mechanical and Chemical Properties of Modified Concrete." *Iranian Journal of Science and Technology, Transactions of Civil Engineering*. doi: 10.1007/s40996-021-00694-9.

Nie, Qingke, Changjun Zhou, Xiang Shu, Qiang He, and Baoshan Huang. 2014. "Chemical, Mechanical, and Durability Properties of Concrete with Local Mineral Admixtures under Sulfate Environment in Northwest China." *Materials*. doi: 10.3390/ma7053772.

Nilimaa, Jonny. 2023. "Smart Materials and Technologies for Sustainable Concrete Construction." *Developments in the Built Environment*. doi: 10.1016/j.dibe.2023.100177.

Niyazbekova, Rimma, Gabit Mukhambetov, Rassul Tlegenov, Saule Aldabergenova, Lazzat Shansharova, Vasiliy Mikhalchenko, and Michał Bembenek. 2023. "The Influence of Addition of Fly Ash from Astana Heat and Power Plants on the Properties of the Polystyrene Concrete." *Energies*. doi: 10.3390/en16104092.

Öhrn Sagrelius, Pär, Godecke Blecken, Annelie Hedström, Richard Ashley, and Maria Viklander. 2022. "Environmental Impacts of Stormwater Bioretention Systems with Various Design and Construction Components." *Journal of Cleaner Production*. doi: 10.1016/j.jclepro.2022.132091.

Osial, Magdalena, Agnieszka Pregowska, Sławomir Wilczewski, Weronika Urbańska, and Michael Giersig. 2022. "Waste Management for Green Concrete Solutions: A Concise Critical Review." *Recycling*.

Otieno, Meshack Oduor, Charles Kabubo, and Zachary Abiero Gariy. 2023. "A Comparative Investigation on Cement Stabilized Lateritic Soil Admixed with Sugarcane Bagasse Ash and Saw Dust Ash for Use in Road Base." *International Journal of Engineering Trends and Technology*. doi: 10.14445/22315381/IJETT-V71I5P211.

Oyebisi, Solomon, Tobit Igba, Akeem Raheem, and Festus Olutoge. 2020. "Predicting the Splitting Tensile Strength of Concrete Incorporating Anacardium Occidentale Nut Shell Ash Using Reactivity Index Concepts and Mix Design Proportions." *Case Studies in Construction Materials*. doi: 10.1016/j.cscm.2020.e00393.

Pan, Lei, Yuanfeng Wang, Kai Li, and Xiaohui Guo. 2023. "Predicting Compressive Strength of Green Concrete Using Hybrid Artificial Neural Network with Genetic Algorithm." *Structural Concrete*. doi: 10.1002/suco.202200034.

Pearce, Annie R., and Jorge A. Vanegas. 2002. "A Parametric Review of the Built Environment Sustainability Literature." *International Journal of Environmental Technology and Management*.

Ramalingam, Chandru, B. Sundarakannan, and R. Vasudevan. 2022. "Investigation on the Behavior of the Plastone Blocks Made with Waste Plastics and Aggregates – Substitute for Cement Concrete Paver Blocks." *SSRN Electronic Journal*. doi: 10.2139/ssrn.4136987.

Rezvan, Sina, Mohammad Javad Moradi, Hamed Dabiri, Kambiz Daneshvar, Moses Karakouzian, and Visar Farhangi. 2023. "Application of Machine Learning to Predict the Mechanical Characteristics of Concrete Containing Recycled Plastic-Based Materials." *Applied Sciences (Switzerland)*. doi: 10.3390/app13042033.

Santhosh, Kumar Gedela, Sk M. Subhani, and A. Bahurudeen. 2022. "Recycling of Palm Oil Fuel Ash and Rice Husk Ash in the Cleaner Production of Concrete." *Journal of Cleaner Production*.

Seghier, Taki Eddine, Mohd Hamdan Ahmad, Lim Yaik Wah, and Williams Opeyemi Samuel. 2018. "Integration Models of Building Information Modelling and Green Building Rating Systems: A Review." *Advanced Science Letters*. doi: 10.1166/asl.2018.11554.

Shao, Xiaoping, Long Wang, Xin Li, Zhiyu Fang, Bingchao Zhao, Yeqing Tao, Lang Liu, Wuliang Sun, and Jianpeng Sun. 2020. "Study on Rheological and Mechanical Properties of Aeolian Sand-Fly Ash-Based Filling Slurry." *Energies*. doi: 10.3390/en13051266.

Shen, Yijuan, Zhi Wei Su, Muhammad Yousaf Malik, Muhammad Umar, Zeeshan Khan, and Mohsin Khan. 2021. "Does Green Investment, Financial Development and Natural Resources Rent Limit Carbon Emissions? A Provincial Panel Analysis of China." *Science of the Total Environment*. doi: 10.1016/j.scitotenv.2020.142538.

Signorini, Cesare, and Valentina Volpini. 2021. "Mechanical Performance of Fiber Reinforced Cement Composites Including Fully-Recycled Plastic Fibers." *Fibers*. doi: 10.3390/fib9030016.

Singh, Siddharth, Mickey Mecon Dalbehera, Soumitra Maiti, Ravindra Singh Bisht, Nagesh Babu Balam, and Soraj Kumar Panigrahi. 2023. "Investigation of Agro-Forestry and Construction Demolition Wastes in Alkali-Activated Fly Ash Bricks as Sustainable Building Materials." *Waste Management*. doi: 10.1016/j.wasman.2023.01.031.

Singh, Yashpal, Jyoti Rani, Jeetesh Kushwaha, Madhumita Priyadarsini, Kailash Pati Pandey, Pratik N. Sheth, Sushil Kumar Yadav, M. S. Mahesh, and Abhishek S. Dhoble. 2023. "Scientific Characterization Methods for Better Utilization of Cattle Dung and Urine: A Concise Review." *Tropical Animal Health and Production*.

Sldozian, Rami Joseph Aghajan, Ali Jihad Hamad, and Hesham Salim Al-Rawe. 2023. "Mechanical Properties of Lightweight Green Concrete Including Nano Calcium Carbonate." *Journal of Building Pathology and Rehabilitation*. doi: 10.1007/s41024-022-00247-1.

Stevulova, Nadezda, Vojtech Vaclavik, Viola Hospodarova, and Tomáš Dvorský. 2021. "Recycled Cellulose Fiber Reinforced Plaster." *Materials*. doi: 10.3390/ma14112986.

Sureshkumar, M. P., B. Sathish Kumar, and J. Ravikanth. 2019. "Green Concrete – A Review." *International Research Journal of Multidisciplinary Technovation*.

Syamsiyah, Nur Rahmawati, Dhani Mutiari, Yayi Arsandrie, Suharyani, and Saidah Aliyatul Himmah. 2020. "Acoustic Performance from a Mixture of Plastic Waste, Wood Dust, and Rice Husk." *Civil Engineering and Architecture*. doi: 10.13189/cea.2020.080412.

Tahir, Furqan, Sami Sbahieh, and Sami G. Al-Ghamdi. 2022. "Environmental Impacts of Using Recycled Plastics in Concrete." *Materials Today: Proceedings*. doi: 10.1016/j.matpr.2022.04.593.

Thiam, Moussa, Mamadou Fall, and M. S. Diarra. 2021. "Mechanical Properties of a Mortar with Melted Plastic Waste as the Only Binder: Influence of Material Composition and Curing Regime, and Application in Bamako." *Case Studies in Construction Materials*. doi: 10.1016/j.cscm.2021.e00634.

Tota-Maharaj, Kiran, Blessing Oluwaseun Adeleke, and Ghassan Nounu. 2022. "Effects of Waste Plastics as Partial Fine-Aggregate Replacement for Reinforced Low-Carbon Concrete Pavements." *International Journal of Sustainable Engineering*. doi: 10.1080/19397038.2022.2108156.

Tramontin, Vittorio, Claudia Loggia, and Martina Basciu. 2010. "Passive Design and Building Renovation in the Mediterranean Area: New Sensitive Approach for Sustainability." *Journal of Civil Engineering and Architecture*.

Tulashie, Samuel Kofi, Daniel Dodoo, Atiiga Abdul Wadud Ibrahim, Stephen Mensah, Sandra Atisey, Raphael Odai, and David Mensah. 2022. "Recycling Plastic Wastes for Production of Sustainable and Decorative Plastic Pavement Bricks." *Innovative Infrastructure Solutions*. doi: 10.1007/s41062-022-00866-0.

Warnock, A. C. C. 1998. "Controlling Sound Transmission through Concrete Block Walls." *Construction Technology Update No. 13*.

Wilson, J. L., and E. Tagaza. 2006. "Green Buildings in Australia: Drivers and Barriers." *Australian Journal of Structural Engineering*. doi: 10.1080/13287982.2006.11464964.

Winkler, Lisa, Drew Pearce, Jenny Nelson, and Oytun Babacan. 2023. "The Effect of Sustainable Mobility Transition Policies on Cumulative Urban Transport Emissions and Energy Demand." *Nature Communications*. doi: 10.1038/s41467-023-37728-x.

Xu, Chao, Hao Hao Liao, You Liang Chen, Xi Du, Bin Peng, and Tomas Manuel Fernandez-Steeger. 2021. "Corrosion Performance of Nano-Tio2-Modified Concrete under a Dry–Wet Sulfate Environment." *Materials*. doi: 10.3390/ma14195900.

Yilmaz, Arin. 2021. "Mechanical and Durability Properties of Cement Mortar Containing Waste Pet Aggregate and Natural Zeolite." *Ceramics - Silikaty*. doi: 10.13168/cs.2021.0001.

Yuan, Yuan, Zhushan Shao, Rujia Qiao, Xuan Guo, and Weitao Wang. 2022. "Crack Damage Evolution in Concrete Coarse Aggregates under Microwave-Induced Thermal Stress." *Archives of Civil and Mechanical Engineering*. doi: 10.1007/s43452-022-00419-3.

Záleská, Martina, Milena Pavlíková, Ondřej Jankovský, Michal Lojka, Filip Antončík, Adam Pivák, and Zbyšek Pavlík. 2019. "Influence of Waste Plastic Aggregate and Water-Repellent Additive on the Properties of Lightweight Magnesium Oxychloride Cement Composite." *Applied Sciences (Switzerland)*. doi: 10.3390/app9245463.

Zerbino, R., G. Giaccio, and G. C. Isaia. 2011. "Concrete Incorporating Rice-Husk Ash without Processing." *Construction and Building Materials*. doi: 10.1016/j.conbuildmat.2010.06.016.

Zeyad, Abdullah M., Megat Azmi Megat Johari, Aref Abadel, Ahmed Abutaleb, M. J. A. Mijarsh, and Ali Almalki. 2022. "Transport Properties of Palm Oil Fuel Ash-Based High-Performance Green Concrete Subjected to Steam Curing Regimes." *Case Studies in Construction Materials*. doi: 10.1016/j.cscm.2022.e01077.

Zhang, Peng, Xiaoyao Sun, Fei Wang, and Juan Wang. 2023. "Mechanical Properties and Durability of Geopolymer Recycled Aggregate Concrete: A Review." *Polymers*.

Zhang, Sherong, Andong Zhu, Kelei Cao, Xiaohua Wang, Chao Wang, and Chao Shang. 2021. "Temperature and Age Effects on Mechanical Behaviour of Phosphorus Slag-Based Concrete." *Advances in Cement Research*. doi: 10.1680/jadcr.18.00230.

Index

For Product Safety Concerns and Information please contact our EU representative GPSR@taylorandfrancis.com
Taylor & Francis Verlag GmbH, Kaufingerstraße 24, 80331 München, Germany

www.ingramcontent.com/pod-product-compliance
Lightning Source LLC
LaVergne TN
LVHW020612110826
845149LV00002B/454

* 9 7 8 1 0 3 2 7 9 9 1 3 1 *